Co-Existence and Co-Release of Classical Neurotransmitters

Rafael Gutierrez
Editor

Co-Existence and Co-Release of Classical Neurotransmitters

Ex uno plures

Springer

Editor
Rafael Gutierrez
Centro de Investigacion y
Estudios Avanzados
Mexico City
grafael@fisio.cinvestav.mx

ISBN: 978-1-4419-3498-7 e-ISBN: 978-0-387-09622-3
DOI 10.1007/978-0-387-09622-3

Printed on acid-free paper

springer.com

Contents

Contributors

Arias-Montaño, J.A. Department of Physiology, Biophysics and Neurosciences, Center for Research and Advanced Studies of the National Polytechnic Institute, Apartado Postal 14-740, México D.F. 07000

Birren, Susan J. Department of Biology and Volen Center for Complex Systems, Brandeis University

Borodinsky, Laura N. Department of Physiology & Membrane Biology, University of California Davis School of Medicine and Institute for Pediatric Regenerative Medicine, Shriners Hospital for Children Northern California, 2425 Stockton Blvd, Sacramento, California 95817 e-mail: lnborodinsky@ucdavis.edu

Cherubini, Enrico Neurobiology Sector, International School for Advanced Studies, Via Beirut 2–4, 34014 Trieste, Italy e-mail: cher@sissa.it

Dal Bo, Grégory Department of Pharmacology, CNS Research Group, Faculty of Medicine, Université de Montréal

Dani, John A. Department of Neuroscience, Menninger Department of Psychiatry and Behavioral Science, Baylor College of Medicine, Houston, TX 77030 e-mail: jdani@bcm.tmc.edu

Diana, Marco Alberto Laboratoire de neurobiology, CNRS UMR8544, Ecole normale supérieure, 46 rue d'Ulm, 75005 Paris e-mail: mdiana@ens.fr

Dieudonné, Stéphane Laboratoire de neurobiologie, CNRS UMR8544, Ecole normale supérieure, 46 rue d'Ulm, 75005 Paris (France) e-mail: dieudon@biologie.ens.fr

Gillespie, Deda C. Department of Psychology, Neuroscience & Behaviour, McMaster University, Hamilton, ON L8S 4K1 Canada e-mail: dgillespie@mcmaster.ca

Gutiérrez, R. Department of Physiology, Biophysics and Neurosciences, Center for Research and Advanced Studies of the National Polytechnic Institute, Post Box 14-740, México D.F. 07000. E-mail: grafael@fisio.cinvestav.mx

Hökfelt, Tomas Department of Neuroscience, Karolinska Institutet, Retzius v. 8, S-171 77 Stockholm, SWEDEN. E-mail: tomas.hokfelt@ki.se

Hugel, S. Université Louis Pasteur, Institut des Neurosciences Cellulaires et Intégratives (INCI), Centre National de la Recherche Scientifique (CNRS), UMR7168, F-67084 Strasbourg, France

Jo, YH Albert Einstein College of Medicine of Yeshiva University, Dept. of Medicine, Div. of Endocrinology, Bronx, NY 10461, New-York, USA, e-mail: schlichter@neurochem.u-strashy.fr

Kandler, Karl Department of Otolaryngology and Neurobiology, School of Medicine, University of Pittsburgh, Pittsburgh, PA 15213 USA e-mail: kkarl@pitt.edu

Li, Wen-Chang School of Biology, Bute Building, University of St Andrews, St Andrews, Fife, KY16, 9TS, UK, e-mail: wl21@st-andrews.ac.uk

Luther, Jason A. Department of Biology and Volen Center for Complex Systems, Brandeis University

Mendez, José Alfredo Department of Pharmacology, CNS Research Group, Faculty of Medicine, Université de Montréal

Miller, Mark W. Institute of Neurobiology and Department of Anatomy & Neurobiology, University of Puerto Rico, 201 Blvd del Valle, San Juan, Puerto Rico 00901

Mohajerani, Majid H. Neurobiology Sector, International School for Advanced Studies, Via Beirut 2–4, 34014 Trieste, Italy

Safiulina, Victoria F. Neurobiology Sector, International School for Advanced Studies, Via Beirut 2–4, 34014 Trieste, Italy

Schlichter, R. Université Louis Pasteur, Institut des Neurosciences Cellulaires et Intégratives (INCI), Centre National de la Recherche Scientifique (CNRS), UMR7168, 21 rue René Decartes, F-67084 Strasbourg, France

Sivakumaran, Sudhir Neurobiology Sector, International School for Advanced Studies, Via Beirut 2–4, 34014 Trieste, Italy

Spitzer, Nicholas C. Neurobiology Section, Division of Biological Sciences and Center for Molecular Genetics, Kavli Institute for Brain and Mind, University of California, San Diego, 9500 Gilman Drive, La Jolla, California 92093. e-mail: nspitzer@ucsd.edu

Trudeau, Louis-Eric Department of Pharmacology, CNS Research Group, Faculty of Medicine, Université de Montréal, Montrĕal, Quévec, Canada

Zhou, Fu-Ming Department of Pharmacology, University of Tennessee College of Medicine Memphis, TN 38163

Chapter 1
Coexistence of Neuromessenger Molecules – A Perspective

Tomas Hökfelt

Abstract Over the last three decades, it has become increasingly clear that multiple messengers synthesized in, and released from, a single nerve ending (or some/dendrite) participate in the chemical transmission process – the one neuron, multiple transmitters concept. The molecules involved encompass a wide variety of chemicals, e.g. aminoacids, monoamines, peptides and others. This first chapter attempts to provide a background to the many novel and interesting aspects on coexistence dealt with in the book.

1.1 Chemical Transmission

Chemical transmission is a fundamental process in nervous system function. The chemicals involved were originally termed neurotransmitters, but other names have subsequently also been used: messenger molecule, signaling/transmitter substance, modulator and more – in Sweden we say "a loved child has many names". Early on, with only few substances around, the term "neurotransmitter" appeared distinct and sufficient. However, as more and more categories of molecules appeared to have a signaling function in the nervous system, and sometimes with additional, even not well-defined functions, the name not rarely became on issue of controversy. For example, in the 1960s some eminent neurophysiologists would not accept the monoamines as neurotransmitters. This discussion is today less intense, perhaps because of the insight that the name really is not the critical issue, but rather to understand under what circumstances this spectrum of molecules is produced and released and what their functional significance is.

T. Hökfelt (✉)
Department of Neuroscience, Karolinska Institutet, Retzius v. 8, S-171 77 Stockholm, SWEDEN
e-mail: tomas.hokfelt@ki.se

R. Gutierrez (ed.), *Co-Existence and Co-Release of Classical Neurotransmitters*,
DOI 10.1007/978-0-387-09622-3_1, © Springer Science+Business Media, LLC 2009

1.2 Identity and Function of Neurotransmitters

What is clear is that messenger molecules not only are involved in different types of transmission, e.g. slow versus fast signaling (the type of receptor being decisive), but many of them also have other effects, e.g. stimulating growth. And the main function of a messenger may vary during the life of a neuron/the nervous system, e.g. early on exerting a role in developmental processes and later on being a regular transmitter; or being postnatally downregulated and then reactivated under certain conditions, e.g. nerve injury. Thus, our view has advanced from the somewhat stereotype view that the function of a transmitter is just to allow axon potential "to jump" from one neuron to another via a chemical message. One could argue, let us only call such a molecule "transmitter" that does exactly that; but in fact there are hardly any messengers with just that function: Even glutamate exerts trophic effects and has both pre- and postsynaptic effects, that is, it also acts as a growth factor and modulator. In summary, molecules released from a nerve ending may have many different functions. If so, we cannot in many cases speak about co-release of transmitters in a strict sense.

1.3 Neurons Only Produce One Transmitter

My upbringing in the Amine Group – established in 1962 by the late Nils-Åke Hillarp (1916–1965) – in the Department of Histology at Karolinska Institutet, taught me that a neuron only has one neurotransmitter. This my view was based on histochemical monoamine research, using the formaldehyde fluorescence (Falck-Hillarp) method developed by Bengt Falck, Nils- Åke Hillarp and collaborators (Falck et al., 1962). With this technique, for the first time, a transmitter could be identified in an individual neuron – if one wants, the first opportunity to approach the coexistence problem. The results clearly showed that dopamine (DA), noradrenaline (NA), 5-hydroxytryptamine (5-HT; serotonin) and (later) adrenaline were synthesized in different systems with their cell bodies distinctly separated along the caudo-cranial axis (Dahlström and Fuxe, 1964; Hökfelt et al., 1974; Hökfelt et al., 1984). Also, early ultrastructural analyses, even when using the highly sensitive potassium permanganate fixation (Richardson, 1966), showed that in the adult animal the peripheral noradrenergic and cholinergic neurons are two separate populations.

Moreover, when it became possible to demonstrate the cellular localization of the large population of inhibitory γ-amino-butyric acid (GABA) neurons, first with ^{3}H-GABA and autoradiography (Hökfelt and Ljungdahl, 1972a, b), and subsequently with immunohistochemistry using antibodies either to the GABA-synthesizing enzyme glutamate decarboxylase (GAD) (Wu et al., 1973; Saito et al., 1974), or to GABA itself (Storm-Mathisen et al., 1983), there was no obvious evidence for overlap and coexistence of GABA with the above-mentioned monoamine neurotransmitter systems (cf. Mugnaini and Oertel, 1985).

1.4 Some Historical Aspects – Dale's Principle

This view was in general agreement with an idea often called the "one neuron-one transmitter" hypothesis. This went back to Sir Henry Dale's statement (Dale, 1935a, b) that a neuron is a metabolic unit and "operates at all its synapses by the same chemical transmission mechanism", one interpretation being that a neuron releases one and the same messenger from all its branches. The concept was then further modified, saying that each nerve cell makes and releases only one transmitter. In the light of the findings of coexistence of messenger molecules described below, this concept was later discussed in some depth (see e.g. Eccles, 1986; Potter et al., 1986). Nevertheless, the "one neuron-one transmitter" idea was not challenged for several decades. But when multiple messengers were shown in neurons (see below), additional interesting findings with bearing on Dale's principle were reported. Thus, in Aplysia two different messengers could be shown to be directed into different processes of the neuron (Sossin et al., 1990), thus not having the same transmission mechanism, in any case not the same transmitter, at all processes. Another interesting concept has been developed by Ludwig and co-workers, showing that dendrite and nerve endings of a neuron can operate separately and independently in releasing a messenger substance (see Ludwig, 2005; Ludwig and Leng, 2006).

1.5 Early Evidence for One Neuron-Multiple Transmitters

In the mid-1970s, the"one neuron-one transmitter" idea came under serious scrutiny. Thus, studies on isolated (large) invertebrate neurons suggested presence/co-release of more than one putative transmitter from a neuron (Kerkut et al., 1967; Brownstein et al., 1974; Hanley et al., 1974; Cottrell, 1976; Osborne, 1984). Also, Jaim-Etcheverry and Zieher (1973), to my knowledge the first ones using the word "coexistence" in this context, reported presence of NA and 5-HT in the same synaptic vesicles in the pineal gland. In this case serotonin had been taken up from the blood, that is not synthesized in the pineal nerves. Nevertheless, when activated these nerve endings presumably release two transmitters, a topic that will be dealt with in this book.

Elegant experiments, initially carried out in mono-neuron cultures (Furshpan et al., 1976; Landis, 1976), showed that there is a developmental switch in autonomic neurons from a noradrenergic to a cholinergic phenotype and that autonomic neurons for a while can synthesize and release both NA and acetylcholine (ACh). Thus, there is coexistence and co-release of two classic transmitters during development, which also occurs in vivo (Francis and Landis, 1999).

Physiological/pharmacological studies on the peripheral nervous system suggested existence of nerves releasing neither NA nor ACh, and this phenomenon was termed NANC (nonadrenergic, noncholinergic) transmission, which could be shown to exist in many peripheral tissues (see Burnstock, 2007). The prime candidate for this type of transmission was ATP, and Geoffrey Burnstock coined the term purinergic transmission. ATP was also one of the early molecules suggested to be involved in cotransmission, and co-release with NA could be demonstrated (Su et al., 1971; Westfall et al., 1978).

1.6 Coexisting Neuropeptides

Meanwhile, many groups had started to analyze the expression and distribution of a further group of neuronal messengers, the neuropeptides. They have now turned out to represent the largest family of signaling molecules in the nervous system, probably more than hundred members (Burbach, 2008), and with a correspondingly large number of receptors (several hundreds), virtually all of the 7-transmembrane, G-protein-coupled type. Radioimmunoassay and immunohistochemistry, sometimes using the same antibodies, clearly showed a very wide distribution in the brain and in all types of peripheral systems, sensory and autonomic neurons and in the gastro-intestinal tract. Geoffrey Burnstock (1976) wrote an influential review article suggesting reexamination of the "one neuron-one transmitter" concept, pointing in particular to the wide distribution of neuropeptides in the nervous system.

The first direct evidence for presence of a peptide and a classic transmitter in the same neuron was then observed in guinea pig sympathetic ganglia, where somatostatin was found in noradrenergic neurons (Hökfelt et al., 1977). Somatostatin, a tridecapeptide and the principal growth hormone release-inhibiting factor, was discovered by Brazeau et al. (1973), and Renaud et al. (1975) rapidly demonstrated a transmitter function for this peptide. Thus, many sympathetic neurons synthesize two transmitters/messenger molecules, NA and somatostatin. Early on efforts went into establishing that the somatostatin-NA was not a single case, and fairly rapidly more and more examples were found, and early reviews often had tables of varying length showing such examples. However, today they are so abundant that it seems useless to produce such a table. In fact, it is likely that every neuropeptide co-exists with a classic transmitter of some kind, as will be discussed below.

1.7 Neurotransmitter Storage

There are some general points that could be discussed in relation to coexistence. First, it may be said that this primarily is an anatomical term, that is, one has to show that two molecules with transmitter function are synthesized and present (preferably transcript and peptide/protein) in the same neuron. In addition to peptide-monoamine coexistence in the same neuron, it was rapidly shown that the peptides have a special storage site, the large dense core vesicles (LDCVs) (see

Pickel, 1985). In fact, it had been recognized in early electron microscopic studies that there are at least two types of storage vesicles in neurons/nerve endings (see Grillo, 1966): (1) synaptic vesicles (diameter around 500 Å); in NA (and other types of monoamine) neurons they can often be shown to have a dense core; and (2) LDCVs (diameter around 1,000 Å); if fixed with glutaraldehyde they have a dense core and are present in most neurons; if fixed with $KMnO_4$ (Richardson, 1966) LDCVs also have a dense core, but only in monoamine neurons (Hökfelt, 1968). So monoamines are stored both in LDCVs and synaptic vesicles.

The evidence for neuropeptide storage in LDCVs was/is based on immuno-histo-chemistry, which showed immunoprecipitate in LDCVs but not in synaptic vesicles (see Pickel, 1985). However, Pelletier et al. (1981) showed with immunohistochemistry that 5-HT is stored in LDCVs, but no precipitate was detected in synaptic vesicles, confirming the old truth that "negative (imuno) histochemistry" is not a final answer. Therefore subcellular fractionation studies were carried out, strongly supporting storage of neuropeptides exclusively in LDCVs (Lundberg et al., 1981; Fried et al., 1985).

An interesting question is whether amino acids are stored not only in synaptic vesicles, but also in LDCVs. Merighi et al. (1991) have triple-stained primary afferent nerve endings in the spinal dorsal horn for glutamate, substance P and calcitonin gene-related peptide (CGRP). Although, often in the same nerve endings, glutamate was never seen in the LDCVs. Thus perhaps one distinct difference between monoamines such as DA, NA and 5-HT on one hand, and aminoacid transmitters on the other hand is that only the former are stored both in synaptic vesicles and in LDCVs.

Taken together, two transmitters present not only in the same neuron but also in the same vesicles suggest co-release, but anatomy does not prove co-release. It is difficult to get direct evidence for co-release, that is, that released molecules indeed are coming from the same neuron/nerve ending(s). This was perhaps first achieved in the mono-neuron cultures discussed above.

1.8 Is the Classic Transmitter Always the Main Messenger?

At one point we considered the interesting question, whether neuropeptides are only present in neurons having a coexisting classic transmitter, and whether the classic transmitter always is the important partner. Here the hypothalamic magno- and parvocellular neurons may provide an answer: Vasopressin, oxytocin and the releasing/inhibitory peptide hormones are of vital importance and, in agreement, these neurons contain, both in cell bodies and nerve endings, large amounts of LDCVs storing the peptides. However, also the earliest electron microscopic studies showed that the nerve endings in the posterior pituitary and the external layer of the median eminence harbor numerous synaptic vesicles in addition, suggesting presence of a classic transmitter. There was early evidence that some CRF neurons produce GABA (Meister et al., 1988), and GHRH-positive neurons have a dopaminergic/GABAergic phenotype (Meister et al., 1986; Meister and Hökfelt, 1988; Hrabovszky et al., 2005a). Moreover, recent studies now clearly demonstrate presence of one of the recently discovered vesicular glutamate transporters, VGLUT-2, transcript and protein, in parvocellular LHRH- and somatostatin-positive hypothalamic neurons (Hrabovszky et al., 2004; Hrabovszky et al., 2005a; Hrabovszky et al., 2005b), as well as in magnocellular vasopressin and oxytocin neurons in the supraoptic and paraventricular nuclei (Hrabovszky et al., 2006).

Thus, even if the peptide hormone certainly is the main messenger molecule in these systems, classical aminoacid transmitters appear to participate in the modulation of these neurons. So we have a reversed situation as compared to most other systems, that is, the peptide is the main signaling molecule and the aminoacid the auxiliary messenger.

1.9 Also Aminoacid Transmitters Coexist

For a while it seemed as if only monoamines and ACh were involved in coexistence situations, that is, the aminoacid transmitters were “single”. However, as mentioned above, GABA and DA coexist in the dorso-medial arcuate neurons (Everitt et al., 1984), but even before that the coexistence of GABA and 5-HT was demonstrated (Belin et al., 1981; Nanopoulos et al., 1981; Belin et al., 1983; Millhorn et al., 1987), and there were indications of glutamate in catecholamine and 5-HT neurons (Kaneko et al., 1990; Nicholas et al., 1990; Minson et al., 1991; Nicholas et al., 1992). Monoamine-glutamate coexistence was also supported by functional studies from single-cell microcultures demonstrating co-release of glutamate and 5-HT (Johnson, 1994; Li and Bayliss, 1998). Evidence for coexistence and co-release of glutamate and dopamine was published by Trudeau and collaborators (Dal Bo et al., 2004), a topic that they will also deal with in this book. Thus, all these studies suggested that neurons can co-release three classes of messengers: Aminoacids, monoamines and neuropeptides. Nevertheless, the glutamatergic nature of these neurons remained somewhat uncertain, because of lack of a truly specific marker. This changed, as indicated above, with the discovery of three vesicular glutamate transporters (see Masson et al., 1999; Fremeau et al., 2004). Thus, using VGLUT3 as a marker final proof was provided for the glutamatergic nature of many serotonin (Gras et al., 2002; Schäfer et al., 2002) and DA (Dal Bo et al., 2004) neurons.

The first example of possible aminoacid–aminoacid coexistence apparent to me was the demonstration by Ottersen et al. (1987) of cerebellar mossy fiber nerve endings characterized by high levels of both GABA and glycine. This type of coexistence, e.g. presence of GABA and glutamate in granule cells/mossy fibers (see Gutierrez, 2003), has during the last years captured increasing interest and is, in fact, the topic of several chapters in this book. It will therefore not be further dealt with here.

1.10 Functional Consequences and Clinical Implications

The insight that neuropeptides coexist with classical transmitters had a major impact, in any case for us, on the functional role of this large family of molecules. This indicated that neuropeptides are auxiliary messengers, not the sole messengers responsible for transmission at synaptic and non-synaptic sites in subpopulations of neurons. In fact, many colleagues are unconvinced about a physiological role of neuropeptides, as reflected in a stimulating and thought-provoking article by Bowers (1994).

The functional consequences of coexistence and cotransmission are manifold and have been explored in many experimental models. In the early days various types of interactions between classic transmitters and neuropeptides were considered. A model used here at Karolinska by Lundberg, Änggård and colleagues was the cat salivary gland exploiting interactions between NA and NPY and between ACh and VIP (Lundberg et al., 1980). A particularly convincing and elegant model was the the frog sympathetic ganglion for studies of interaction between ACh and LHRH-like peptides explored by Yuh Nung Jan and Lily Jan working in the legendary Stephen Kuffler's laboratory at Harvard Medical School (Jan and Jan, 1983).

We hypothesize that coexistence and cotransmission also has clinical implications. For example, in the rat the 29-aminoacid peptide galanin (Tatemoto et al., 1983) is expressed both in NA and 5-HT neurons (Melander et al., 1986), and NA-galanin coexistence has also been demonstrated in the human LC (Chan-Palay et al., 1990; Fodor et al., 1992; Kordower et al., 1992). Many NA neurons in the human LC also contain substance P (Baker et al., 1991; Sergeyev et al., 1999). Both NA and 5-HT neurons are targets for so called selective 5-HT and NA uptake inhibitors (SSRIs, SNRIs) for treatment of major (unipolar) depression. And also NK1 (substance P) antagonists have been reported to have antidepressant activity (Kramer et al., 1998; Kramer et al., 2004). This opens up interesting possibilities that coexisting molecules and their receptors can be target for development of novel treatment strategies for various disorders.

1.11 Concluding Remarks

We are witnessing an exciting development in our understanding of chemical transmission in the nervous system, characterized by an amazing complexity, at least as compared to the situation when I started in research some four decades ago. It was difficult enough to explain how a motoneuron in the ventral horn is controlled by some 10.000 boutons releasing one transmitter, but additional messengers in each bouton certainly does not make it easier to understand the functional execution.

Numerous papers have dealt with coexistence: More than 900 hits in PubMed under the terms "neurotransmitter, coexistence", more than 800 on "neuropeptide, coexistence" and more than 100 reviews with the two latter terms in the beginning of February 2008. So much focus has been on the neuropeptides. Early results on coexistence have been summarized in review articles (e.g. Hökfelt et al., 1980; Lundberg and Hökfelt, 1983; Furness et al., 1989; Burnstock, 1990; Lundberg, 1996; Merighi, 2002; Burnstock, 2004) and in books (Cuello, 1982; Osborne, 1983; Chan-Palay and Palay, 1984; Hökfelt et al., 1986). There are also important studies on the evertebrate nervous system that lends itself in an ideal way to coexistence/co-release studies, as also mentioned in the Introduction (for review see Osborne, 1984; Kupfermann, 1991; Nusbaum et al., 2001).

In the present book another chapter in the history of transmitter coexistence is written in a series of exciting chapters dealing with topics so far not summarized. They include novel aspects on transmitter combinations, such as NA-ACh, monoamines-glutamate, ACh-glutamate, cross-talk between monoamines, GABA and ATP, and especially various combinations of aminoacid transmitters, actually coexistence of excitatory and inhibitory ones. There are also chapters on synapse formation and invertebrates.

I thank the Editor Dr. Rafael Gutierrez for asking me to write this introduction. Many colleagues I am sure, and I for certain, look forward to read the final "product".

Acknowledgments Our work on transmitter coexistence for more than three decades has been supported in particular by the Swedish (Medical) Research Council, the Marianne and Marcus Wallenberg and the Knut and Alice Wallenberg Foundations, an Unrestricted Bristol-Myers-Squibb Neuroscience Grant, AFA Insurance, The Swedish Brare Foundation, The Swedish Foundation for International cooperation in Research and Higher Education (STINT) and EC grants (NEWMOOD; LHSM-CT-2003-503474).

References

Baker KG, Halliday GM, Hornung JP, Geffen LB, Cotton RG, Tork I (1991) Distribution, morphology and number of monoamine-synthesizing and substance P–containing neurons in the human dorsal raphe nucleus. Neuroscience 42:757–775

Belin MF, Nanopoulos D, Didier M, Aguera M, Steinbusch H, Verhofstad A, Maitre M, Pujol JF (1983) Immunohistochemical evidence for the presence of gamma-aminobutyric acid and serotonin in one nerve cell. A study on the raphe nuclei of the rat using antibodies to glutamate decarboxylase and serotonin. Brain Res 275:329–339

Belin MF, Weisman-Nanopoulos D, Steinbusch H, Verhofstad A, Maitre M, Jouvet M, Pujol JF (1981) [Demonstration of glutamate decarboxylase and serotonin in the same neuron of the nucleus raphe dorsalis of the rat by the methods of immunocytochemical doubling labelling]. C R Seances Acad Sci III 293:337–341

Bowers CW (1994) Superfluous neurotransmitters? Trends Neurosci 17:315–320

Brazeau P, Vale W, Burgus R, Ling N, Butcher M, Rivier J, Guillemin R (1973) Hypothalamic polypeptide that inhibits the secretion of immunoreactive pituitary growth hormone. Science 179:77–79

Brownstein MJ, Saavedra JM, Axelrod J, Zeman GH, Carpenter DO (1974) Coexistence of several putative neurotransmitters in single identified neurons of Aplysia. Proc Natl Acad Sci U S A 71:4662–4665
Burbach JPH (2008) Mammalian Neuropeptide Families. New Neuroscience Encyclopedia, in press
Burnstock G (1976) Do some nerve cells release more than one transmitter? Neuroscience 1:239–248
Burnstock G (1990) The fifth Heymans memorial lecture-Ghent, February 17, 1990. Cotransmission. Arch Int Pharmacodyn Ther 304:7–33
Burnstock G (2004) Cotransmission. Curr Opin Pharmacol 4:47–52
Burnstock G (2007) Physiology and pathophysiology of purinergic neurotransmission. Physiol Rev 87:659–797
Chan-Palay V, Jentsch B, Lang W, Höchli M, Asan E (1990) Distribution of neuropeptide Y, C-terminal flanking peptide of NPY and galanin coexistence with catecholamine in the locus coeruleus of normal human, Alzheimeräs dementia and Parkinson's disease brains. Dementia 1:18–31
Chan-Palay V, Palay SL (1984) Co-existence of Neuroactive Substances in Neurones. Wiley, New York
Cottrell GA (1976) Proceedings: Does the giant cerebral neurone of Helix release two transmitters: ACh and serotonin? J Physiol 259:44P–45P
Cuello AC (ed), 1982. CoTransmission. MacMillan, London
Dahlström A, Fuxe K (1964) Evidence for the existence of monoamine neurons in the central nervous system. I. Demonstration of monoamines in the cell bodies of brainstem neurons. Acta Physiol Scand 62, Suppl. 232:1–55
Dal Bo G, St-Gelais F, Danik M, Williams S, Cotton M, Trudeau LE (2004) Dopamine neurons in culture express VGLUT2 explaining their capacity to release glutamate at synapses in addition to dopamine. J Neurochem 88:1398–1405
Dale HH (1935a) Pharmacology and nerve endings. Proc Roy Soc Med 28:319–332
Dale HH (1935b) Reizübertragung durch chemische Mittel im peripheren Nervensystem. Sammlung von der Nathnagel-Stiftung veranstalteten Vorträge, Heft 4. Urban & Schwarzenberg, Berlin, Wien, pp 1–23
Eccles JC (1986) Chemical transmission and Dale's principle. In: Hökfelt, T et al. (eds), Progress in Brain Research. Coexistence of neuronal messengers: A new principle in chemical transmission, vol 68. Elsevier, Amsterdam, pp 3–13
Everitt BJ, Hökfelt T, Wu JY, Goldstein M (1984) Coexistence of tyrosine hydroxylase-like and gamma-aminobutyric acid-like immunoreactivities in neurons of the arcuate nucleus. Neuroendocrinology 39:189–191
Falck B, Hillarp N-Å, Thieme G, Torp A (1962) Fluoresence of catecholamines and related compounds with formaldehyde. J Histochem Cytochem 10:348–354
Fodor M, Gorcs TJ, Palkovits M (1992) Immunohistochemical study on the distribution of neuropeptides within the pontine tegmentum—particularly the parabrachial nuclei and the locus coeruleus of the human brain. Neuroscience 46:891–908
Francis NJ, Landis SC (1999) Cellular and molecular determinants of sympathetic neuron development. Annu Rev Neurosci 22:541–566
Fremeau RT, Jr., Voglmaier S, Seal RP, Edwards RH (2004) VGLUTs define subsets of excitatory neurons and suggest novel roles for glutamate. Trends Neurosci 27:98–103
Fried G, Terenius L, Hökfelt T, Goldstein M (1985) Evidence for differential localization of noradrenaline and neuropeptide Y in neuronal storage vesicles isolated from rat vas deferens. J Neurosci 5:450–458
Furness JB, Morris JL, Gibbins IL, Costa M (1989) Chemical coding of neurons and plurichemical transmission. Annu Rev Pharmacol Toxicol 29:289–306
Furshpan EJ, MacLeish PR, O'Lague PH, Potter DD (1976) Chemical transmission between rat sympathetic neurons and cardiac myocytes developing in microcultures: evidence for

cholinergic, adrenergic, and dual-function neurons. Proc Natl Acad Sci U S A 73:4225–4229
Gras C, Herzog E, Bellenchi GC, Bernard V, Ravassard P, Pohl M, Gasnier B, Giros B, El Mestikawy S (2002) A third vesicular glutamate transporter expressed by cholinergic and serotoninergic neurons. J Neurosci 22:5442–5451
Grillo MA (1966) Electron microscopy of sympathetic tissues. Pharmacol Rev 18:387–399
Gutierrez R (2003) The GABAergic phenotype of the "glutamatergic" granule cells of the dentate gyrus. Prog Neurobiol 71:337–358
Hanley MR, Cottrell GA, Emson PC, Fonnum F (1974) Enzymatic synthesis of acetylcholine by a serotonin-containing neurone from Helix. Nature 251:631–633
Hrabovszky E, Csapo AK, Kallo I, Wilheim T, Turi GF, Liposits Z (2006) Localization and osmotic regulation of vesicular glutamate transporter-2 in magnocellular neurons of the rat hypothalamus. Neurochem Int 48:753–761
Hrabovszky E, Turi GF, Kallo I, Liposits Z (2004) Expression of vesicular glutamate transporter-2 in gonadotropin-releasing hormone neurons of the adult male rat. Endocrinology 145:4018–4021
Hrabovszky E, Turi GF, Liposits Z (2005a) Presence of vesicular glutamate transporter-2 in hypophysiotropic somatostatin but not growth hormone-releasing hormone neurons of the male rat. Eur J Neurosci 21:2120–2126
Hrabovszky E, Wittmann G, Turi GF, Liposits Z, Fekete C (2005b) Hypophysiotropic thyrotropin-releasing hormone and corticotropin-releasing hormone neurons of the rat contain vesicular glutamate transporter-2. Endocrinology 146:341–347
Hökfelt T (1968) In vitro studies on central and peripheral monoamine neurons at the ultrastructural level. Z Zellforsch Mikrosk Anat 91:1–74
Hökfelt T, Elfvin LG, Elde R, Schultzberg M, Goldstein M, Luft R (1977) Occurrence of somatostatin-like immunoreactivity in some peripheral sympathetic noradrenergic neurons. Proc Natl Acad Sci U S A 74:3587–3591
Hökfelt T, Fuxe K, Goldstein M, Johansson O (1974) Immunohistochemical evidence for the existence of adrenaline neurons in the rat brain. Brain Res 66:235–251
Hökfelt T, Fuxe K, Pernow B (eds), (1986) Coexistence of Neuronal Messengers: A New Principle in Chemical Transmission. Progress in Brain Research, vol 68. Elsevier, Amsterdam
Hökfelt T, Johansson O, Goldstein M (1984) Chemical anatomy of the brain. Science 225:1326–1334
Hökfelt T, Johansson O, Ljungdahl A, Lundbeg JM, Schultzberg M (1980) Peptidergic neurones. Nature 284:515–521
Hökfelt T, Ljungdahl Å (1972a) Application of cytochemical techniques to the study of suspected transmitter substances in the nervous system. Adv Biochem Psychopharmacol 6:1–36
Hökfelt T, Ljungdahl Å (1972b) Autoradiographic identification of cerebral and cerebellar cortical neurons accumulating labeled gamma-aminobutyric acid (3 H-GABA). Exp Brain Res 14:354–362
Jaim-Etcheverry G, Zieher LM (1973) Proceedings: Coexistence of monoamines in adrenergic synaptic vesicles. Acta Physiol Lat Am 23:616–618
Jan YN, Jan LY (1983) Coexistence and corelease of cholinergic and peptidergic transmitters in frog sympathetic ganglia. Fed Proc 42:2929–2933
Johnson MD (1994) Synaptic glutamate release by postnatal rat serotonergic neurons in microculture. Neuron 12:433–442
Kaneko T, Akiyama H, Nagatsu I, Mizuno N (1990) Immunohistochemical demonstration of glutaminase in catecholaminergic and serotonergic neurons of rat brain. Brain Res 507:141–154
Kerkut GA, Sedden CB, Walker RJ (1967) Uptake of DOPA and 5-hydroxytryptophan by monoamine-forming neurones in the brain of Helix aspersa. Comp Biochem Physiol 23:159–162

Kordower JH, Le HK, Mufson EJ (1992) Galanin immunoreactivity in the primate central nervous system. J Comp Neurol 319:479–500

Kramer MS, Cutler N, Feighner J, Shrivastava R, Carman J, Sramek JJ, Reines SA, Liu G, Snavely D, Wyatt-Knowles E, Hale JJ, Milss SG, MacCoss M, Swain CJ, Harrison T, Hill RG, Hefti F, Scolnick EM, Cascieri MA, Chicchi GG, Sadowski S, Williams AR, Hewson L, Smith D, Carlson E, Hargreaves RJ, Rupniak NMJ (1998) Distinct mechanism for antidepressant activity by blockade of central substance P receptors. Science 281:1640–1645

Kramer MS, Winokur A, Kelsey J, Preskorn SH, Rothschild AJ, Snavely D, Ghosh K, Ball WA, Reines SA, Munjack D, Apter JT, Cunningham L, Kling M, Bari M, Getson A, Lee Y (2004) Demonstration of the efficacy and safety of a novel substance P (NK1) receptor antagonist in major depression. Neuropsychopharmacology 29:385–392

Kupfermann I (1991) Functional studies of cotransmission. Physiol Rev 71:683–732

Landis SC (1976) Rat sympathetic neurons and cardiac myocytes developing in microcultures: correlation of the fine structure of endings with neurotransmitter function in single neurons. Proc Natl Acad Sci U S A 73:4220–4224

Li YW, Bayliss DA (1998) Presynaptic inhibition by 5-HT1B receptors of glutamatergic synaptic inputs onto serotonergic caudal raphe neurones in rat. J Physiol 510 (Pt 1):121–134

Ludwig M, Leng G (2006) Dendritic peptide release and peptide-dependent behaviours. Nat Rev Neurosci 7:126–136

Ludwig ML (ed) (2005) Dendritic Neurotransmitter Release. Springer, New York

Lundberg JM (1996) Pharmacology of cotransmission in the autonomic nervous system: integrative aspects on amines, neuropeptides, adenosine triphosphate, amino acids and nitric oxide. Pharmacol Rev 48:113–178

Lundberg JM, Anggard A, Fahrenkrug J, Hökfelt T, Mutt V (1980) Vasoactive intestinal polypeptide in cholinergic neurons of exocrine glands: functional significance of coexisting transmitters for vasodilation and secretion. Proc Natl Acad Sci U S A 77:1651–1655

Lundberg JM, Fried G, Fahrenkrug J, Holmstedt B, Hökfelt T, Lagercrantz H, Lundgren G, Anggard A (1981) Subcellular fractionation of cat submandibular gland: comparative studies on the distribution of acetylcholine and vasoactive intestinal polypeptide (VIP). Neuroscience 6:1001–1010

Lundberg JM, Hökfelt T (1983) Coexistence of peptides and classical neurotransmitters. TINS 6:325–333

Masson J, Sagne C, Hamon M, El Mestikawy S (1999) Neurotransmitter transporters in the central nervous system. Pharmacol Rev 51:439–464

Meister B, Hökfelt T (1988) Peptide- and transmitter-containing neurons in the mediobasal hypothalamus and their relation to GABAergic systems: possible roles in control of prolactin and growth hormone secretion. Synapse 2:585–605

Meister B, Hökfelt T, Geffard M, Oertel W (1988) Glutamic acid decarboxylase- and gamma-aminobutyric acid-like immunoreactivities in corticotropin-releasing factor-containing parvocellular neurons of the hypothalamic paraventricular nucleus. Neuroendocrinology 48:516–526

Meister B, Hökfelt T, Vale WW, Sawchenko PE, Swanson L, Goldstein M (1986) Coexistence of tyrosine hydroxylase and growth hormone-releasing factor in a subpopulation of tubero-infundibular neurons of the rat. Neuroendocrinology 42:237–247

Melander T, Hökfelt T, Rökaeus Å, Cuello AC, Oertel WH, Verhofstad A, Goldstein M (1986) Coexistence of galanin-like immunoreactivity with catecholamines, 5-hydroxytryptamine, GABA and neuropeptices in the rat CNS. J Neurosci 6:3640–3654

Merighi A (2002) Costorage and coexistence of neuropeptides in the mammalian CNS. Prog Neurobiol 66:161–190

Merighi A, Polak JM, Theodosis DT (1991) Ultrastructural visualization of glutamate and aspartate immunoreactivities in the rat dorsal horn, with special reference to the

co-localization of glutamate, substance P and calcitonin-gene related peptide. Neuroscience 40:67–80
Millhorn DE, Hökfelt T, Seroogy K, Oertel W, Verhofstad AA, Wu JY (1987) Immunohistochemical evidence for colocalization of gamma-aminobutyric acid and serotonin in neurons of the ventral medulla oblongata projecting to the spinal cord. Brain Res 410:179–185
Minson J, Pilowsky P, Llewellyn-Smith I, Kaneko T, Kapoor V, Chalmers J (1991) Glutamate in spinally projecting neurons of the rostral ventral medulla. Brain Res 555:326–331
Mugnaini E, Oertel WH (1985) An atlas of the distribution of GABAergic neurons and terminals in the rat CNS. Part I. In: Björklund, A and Hökfelt T (eds), Handbook of Chemical Neuroanatomy. GABA and Neuropeptides in the CNS, vol 4. Elsevier, Amsterdam, pp 436–608
Nanopoulos D, Maitre M, Belin MF, Aguera M, Pujol JF, Gamrani H, Calas A (1981) Autoradiographic and immunocytochemical evidence for the existence of GABAergic neurons in the nucleus raphe dorsalis—possible existence of neurons containing 5HT and glutamate decarboxylase. Adv Biochem Psychopharmacol 29:519–525
Nicholas AP, Cuello AC, Goldstein M, Hökfelt T (1990) Glutamate-like immunoreactivity in medulla oblongata catecholamine/substance P neurons. NeuroReport 1:235–238
Nicholas AP, Pieribone VA, Arvidsson U, Hökfelt T (1992) Serotonin-, substance P- and glutamate/aspartate-like immunoreactivities in medullo-spinal pathways of rat and primate. Neuroscience 48:545–559
Nusbaum MP, Blitz DM, Swensen AM, Wood D, Marder E (2001) The roles of cotransmission in neural network modulation. Trends Neurosci 24:146–154
Osborne NN (1983) Dale's Principle and Communication Between Neurones. Pergamon Press, Oxford
Osborne NN (1984) Putative neurotransmitters and their coexistence in gastropod mollusks. In: Chan-Palay, V and Palay SL (eds), Coexistence of neuroactive substances in neurons. John Wiley, New York, pp 395–409
Ottersen OP, Davanger S, Storm-Mathisen J (1987) Glycine-like immunoreactivity in the cerebellum of rat and Senegalese baboon, Papio papio: a comparison with the distribution of GABA-like immunoreactivity and with [3H]glycine and [3H]GABA uptake. Exp Brain Res 66:211–221
Pelletier G, Steinbusch HW, Verhofstad AA (1981) Immunoreactive substance P and serotonin present in the same dense-core vesicles. Nature 293:71–72
Pickel VM (1985) General morphological features of peptidergic neurons. In: Hökfelt, T and Björklund A (eds), Handbook of Chemical Anatomy (IV): GABA and neuropeptides in the CNS. Elsevier, Amsterdam, pp 72–92
Potter DD, Matsumoto SG, Landis SC, Sah DWY, Furshpan EJ (1986) Transmitter status in cultured sympathetic principal neurons: plasticity, graded expression and diversity. In: Hökfelt, T et al. (eds), Progress in Brain Research. Coexistence of neuronal messengers: A new principle in chemical transmission. vol 68. Elsevier, Amsterdam, pp 103–120
Renaud LP, Martin JB, Brazeau P (1975) Depressant action of TRH, LH-RH and somatostatin on activity of central neurones. Nature 255:233–235
Richardson KC (1966) Electron microscopic identification of autonomic nerve endings. Nature 210:756
Saito K, Barber R, Wu J, Matsuda T, Roberts E, Vaughn JE (1974) Immunohistochemical localization of glutamate decarboxylase in rat cerebellum. Proc Natl Acad Sci U S A 71:269–273
Schäfer MK, Varoqui H, Defamie N, Weihe E, Erickson JD (2002) Molecular cloning and functional identification of mouse vesicular glutamate transporter 3 and its expression in subsets of novel excitatory neurons. J Biol Chem 277:50734–50748
Sergeyev V, Hökfelt T, Hurd Y (1999) Serotonin and substance P coexist in dorsal raphe neurons of the human brain. NeuroReport 10:3967–3970

Sossin WS, Sweet-Cordero A, Scheller RH (1990) Dale's hypothesis revisited: different neuropeptides derived from a common prohormone are targeted to different processes. Proc Natl Acad Sci U S A 87:4845–4848
Storm-Mathisen J, Leknes AK, Bore AT, Vaaland JL, Edminson P, Haug FM, Ottersen OP (1983) First visualization of glutamate and GABA in neurones by immunocytochemistry. Nature 301:517–520
Su C, Bevan JA, Burnstock G (1971) [3H]adenosine triphosphate: release during stimulation of enteric nerves. Science 173:336–338
Tatemoto K, Rökaeus Å, Jörnvall H, McDonald TJ, Mutt V (1983) Galanin- a novel biologically active peptide from porcine intestine. FEBS 164:124–128
Westfall DP, Stitzel RE, Rowe JN (1978) The postjunctional effects and neural release of purine compounds in the guinea-pig vas deferens. Eur J Pharmacol 50:27–38
Wu JY, Matsuda T, Roberts E (1973) Purification and characterization of glutamate decarboxylase from mouse brain. J Biol Chem 248:3029–3034

Chapter 2
Ex uno plures: Out of One, Many

R. Gutiérrez

Abstract Ex uno plures, out of one (cell) many (neurotransmitters), seems to be a principle that applies to many, if not all, neuronal types. The co-release of signaling molecules has been long recognized and the terms "classical neuro-transmitter" and "neuromodulator" have been used to label the co-released substances, often being the former of low molecular weight and the latter of high molecular weight. Indeed, the use of these terms confers a distinctive function for each substance. However, the co-release of two or more low weight, fast-acting "classical neurotransmitters" is until recently subject of intense investigation. Initially, the co-existence of classical neurotransmitters in a given cell or its terminals was a curious observation, and the possibility of they being released was not directly approached as it contradicted a dogma: "one cell, one neurotransmitter". Presently, the co-existence and co-release of classical neurotransmitters is known to occur in different animal species and neuronal systems, from invertebrates to human. Moreover, the specification of the neurotransmitter phenotype of neurons has been shown to be plastic. In some cases this plasticity follows a developmental program and, in others, it depends on activity-dependent and even on pathological processes. Therefore, the listener cell should already have the receptors in the postsynaptic site or should actively put them in place to interpret a compound message, carried by two or more neurotransmitters, and integrate it to display a response. The time-locked release and thus, the action of two or more classical neurotransmitter provide the central nervous system with a powerful communication and computational tool.

It is well accepted that neurons can release both a "classical" neurotransmitter together with a modulatory transmitter. However, the recognition that they contain and co-release two or more "classical" neurotransmitters is a new

R. Gutiérrez (✉)
Department of Physiology, Biophysics and Neurosciences, Center for Research and Advanced Studies of the National Polytechnic Institute, Post Box 14-740, México D.F. 07000
e-mail: grafael@fisio.cinvestav.mx

R. Gutierrez (ed.), *Co-Existence and Co-Release of Classical Neurotransmitters*, 15
DOI 10.1007/978-0-387-09622-3_2, © Springer Science+Business Media, LLC 2009

avenue in the study of neurotransmission that has started recently to be explored. Colocalization of classical neurotransmitters within single terminals was initially perceived as a curiosity and their potential co-release was viewed with skepticism. The molecular explanation, as well as the physiological and physiopathological relevance of the colocalization and co-release of classical neurotransmitters has recently been the subject of intense research. Indeed, many examples of coexistence and co-release of classical transmitters have been described in invertebrate and vertebrate organisms, and within the mammalian central and peripheral nervous system.

The idea that two or more classic neurotransmitters might coexist in an individual neuron and hence, be co-released by it was not derived from a direct experimental approach. Perhaps with good reason, why would a neuron have and use two chemical neurotransmitters? Indeed, why would it convey more than one message? Unfortunately, the hypothesis of co-release was also often considered as heretical because it contradicted the commonly understood idea behind Dale's postulate that "a single cell releases only one neurotransmitter." Serendipitous observations of neurotransmitter colocalization, as well as of their synthetic enzymes and vesicular transporters, as well as the observation that postsynaptic responses did not match the supposed activity of the fibers stimulated, led researchers to seriously test the co-release hypothesis. Although still questioned in some cases, the overwhelming evidence in favor of this phenomenon has opened new possibilities to understand neural communication and in particular, to address synaptic physiology.

Moreover, it now appears that the neurotransmitter phenotype of neurons may be very plastic. Evidence of activity-dependent plasticity of neurotransmitter phenotype, even after brain insult, has drawn the attention of many investigators to the possibility of finding more cells that express multiple neurotransmitters in response to environmental changes. Specification of neurotransmitter phenotype is a process that initially takes place during development. It is a mechanism by which the genetic program and the environmental signals received by a neuron fine tunes the expression of a series of proteins, to define a given neurotransmitter phenotype. The silencing and the turning on of positive or negative signals are complementary and active processes during development. However, while some genes are definitively turned off by the end of development, others seem to remain latent and can be turned on later in life upon specific demand (e.g., increase or decrease of electrical activity, the action of trophic factors, hormones, etc.). This type of plasticity provides the nervous system with a powerful communicational tool.

2.1 What Is a Classical Neurotransmitter?

A "classical neurotransmitter" has been defined as a chemical substance that (1) is synthesized in the cell (indicating that the synthetic machinery has to be present in the cell); (2) is present in the presynaptic terminal and is released from a specific zone during activity; (3) acts directly on receptors that are present in the postsynaptic cell, altering their activity; (4) if applied exogenously, it mimics the effects of the endogenously released substance by activating the same receptors/channels; and (5) is removed from the extracellular space by defined catalytic or transport mechanisms. Classical neurotransmitter substances are: acetylcholine (ACh); the biogenic amines dopamine (DA), epinephrin (E) and norepinephrine (NE); serotonin (5-HT); histamine (H); and the amino acids glycine (gly), γ-aminobutyric acid (GABA), glutamate (Glu) and aspartate (asp). Some of these classical transmitters activate receptors directly coupled to ion channels, while others activate metabotropic receptors that initiate intracellular signaling cascades that produce the opening or closing of ion channels. Indeed, some transmitters may activate both types of receptors. Accordingly, it is the receptor and not the transmitter that determines whether the action of the released substance will be excitatory or inhibitory.

2.2 What Is a Modulatory Transmitter?

In contrast, other substances that are also released from nerve cells, not necessarily from an active zone where the classical transmitters are released, may exert a more diffuse effect and modulate the signal that the primary ("classical") neurotransmitter conveys. In general, these substances activate intracellular signaling cascades that modify the action of the primary transmitter. Therefore, the usual concept of "co-release" most often implies the simultaneous release of a "classical transmitter" (usually of low molecular weight) and a "modulator" (usually of high molecular weight), which seems to be a general phenomenon in neuronal cells. Among the modulatory substances are peptides, nucleotide (e.g., ATP), Zn^{2+}, neurotrophic factors, nitric oxide and endogenous cannabinoids. In all these cases the co-released factors are synthesized in the cell (except for Zn^{2+}), they are released in a selectively regulated fashion, they act on receptors, and they are removed or inactivated by specific mechanisms. Classical transmitters may also have modulatory roles acting both postsynaptically and/or presynaptically.

2.3 Dale's Principle

Based on the ideas from the studies of Henry Dale on cholinergic and adrenergic neurons in the spinal cord in the early 1930s, John Eccles formulated a functional principle, which was then further transformed to imply that "a single cell releases only one neurotransmitter."

In fact, what Dale stated was that a neuron functions as a metabolic unity, whereby a single process in a cell can influence all the compartments of the same neuron. From this, Eccles claimed that all the terminals of a given neuron should release the same neurotransmitter. However, in the light of our current knowledge about the coexistence of transmitter substances, a more correct interpretation of Dale's principle would be: "A neuron *normally* releases the same chemical messengers from all of its synapses." As new evidence for the segregation and compartmentalization in neurotransmission has been disclosed, this can be further adapted to suggest "that the same chemical messengers *could* be released from all the terminals of a given neuron."

After the discovery of the coexistence of classical and modulatory transmitters in single neurons and their co-release, it was thought that co-transmission only permitted this combination of transmitter substances. But what about the coexistence and co-release of two or more classical neurotransmitters?

2.4 Coexistence and Co-release of Classical Neurotransmitters

The transformed Dale's principle that "a single cell releases only one neurotransmitter" has influenced many neuroscientists to consider with skepticism the idea that two or more classical neurotransmitters could coexist and hence, be co-released by neurons.

The coexistence and co-release of two or more classical neurotransmitters, each conveying a "principal message," has therefore been studied less extensively than that of classical transmitters and peptide modulators (see Hökfelt in this volume). Indeed, in some ways the initial descriptions of these events still remain a curiosity. However, during recent years, ample evidence has shown that co-release is in fact not that uncommon, and it is now known to occur in a variety of neural systems. In addition, although many populations of adult neurons may not release two classical neurotransmitters under basal conditions, many appear to do so transiently following an established program during early development, or in response to a variety of physiological and pathophysiological stimuli.

A fundamental question in neuroscience is how the remarkable cellular diversity is established during development. The influence of external signals and transcription factors determines when the distinct types of neurons are formed at specific sites. It is equally important to unravel how the specification of their morphological and functional traits takes place, particularly how neurotransmitter phenotypes are specified. The expression of the neurotransmitter phenotype of a given neuron follows a specific program, whereby transcription factors exert a strong regulatory role. However, electrical activity modulated by external signals seems to trigger the expression or repression of a given phenotype by virtue of calcium entering the cell. The only way through which the environmental demands can affect this expression is to activate the existing capability of the neuron to express a given phenotype. Thus, a neuron can possess a latent or potential phenotype, which can be activated during development or in adult life. This also implies that a cell must possess a message that restricts its expression at a given point in time. The effect of its activity and thus, of calcium influx on transmitter selection suggests that the number of neurons expressing a particular transmitter may be regulated so as to maintain a steady level of excitability in the nervous system (see Borodinsky and Spitzer, this volume; see also Spitzer et al., 2005).

2.5 Colocalization of Receptors for Classical Neurotransmitters

In order for a neuron to respond to a given chemical signal, it must have the appropriate receptors at the correct site. Thus, there must be adequate apposition of the presynaptic and postsynaptic elements. In other words, the neurotransmitter released by the presynaptic terminal must precisely match the postsynaptic receptors for communication to occur across the synapse. This implies that when co-release of classical neurotransmitters takes place, the receptors for the different neurotransmitters must be colocalized in the postsynaptic element, specifically in the subsynaptic zone apposing the site of release. Another possibility is that the receptors for one neurotransmitter are in the subsynaptic zone, while the receptors for the other neurotransmitter are in the perisynaptic or extrajunctional membrane. Such distributions would produce different types of postsynaptic effects. Moreover, the phenotypic plasticity of the presynaptic neuron must be paralleled by plastic changes of the postsynaptic neuron, probably by triggering receptor motility of the matching receptor in the vicinity of the release site. Indeed, receptor selection during development parallels changes in neurotransmitter phenotype (i.e., activity regulates the matching of transmitters and their receptors in the assembly of functional synapses). On the other hand, clusters of different neurotransmitter receptors have been identified, albeit in culture preparations, apposing terminals that apparently release only one type of neurotransmitter. Whether changes in the presynaptic neuron are triggered to make use of these mismatched receptors is still to be determined. Therefore, it would be also important to determine whether a constitutive, redundant co-expression of receptors is a general phenomenon.

Not only do the postsynaptic neurons contain the receptors to the co-released neurotransmitters but also, the terminals or axons that co-release the neurotransmitters can possess receptors that could be readily activated, producing presynaptic auto-modulation or collateral presynaptic actions. This mechanism seems especially relevant for activity-dependent plasticity and can be of marked functional significance in situations of enhanced excitability.

2.6 Consequences and Functional Advantages of Classical Neurotransmitter Co-release

The site of synthesis within the neuron, the rate of synthesis and finally, the type of activity needed to release small-molecule neurotransmitters and peptides differ. While small-molecule neurotransmitters can be synthesized locally in the nerve terminals where they are to be released, peptides are synthesized in the soma and then transported to the terminals for release. In this regard, the rate at which the chemical substances are produced and prepared for release can produce marked functional differences. Another difference between the release of peptides and "classical" neurotransmitters is that peptides are only released by repetitive action potentials, whereas classical transmitters can be released by single action potentials, providing them with distinct functional characteristics. Thus, most modulators are released in an activity-dependent manner.

This difference underlies the effectiveness of co-released classical neurotransmitters in producing fast signaling and therefore, fast modulatory interactions. In some cases, these interactions are synergistic, for instance, if two inhibitory or two excitatory transmitters are co-released. However, in other cases, two neurotransmitters may exert opposing effects. Of course, the specific receptors have to be in the right postsynaptic site for this interaction of the neurotransmitters to occur. Therefore, because the receptors determine the way in which a cell responds to its inputs, the coexistence of receptors to different neurotransmitters apposed to presynaptic terminals expands the possible responses of the neuron to a variety of signals. In this way fast and efficient modulation of incoming signals can be established at very restricted membrane sites. Therefore, neuronal activity can lead to a high concentration of (classical) neurotransmitters in a very narrow time-window. Accordingly, neurotransmitter-mediated modulation could occur more rapidly than peptide-to-neurotransmitter mediated modulation, and it may take place right at the synapse involved, spatially restricting the modulation. The last chapter of this book will deal with the integration of two or more signals conveyed by classical neurotransmitters: E pluribus unum, out of many (signals), one (response).

2.7 What Do We Still Need to Know?

The response is simple, more than we already know.

A list of questions might be helpful in illustrating the aspects related to the coexistence and co-release of classical neurotransmitters that need to be investigated. This might include questions such as:

How are two or more neurotransmitters synthesized and packaged within single boutons and in some cases, within single vesicles? For example, glutamic acid is the precursor of GABA and although essentially all GABAergic cells must contain glutamate for its conversion to GABA, they do not release glutamate. If a cell were to have and to release both amino acids, how does it control the passage of glutamate to GABA without running out of glutamate to be used directly? Indeed, if glutamate is used for GABA synthesis, is there a compensatory mechanism to replenish the glutamate needed for release?

This example considers two amino acids that are closely related in metabolic terms. But in neurons containing other transmitters that are not that related, how does the cell cope with the synthetic machinery for two neurotransmitters? Are the vesicles segregated? How is "access" to the releasable and reserve pools organized? How are vesicles distributed and docked in the membrane, and then secreted from the presynaptic terminal? How is the trafficking of vesicles destined to package different neurotransmitters? Are they tagged?

How does the secretory mechanism work when there are two or more neurotransmitters to be released? Are there different release sites for each neurotransmitter? Is the release machinery the same for the different populations of vesicles within single boutons? Is the release machinery (considering the membrane scaffold proteins that anchor the vesicles to the membrane) of one family of vesicles the same as that for a second family? Can the release of either neurotransmitter be differentially controlled? Are the kinetics the same for the release of all the neurotransmitters a neuron contains? What mechanism controls the release of one neurotransmitter at a given time but not at another?

How do the transporters for different neurotransmitters work? How do vesicles "select" the neurotransmitter transporters (or vice versa) that will enable them to capture certain neurotransmitters?

Are all neurotransmitters within a cell released from all terminals? Does segregation occur and if so, how is segregation determined? Is there a postsynaptic retrograde signal (NO, cannabinoids, trophic factors, others) that controls release in a differential manner? Does the postsynaptic target cell have a say in all this? Are the plastic properties the same for all substances released by the neuron?

How is the expression of the neurotransmitter phenotype controlled? Does the same intracellular signal control the expression of a given phenotype and the suppression of the other?

Although some of these questions are currently being addressed by different research groups, the "initial", more basic questions remain on the table, i.e., which are the neurotransmitters that are co-released? Where from? How? What do their signals do and how are they integrated? etc.

The problem we are dealing with has been clearly presented in this chapter: *Ex uno, plures*, from one (cell) many (neurotransmitters). The chapters that follow will review the specific issues surrounding the coexistence and co-release of these neurotransmitters. Finally, the last chapter will focus on what we are looking for: the meaning of co-expression and co-release of classical neurotransmitters, i.e., *E pluribus unum*, out of many (signals), one (response).

We are sure that this volume will grow rapidly and we will make efforts to update it as new and exciting advances appear.

Acknowledgments I thank Dr. Peter Somogyi for critical review of this chapter. Supported by the Consejo Nacional de Ciencia y Tecnología.

References

Eccles JC (1976) From electrical to chemical transmission in the central nervous system. The closing address of the Sir Henry Dale Centennial Symposium. Notes Rec R Soc London 30:219–230

Dale H (1935) Pharmacology and nerve-endings. Proc R Soc Med London 28:319–332

Spitzer NC, Borodinsky LN, Root CM (2005) Homeostatic activity-dependent paradigm for neurotransmitter specification. Cell Calcium 37:417–423

Chapter 3
Mechanisms of Synapse Formation: Activity-Dependent Selection of Neurotransmitters and Receptors

Laura N. Borodinsky and Nicholas C. Spitzer

Abstract Specification of the neurotransmitters and receptors involved in a synapse is a key feature of development of the nervous system. Multiple mechanisms govern these aspects of neuronal differentiation. Among them electrical activity appears to be an important factor, engaging the functionality of the nervous system right from the beginning of its formation. In the developing *Xenopus* neuromuscular junction, the specification of neurotransmitters and receptors responds to a selection process that depends on calcium-mediated neuronal activity. This mechanism may provide a level of plasticity that is responsive to changes in environmental cues at very early stages of development.

3.1 Introduction

Chemical synapses represent the main form of communication in the nervous system. They require the delicate apposition of presynaptic structures containing neurotransmitter-filled vesicles and neurotransmitter receptors exposed in postsynaptic cell membranes. Formation of synapses occurs during development when presynaptic axons are guided to target postsynaptic cells. Although this process is already complex, if we add to this description the fact that more than fifty different neurotransmitters and multiple neurotransmitter receptor types for each of them can be expressed by cells at the synapse, the level of complexity acquires a new dimension. The process requires both the correct targeting of innervating axons to appropriate cells and the precise matching between neurotransmitter/s and receptors on each side of the synapse. Moreover, in contrast to the idea that neurons express a single classical neurotransmitter, recent work suggests that this may be the exception rather than the rule.

L.N. Borodinsky (✉)
Department of Physiology & Membrane Biology, University of California Davis School of Medicine and Institute for Pediatric Regenerative Medicine, Shriners Hospital for Children Northern California, 2425 Stockton Blvd, Sacramento, California 95817
e-mail: lnborodinsky@ucdavis.edu

R. Gutierrez (ed.), *Co-Existence and Co-Release of Classical Neurotransmitters*,
DOI 10.1007/978-0-387-09622-3_3,

Multiple neurotransmitters can be coexpressed and released from single presynaptic neurons.

In this chapter we review how neurotransmitters and their receptors are specified in cells and how the specification of one side of the synapse could influence the other. We propose that neuronal activity is crucial in neurotransmitter-receptor matching.

3.2 Mechanisms of Neurotransmitter Specification

The mechanisms by which neurons make the choice of the neurotransmitter/s they will express are not completely understood. Different lines of study have identified several factors necessary in this process.

Transcription factors are direct regulators of the expression of specific neurotransmitter phenotypes in many structures and for different neuronal populations (Goridis and Brunet, 1999). A combinatorial code of transcription factors expression allows the differentiation of distinct neuronal phenotypes along the dorso-ventral axis of the developing spinal cord (Briscoe et al., 2000). For instance, Lbx1 and Tlx3 are opposing switches in determining glutamatergic over GABAergic phenotypes in mouse dorsal horn neurons (Cheng et al., 2005). What controls transcription factor expression? Gradients of morphogenetic proteins such as Sonic Hedgehog, bone morphogenetic proteins (BMPs) and Wnts are generated from dorsal and ventral secretory centers established early during development. These gradients establish a positional code of transcription factor expression that shapes differentiation of spinal neurons (Ericson et al., 1996; Ericson et al., 1997). In addition, neurotrophins and cytokines are also regulators of neurotransmitter phenotype (Habecker et al., 1997; Daadi and Weiss, 1999; Francis and Landis, 1999; Zweifel et al., 2005), likely through the regulation of gene expression as well.

Neuronal activity has been shown to be relevant for transmitter specification in developing and mature nervous system structures in different species. It is not restricted to a particular neurotransmitter phenotype. The influence of electrical activity in neurotransmitter specification applies to GABA (Hendry and Jones, 1986, 1988; Akbarian et al., 1995; Gu and Spitzer, 1995; Watt et al., 2000; Ramirez and Gutierrez, 2001; Gutierrez, 2002; Ciccolini et al., 2003; Gutierrez et al., 2003; Patz et al., 2003; Borodinsky et al., 2004; Rosselet et al., 2006), acetylcholine (Walicke et al., 1977; Walicke and Patterson, 1981; Belousov et al., 2002; Borodinsky et al., 2004; Slonimsky et al., 2006), glycine (Borodinsky et al., 2004), glutamate (Carder and Hendry, 1994; Gutierrez, 2000, 2002; Gutierrez et al., 2003; Borodinsky et al., 2004), dopamine (Baker, 1990; Hertzberg et al., 1995; Brosenitsch and Katz, 2001, 2002), norepinephrine (Walicke and Patterson, 1981; Slonimsky et al., 2006), somatostatin (Marty and Onteniente, 1997), pitutitary adenylate cyclase activating-polypeptide (Fukuchi et al., 2004), proenkephalin (Hahm et al., 2003), vasoactive intestinal peptide (Agoston et al., 1991b), enkephalin (Agoston et al., 1991a) and substance P (Tu and Debski, 1999;

Table 3.1 Activity-dependent neurotransmitter expression

Tissue / Species	Neurotransmitter	Citations
	Developing Nervous System	
Spinal cord / Xenopus laevis	GABA, acetylcholine, glycine, glutamate	Gu & Spitzer, 1995; Watt et al., 2000; Borodinsky et al., 2004
Neural precursors / mice	GABA	Ciccolini et al., 2003
Visual cortex / rat	GABA	Patz et al., 2003
Sympathetic ganglion / rat	acetylcholine, norepinephrine	Walicke et al., 1977; Walicke & Patterson, 1981
Hypothalamus / rat	acetylcholine	Belousov et al., 2002
Sensory ganglion / rat	dopamine	Hertzberg et al., 1995; Brosenitsch & Katz, 2001, 2002
Hippocampus / rat	somatostatin	Marty & Onteniente, 1997
Optic tectum / Rana pipiens	substance P	Tu & Debski, 1999
Superior cervical ganglion neurons / rat	acetylcholine, norepinephrine	Slonimsky et al., 2006
Cerebral cortex / rat	pituitary adenylate cyclase activating-polypeptide	Fukuchi et al., 2004
Spinal cord / mice	enkephalin	Agoston et al., 1991a
	Mature Nervous System	
Visual cortex / monkey	GABA	Hendry & Jones, 1986, 1988
Somatosensory cortex / rat	GABA	Rosselet et al., 2006
Prefrontal cortex / human	GABA	Akbarian et al., 1995
Optic tectum / Rana Pipiens	substance P	Tu & Debski, 2000
Hippocampus / rat	GABA, glutamate	Ramirez & Gutierrez, 2001, Gutierrez 2002; Gutierrez et al., 2003
Striate cortex / monkey	glutamate	Carder & Hendry, 1994
Olfactory bulb / rat	dopamine	Baker, 1990

Tu et al., 2000) (Table 3.1). In many cases, the activity-dependent regulation of neurotransmitter and neuropeptide expression follows a homeostatic paradigm. This means that expression of excitatory neuroactive molecules is enhanced when activity is suppressed and decreased when activity is enhanced, and vice versa for inhibitory neuroactive molecules (Spitzer et al., 2005). In particular, we found that in developing *Xenopus* spinal neurons Ca-dependent patterns of activity control specification of GABA, glycine, glutamate and acetylcholine. In addition, perturbations of spontaneous patterns of Ca spike activity lead to co-expression of more than one neurotransmitter (Borodinsky et al., 2004).

3.3 Mechanisms of Neurotransmitter Receptor Specification

Functional expression of neurotransmitter receptors involves not only transcription and translation of the appropriate genes and transcripts but also the localization of the assembled receptors in the cell membrane. All these processes have been shown to depend on electrical activity for many different receptor subunits in diverse postsynaptic cells. $GABA_A$ receptor subunit expression is regulated by levels of activity in specific layers of the monkey visual cortex (Huntsman et al., 1994) and in rat hippocampus (Marty et al., 2004). $GABA_B$ receptor expression in the monkey thalamus dorsal lateral geniculate is decreased after monocular deprivation (Munoz et al., 1998). Surface expression of NMDAR is controlled by activity-dependent phosphorylation in cultured rat cortical neurons (Chung et al., 2004). NMDAR subunit composition varies with activity-dependent induction of calcineurin in superficial visual layers of the rat superior colliculus (Shi et al., 2000). AMPAR expression decreases when synaptic levels of activity increase in cultured hippocampal neurons (Lissin et al., 1998). Even metabotropic glutamate receptor surface expression is regulated by neuronal activity (Ango et al., 2000). Many studies have analyzed activity-dependent regulation of acetylcholine receptors, starting with investigations of muscle receptor expression, distribution and localization (Merlie and Sanes, 1985; Goldman et al., 1988; Goldman and Staple, 1989; Bessereau et al., 1994), and more recently their expression in neurons (Levey et al., 1995; Brumwell et al., 2002) (Table 3.2).

Table 3.2 Activity-dependent neurotransmitter receptor expression

Tissue / Species	Neurotransmitter receptor	Citations
	Developing Nervous System	
Hippocampus / rat	$GABA_AR$: $\alpha1$, $\alpha2$	Marty et al., 2004
Cortical neurons / rat	NMDAR: NR2B	Chung et al., 2004
Superior colliculus / rat	NMDAR: NR1-NR2B, NR1-NR2A	Shi et al., 2000
Hippocampus / rat	AMPAR: GluR1	Lissin et al., 1998
Cerebellar granule cells / mouse	Group I mGluR: mGluR5a	Ango et al., 2000
Ciliary ganglion neurons / chicken	nAChR: $\alpha3$, $\alpha5$, $\beta4$	Levey et al., 1995; Brumwell et al., 2002
Skeletal muscle / Xenopus laevis	nAChR; NMDAR: NR1; AMPAR: GluR1; $GABA_AR$; GlyR: $\alpha1$	Borodinsky & Spitzer, 2007
	Mature Nervous System	
Visual cortex / monkey	$GABA_AR$: $\alpha1$, $\alpha5$, $\beta2$, $\gamma2$	Huntsman et al., 1994
Thalamus / monkey	$GABA_BR$	Muñoz et al., 1998
Skeletal muscle / mouse	nAChR: α, δ	Merlie & Sanes, 1985
Skeletal muscle / rat	nAChR: α, β, γ, δ	Goldman et al., 1988; Goldman & Staple, 1989
Skeletal muscle / chicken	nAChR: $\alpha1$	Bessereau et al., 1994

On the other hand, neurotransmitter receptor specification can be driven by intrinsic genetically predetermined programs. Expression of the $GABA_AR$ alpha 6 subunit in cerebellar granule cells seems to be governed by a specific and intrinsic mechanism that is independent of the environment in which these cells are grown (Bahn et al., 1999; Wang et al., 2004). However, electrical activity, brain-derived neurotrophic factor and cAMP all affect $\alpha 6$ expression level (Thompson et al., 1996; Ghose et al., 1997; Mellor et al., 1998; Thompson et al., 1998). In addition, many studies have shown that neurotransmitter receptors are already expressed prior to innervation. For instance, mRNA for alpha, delta and gamma skeletal muscle acetylcholine receptor subunits start to be expressed in *Xenopus* somites at late gastrula stages, long before synaptic connections are formed (Baldwin et al., 1988).

How can these two pathways for neurotransmitter receptor expression be reconciled?

In a recent study we found that embryonic muscle cells express a surprising variety of neurotransmitter receptors: nAChR, AMPAR, NMDAR, $GABA_AR$ and GlyR. Detection of mRNA and immunohistochemistry for subunits of these receptors in developing myotomes, in addition to neurotransmitter-induced responses in isolated embryonic muscle cells, support the idea that *Xenopus* embryonic skeletal muscle is receptive not only to ACh but also to glutamate, GABA and glycine. As maturation and innervation progress, the cholinergic phenotype prevails (Borodinsky and Spitzer, 2007).

The scenario implied by these results is one in which "default" mechanisms trigger the expression of different types of neurotransmitter receptors early during development. The receptors that respond to specific neurotransmitters released by the innervating axons remain, while the others disappear.

3.4 Matching of Neurotransmitters and Their Receptors: Perfect Encounter or a Selection Process?

In studying neurotransmitter-receptor matching at the frog neuromuscular junction during development, we found that changing the level of activity changed the neurotransmitters and transmitter receptors functionally expressed at this synapse. When early neuronal calcium-dependent activity was suppressed, a condition that increases the number of glutamatergic and cholinergic spinal neurons, glutamatergic along with cholinergic synapses were established in the skeletal muscle. Conversely, when early neuronal calcium-dependent activity was enhanced a condition that increases the number of GABAergic and glycinergic spinal neurons, inhibitory synapses as well as cholinergic synapses were detected on the skeletal muscle (Borodinsky and Spitzer, 2007).

This activity-dependent matching of pre- and postsynaptic players in the developing neuromuscular junction resembles the process of synapse elimination during development (Purves and Lichtman, 1980). It is an activity-dependent event in which only synapses that are used and reinforced become stabilized and the others are eliminated (Nguyen and Lichtman, 1996).

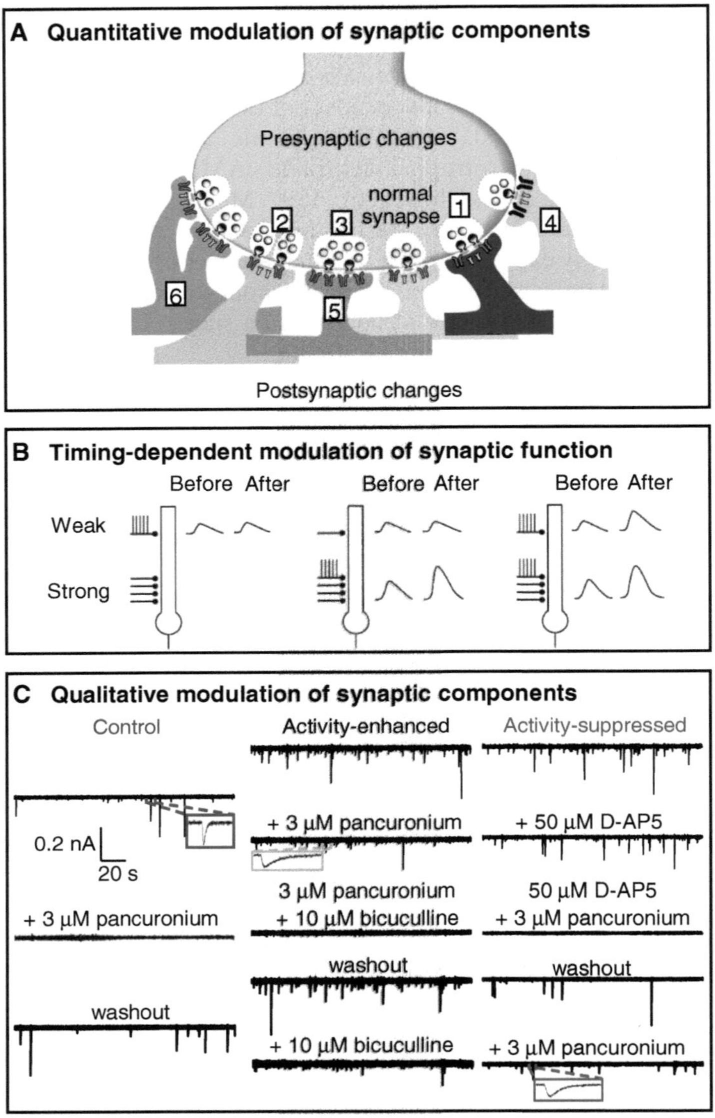

Fig. 3.1 Different levels of modulation of synaptic activity. (**a**) Synaptic activity can be enhanced by presynaptic (1–3), postsynaptic (4–5) or coordinated (6) changes: 1. Increase in release probability; 2. Increase in the number of release sites; 3. Increase in the number of vesicles available for release; 4. Increase in the sensitivity of preexisting neurotransmitter receptors; 5. Increase in the number of functional receptors; 6. Growth of new synaptic contacts. (**b**) Synchronization of stimuli at different synapses can cause changes in synaptic activity. Left, a tetanic stimulation of the weak input on a pyramidal cell does not cause long-term potentiation (LTP) in the pathway (same EPSPs); middle, tetanic stimulation of the strong input alone causes LTP in the strong pathway but not in the weak one; right, tetanic stimulation of both inputs together causes LTP in both pathways. (**c**) Perturbation of electrical

What are the advantages, if any, of having a pool of different neurotransmitter receptors expressed prior to innervation, followed by the selection of the one that matches the transmitter of the innervating axon? Wouldn't it be more "cost-effective" to synthesize the putative neurotransmitter and neurotransmitter receptor from the beginning?

Perhaps the selection of this mechanism was not based on economics but on plasticity. Having a palette of neurotransmitter receptors to choose from allows the system to respond quickly to changes in levels of activity and consequent changes in neurotransmitters released by innervating axons. It ensures formation of functional connections that serve to respond to the underlying electrical activity of the developing nervous system. Genetic hardwiring of this aspect of circuit assembly would be less responsive to changes in the environment.

Matching of neurotransmitters and receptors has also been studied from a more structural point of view. What are the molecules necessary to establish inhibitory and excitatory synapses? The neurexins and their ligands, the neuroligins (NLs), form a cell adhesion complex and promote the initial establishment of synapses (Cantallops and Cline, 2000; Levinson and El-Husseini, 2005). However, whether different types of NLs regulate specifically the establishment of excitatory versus inhibitory synapses has been controversial (Song et al., 1999; Graf et al., 2004; Prange et al., 2004; Varoqueaux et al., 2004; Chih et al., 2005; Levinson et al., 2005; Nam and Chen, 2005; Chih et al., 2006). A detailed recent analysis attempting to solve this controversy shows that NL1 and NL2 specifically increase the number of excitatory and inhibitory synapses, respectively. Interestingly, the effects of both NLs are dependent on synaptic activity, indicating that neuroligins do not establish but specify and validate synapses by an activity-dependent mechanism (Chubykin et al., 2007).

3.5 Qualitative Changes in Transmission and Cotransmission of Classical Neurotransmitters: What for?

Several distinct mechanisms of modulation of synaptic activity are now recognized (Fig. 3.1). The first involves quantitative changes in the levels of different synaptic components. Specific synapses need to be strengthened or weakened.

Fig. 3.1 (continued) activity induces co-transmission and homeostatic changes in types of neurotransmitter and receptors involved in the synapse. Recordings of skeletal muscle miniature postsynaptic currents (mpscs) are shown for control (left), activity-enhanced (middle) and activity-suppressed (right) *Xenopus* larvae. While in control embryos mpscs are completely blocked by a nicotinic acetylcholine receptor antagonist (pancuronium), in activity-suppressed and –enhanced embryos NMDAR- (D-AP5-sensitive) and $GABA_AR$-mediated currents (bicuculline-sensitive) can be recorded along with cholinergic mpscs (pancuronium-sensitive). (a) *Adapted from Wang et al., 1997 and Fundamental Neuroscience*, 2nd edition. (b) *Adapted from Nicoll et al., 1988*. (c) *Adapted from Borodinsky and Spitzer, 2007*. (*See* Color Plate 1)

To respond to this demand, neurons quantitatively change presynaptic neurotransmitter release and postsynaptic neurotransmitter receptor expression and distribution (Wang et al., 1997; Nicoll and Schmitz, 2005). A second mechanism involves timing of synaptic activity. Synchronized synapses can encode a completely different type of information than unsynchronized synapses. Plastic events such as LTP and LTD are generated only when timely concurrent synaptic inputs coincide in a given neuron (Nicoll et al., 1988; Malenka and Bear, 2004). Moreover, synaptic efficacy can change depending on the sequence and timing of stimuli in a phenomenon known as spike timing-dependent plasticity (Dan and Poo, 2004). The third mechanism has been appreciated more recently, through work reviewed in this chapter, and entails qualitative changes in synaptic components. This form of modulation may arise when the synapse must change the qualitative response triggered in the postsynaptic cell. For instance, the target cell may need to be inhibited, instead of excited, or the kinetics and signaling pathways in the postsynaptic cell may need to be different in order to respond appropriately. This third level of modulation could be exerted by cotransmission of neurotransmitters with different modes of action. Our recent work is supportive of this scenario. When electrical activity is enhanced, cotransmission of excitatory and inhibitory neurotransmitters becomes apparent. When electrical activity is suppressed cotransmission of acetylcholine and glutamate can be recorded from a single muscle cell (Borodinsky and Spitzer, 2007). This form of synaptic plasticity is manifest at many synapses in the developing nervous system. It will be of interest to learn the extent to which it operates in the adult.

References

Agoston DV, Eiden LE, Brenneman DE (1991a) Calcium-dependent regulation of the enkephalin phenotype by neuronal activity during early ontogeny. J Neurosci Res 28:140–148

Agoston DV, Eiden LE, Brenneman DE, Gozes I (1991b) Spontaneous electrical activity regulates vasoactive intestinal peptide expression in dissociated spinal cord cell cultures. Brain Res Mol Brain Res 10:235–240

Akbarian S, Kim JJ, Potkin SG, Hagman JO, Tafazzoli A, Bunney WE, Jr., Jones EG (1995) Gene expression for glutamic acid decarboxylase is reduced without loss of neurons in prefrontal cortex of schizophrenics. Arch Gen Psychiatry 52:258–266

Ango F, Pin JP, Tu JC, Xiao B, Worley PF, Bockaert J, Fagni L (2000) Dendritic and axonal targeting of type 5 metabotropic glutamate receptor is regulated by homer1 proteins and neuronal excitation. J Neurosci 20:8710–8716

Bahn S, Wisden W, Dunnett SB, Svendsen C (1999) The intrinsic specification of gamma-aminobutyric acid type A receptor alpha6 subunit gene expression in cerebellar granule cells. Eur J Neurosci 11:2194–2198

Baker H (1990) Unilateral, neonatal olfactory deprivation alters tyrosine hydroxylase expression but not aromatic amino acid decarboxylase or GABA immunoreactivity. Neuroscience 36:761–771

Baldwin TJ, Yoshihara CM, Blackmer K, Kintner CR, Burden SJ (1988) Regulation of acetylcholine receptor transcript expression during development in Xenopus laevis. J Cell Biol 106:469–478

Belousov AB, Hunt ND, Raju RP, Denisova JV (2002) Calcium-dependent regulation of cholinergic cell phenotype in the hypothalamus in vitro. J Neurophysiol 88:1352–1362
Bessereau JL, Stratford-Perricaudet LD, Piette J, Le Poupon C, Changeux JP (1994) In vivo and in vitro analysis of electrical activity-dependent expression of muscle acetylcholine receptor genes using adenovirus. Proc Natl Acad Sci U S A 91:1304–1308
Borodinsky LN, Spitzer NC (2007) Activity-dependent neurotransmitter-receptor matching at the neuromuscular junction. Proc Natl Acad Sci U S A 104:335–340
Borodinsky LN, Root CM, Cronin JA, Sann SB, Gu X, Spitzer NC (2004) Activity-dependent homeostatic specification of transmitter expression in embryonic neurons. Nature 429:523–530
Briscoe J, Pierani A, Jessell TM, Ericson J (2000) A homeodomain protein code specifies progenitor cell identity and neuronal fate in the ventral neural tube. Cell 101:435–445
Brosenitsch TA, Katz DM (2001) Physiological patterns of electrical stimulation can induce neuronal gene expression by activating N-type calcium channels. J Neurosci 21:2571–2579
Brosenitsch TA, Katz DM (2002) Expression of Phox2 transcription factors and induction of the dopaminergic phenotype in primary sensory neurons. Mol Cell Neurosci 20:447–457
Brumwell CL, Johnson JL, Jacob MH (2002) Extrasynaptic alpha 7-nicotinic acetylcholine receptor expression in developing neurons is regulated by inputs, targets, and activity. J Neurosci 22:8101–8109
Cantallops I, Cline HT (2000) Synapse formation: if it looks like a duck and quacks like a duck. Curr Biol 10:R620–623
Carder RK, Hendry SH (1994) Neuronal characterization, compartmental distribution, and activity-dependent regulation of glutamate immunoreactivity in adult monkey striate cortex. J Neurosci 14:242–262
Cheng L, Samad OA, Xu Y, Mizuguchi R, Luo P, Shirasawa S, Goulding M, Ma Q (2005) Lbx1 and Tlx3 are opposing switches in determining GABAergic versus glutamatergic transmitter phenotypes. Nat Neurosci 8:1510–1515
Chih B, Engelman H, Scheiffele P (2005) Control of excitatory and inhibitory synapse formation by neuroligins. Science 307:1324–1328
Chih B, Gollan L, Scheiffele P (2006) Alternative splicing controls selective trans-synaptic interactions of the neuroligin-neurexin complex. Neuron 51:171–178
Chubykin AA, Atasoy D, Etherton MR, Brose N, Kavalali ET, Gibson JR, Sudhof TC (2007) Activity-dependent validation of excitatory versus inhibitory synapses by neuroligin-1 versus neuroligin-2. Neuron 54:919–931
Chung HJ, Huang YH, Lau LF, Huganir RL (2004) Regulation of the NMDA receptor complex and trafficking by activity-dependent phosphorylation of the NR2B subunit PDZ ligand. J Neurosci 24:10248–10259
Ciccolini F, Collins TJ, Sudhoelter J, Lipp P, Berridge MJ, Bootman MD (2003) Local and global spontaneous calcium events regulate neurite outgrowth and onset of GABAergic phenotype during neural precursor differentiation. J Neurosci 23:103–111
Daadi MM, Weiss S (1999) Generation of tyrosine hydroxylase-producing neurons from precursors of the embryonic and adult forebrain. J Neurosci 19:4484–4497
Dan Y, Poo MM (2004) Spike timing-dependent plasticity of neural circuits. Neuron 44:23–30
Ericson J, Morton S, Kawakami A, Roelink H, Jessell TM (1996) Two critical periods of Sonic Hedgehog signaling required for the specification of motor neuron identity. Cell 87:661–673
Ericson J, Rashbass P, Schedl A, Brenner-Morton S, Kawakami A, van Heyningen V, Jessell TM, Briscoe J (1997) Pax6 controls progenitor cell identity and neuronal fate in response to graded Shh signaling. Cell 90:169–180
Francis NJ, Landis SC (1999) Cellular and molecular determinants of sympathetic neuron development. Annu Rev Neurosci 22:541–566
Fukuchi M, Tabuchi A, Tsuda M (2004) Activity-dependent transcriptional activation and mRNA stabilization for cumulative expression of pituitary adenylate cyclase-activating polypeptide mRNA controlled by calcium and cAMP signals in neurons. J Biol Chem 279:47856–47865

Ghose S, Wroblewska B, Corsi L, Grayson DR, De Blas AL, Vicini S, Neale JH (1997) N-acetylaspartylglutamate stimulates metabotropic glutamate receptor 3 to regulate expression of the GABA(A) alpha6 subunit in cerebellar granule cells. J Neurochem 69:2326–2335
Goldman D, Staple J (1989) Spatial and temporal expression of acetylcholine receptor RNAs in innervated and denervated rat soleus muscle. Neuron 3:219–228
Goldman D, Brenner HR, Heinemann S (1988) Acetylcholine receptor alpha-, beta-, gamma-, and delta-subunit mRNA levels are regulated by muscle activity. Neuron 1:329–333
Goridis C, Brunet JF (1999) Transcriptional control of neurotransmitter phenotype. Curr Opin Neurobiol 9:47–53
Graf ER, Zhang X, Jin SX, Linhoff MW, Craig AM (2004) Neurexins induce differentiation of GABA and glutamate postsynaptic specializations via neuroligins. Cell 119:1013–1026
Gu X, Spitzer NC (1995) Distinct aspects of neuronal differentiation encoded by frequency of spontaneous Ca2+ transients. Nature 375:784–787
Gutierrez R (2000) Seizures induce simultaneous GABAergic and glutamatergic transmission in the dentate gyrus-CA3 system. J Neurophysiol 84:3088–3090
Gutierrez R (2002) Activity-dependent expression of simultaneous glutamatergic and GABAergic neurotransmission from the mossy fibers in vitro. J Neurophysiol 87:2562–2570
Gutierrez R, Romo-Parra H, Maqueda J, Vivar C, Ramirez M, Morales MA, Lamas M (2003) Plasticity of the GABAergic phenotype of the "glutamatergic" granule cells of the rat dentate gyrus. J Neurosci 23:5594–5598
Habecker BA, Asmus SA, Francis N, Landis SC (1997) Target regulation of VIP expression in sympathetic neurons. Ann N Y Acad Sci 814:198–208
Hahm SH, Chen Y, Vinson C, Eiden LE (2003) A calcium-initiated signaling pathway propagated through calcineurin and cAMP response element-binding protein activates proenkephalin gene transcription after depolarization. Mol Pharmacol 64:1503–1511
Hendry SH, Jones EG (1986) Reduction in number of immunostained GABAergic neurones in deprived-eye dominance columns of monkey area 17. Nature 320:750–753
Hendry SH, Jones EG (1988) Activity-dependent regulation of GABA expression in the visual cortex of adult monkeys. Neuron 1:701–712
Hertzberg T, Brosenitsch T, Katz DM (1995) Depolarizing stimuli induce high levels of dopamine synthesis in fetal rat sensory neurons. Neuroreport 7:233–237
Huntsman MM, Isackson PJ, Jones EG (1994) Lamina-specific expression and activity-dependent regulation of seven GABAA receptor subunit mRNAs in monkey visual cortex. J Neurosci 14:2236–2259
Levey MS, Brumwell CL, Dryer SE, Jacob MH (1995) Innervation and target tissue interactions differentially regulate acetylcholine receptor subunit mRNA levels in developing neurons in situ. Neuron 14:153–162
Levinson JN, El-Husseini A (2005) Building excitatory and inhibitory synapses: balancing neuroligin partnerships. Neuron 48:171–174
Levinson JN, Chery N, Huang K, Wong TP, Gerrow K, Kang R, Prange O, Wang YT, El-Husseini A (2005) Neuroligins mediate excitatory and inhibitory synapse formation: involvement of PSD–95 and neurexin-1beta in neuroligin-induced synaptic specificity. J Biol Chem 280:17312–17319
Lissin DV, Gomperts SN, Carroll RC, Christine CW, Kalman D, Kitamura M, Hardy S, Nicoll RA, Malenka RC, von Zastrow M (1998) Activity differentially regulates the surface expression of synaptic AMPA and NMDA glutamate receptors. Proc Natl Acad Sci U S A 95:7097–7102
Malenka RC, Bear MF (2004) LTP and LTD: an embarrassment of riches. Neuron 44:5–21
Marty S, Onteniente B (1997) The expression pattern of somatostatin and calretinin by postnatal hippocampal interneurons is regulated by activity-dependent and -independent determinants. Neuroscience 80:79–88

Marty S, Wehrle R, Fritschy JM, Sotelo C (2004) Quantitative effects produced by modifications of neuronal activity on the size of GABAA receptor clusters in hippocampal slice cultures. Eur J Neurosci 20:427–440
Mellor JR, Merlo D, Jones A, Wisden W, Randall AD (1998) Mouse cerebellar granule cell differentiation: electrical activity regulates the GABAA receptor alpha 6 subunit gene. J Neurosci 18:2822–2833
Merlie JP, Sanes JR (1985) Concentration of acetylcholine receptor mRNA in synaptic regions of adult muscle fibres. Nature 317:66–68
Munoz A, Huntsman MM, Jones EG (1998) GABA(B) receptor gene expression in monkey thalamus. J Comp Neurol 394:118–126
Nam CI, Chen L (2005) Postsynaptic assembly induced by neurexin-neuroligin interaction and neurotransmitter. Proc Natl Acad Sci U S A 102:6137–6142
Nguyen QT, Lichtman JW (1996) Mechanism of synapse disassembly at the developing neuromuscular junction. Curr Opin Neurobiol 6:104–112
Nicoll RA, Schmitz D (2005) Synaptic plasticity at hippocampal mossy fibre synapses. Nat Rev Neurosci 6:863–876
Nicoll RA, Kauer JA, Malenka RC (1988) The current excitement in long-term potentiation. Neuron 1:97–103
Patz S, Wirth MJ, Gorba T, Klostermann O, Wahle P (2003) Neuronal activity and neurotrophic factors regulate GAD-65/67 mRNA and protein expression in organotypic cultures of rat visual cortex. Eur J Neurosci 18:1–12
Prange O, Wong TP, Gerrow K, Wang YT, El-Husseini A (2004) A balance between excitatory and inhibitory synapses is controlled by PSD-95 and neuroligin. Proc Natl Acad Sci U S A 101:13915–13920
Purves D, Lichtman JW (1980) Elimination of synapses in the developing nervous system. Science 210:153–157
Ramirez M, Gutierrez R (2001) Activity-dependent expression of GAD67 in the granule cells of the rat hippocampus. Brain Res 917:139–146
Rosselet C, Zennou-Azogui Y, Xerri C (2006) Nursing-induced somatosensory cortex plasticity: temporally decoupled changes in neuronal receptive field properties are accompanied by modifications in activity-dependent protein expression. J Neurosci 26: 10667–10676
Shi J, Townsend M, Constantine-Paton M (2000) Activity-dependent induction of tonic calcineurin activity mediates a rapid developmental downregulation of NMDA receptor currents. Neuron 28:103–114
Slonimsky JD, Mattaliano MD, Moon JI, Griffith LC, Birren SJ (2006) Role for calcium/calmodulin-dependent protein kinase II in the p75-mediated regulation of sympathetic cholinergic transmission. Proc Natl Acad Sci U S A 103:2915–2919
Song JY, Ichtchenko K, Sudhof TC, Brose N (1999) Neuroligin 1 is a postsynaptic cell-adhesion molecule of excitatory synapses. Proc Natl Acad Sci U S A 96:1100–1105
Spitzer NC, Borodinsky LN, Root CM (2005) Homeostatic activity-dependent paradigm for neurotransmitter specification. Cell Calcium 37:417–423
Thompson CL, Pollard S, Stephenson FA (1996) Bidirectional regulation of GABAA receptor alpha1 and alpha6 subunit expression by a cyclic AMP-mediated signalling mechanism in cerebellar granule cells in primary culture. J Neurochem 67:434–437
Thompson CL, Tehrani MH, Barnes EM, Jr., Stephenson FA (1998) Decreased expression of GABAA receptor alpha6 and beta3 subunits in stargazer mutant mice: a possible role for brain-derived neurotrophic factor in the regulation of cerebellar GABAA receptor expression? Brain Res Mol Brain Res 60:282–290
Tu S, Debski EA (1999) Development and regulation of substance P expression in neurons of the tadpole optic tectum. Vis Neurosci 16:695–705
Tu S, Butt CM, Pauly JR, Debski EA (2000) Activity-dependent regulation of substance P expression and topographic map maintenance by a cholinergic pathway. J Neurosci 20:5346–5357

Varoqueaux F, Jamain S, Brose N (2004) Neuroligin 2 is exclusively localized to inhibitory synapses. Eur J Cell Biol 83:449–456
Walicke PA, Patterson PH (1981) On the role of Ca2+ in the transmitter choice made by cultured sympathetic neurons. J Neurosci 1:343–350
Walicke PA, Campenot RB, Patterson PH (1977) Determination of transmitter function by neuronal activity. Proc Natl Acad Sci U S A 74:5767–5771
Wang JH, Ko GY, Kelly PT (1997) Cellular and molecular bases of memory: synaptic and neuronal plasticity. J Clin Neurophysiol 14:264–293
Wang W, Stock RE, Gronostajski RM, Wong YW, Schachner M, Kilpatrick DL (2004) A role for nuclear factor I in the intrinsic control of cerebellar granule neuron gene expression. J Biol Chem 279:53491–53497
Watt SD, Gu X, Smith RD, Spitzer NC (2000) Specific frequencies of spontaneous Ca2+ transients upregulate GAD 67 transcripts in embryonic spinal neurons. Mol Cell Neurosci 16:376–387
Zweifel LS, Kuruvilla R, Ginty DD (2005) Functions and mechanisms of retrograde neurotrophin signalling. Nat Rev Neurosci 6:615–625

Chapter 4
Co-Release of Norepinephrine and Acetylcholine by Mammalian Sympathetic Neurons: Regulation by Target-Derived Signaling

Jason A. Luther and Susan J. Birren

Abstract Neurotransmitter expression has long been thought of as a defining phenotypic property of adult neurons. However, it has now been shown that most neurons co-release multiple signaling molecules. Many examples of neurons that co-release a classical transmitter (e.g., acetylcholine, norepinephrine or glutamate) and neuromodulators have been demonstrated, but neurons can also co-release more than one classical transmitter. Defining the mechanisms that determine released transmitter(s) is important for understanding neural function since this largely determines the influence of neural activity. This chapter details evidence showing that mammalian sympathetic neurons co-release acetylcholine (ACh) and norepinephrine (NE). Sympathetic neurons project to body tissues including blood vessels and heart to control functions such as regulation of blood pressure and cardiac output. Transmitter choice in sympathetic neurons is controlled by target-derived, soluble growth factors. Current data suggests that these factors may operate to regulate the relative amounts of ACh and NE released by sympathetic neurons, which may play an important role in homeostatic regulation of essential physiological processes.

4.1 Introduction

It has historically been held that the identity of a released neurotransmitter is a defining and unchanging characteristic of differentiated adult neurons (Dale 1935); Eccles 1976). However, more recent evidence suggests that neurons can release different classical transmitters under different physiological situations and even, in some cases, co-release multiple transmitters. The existence of co-transmission of classical transmitters (e.g., acetylcholine, norepinephrine, glutamate and GABA) with neuropeptides or neuromodulators is well established in many types of neurons (Kupfermann 1991; Elfvin et al. 1993).

J.A. Luther (✉)
Department of Biology and Volen Center for Complex Systems, Brandeis University, MS 008, 415 South Street, Waltham, MA 02454
e-mail: jluther@brandeis.edu.

R. Gutierrez (ed.), *Co-Existence and Co-Release of Classical Neurotransmitters*,
DOI 10.1007/978-0-387-09622-3_4, © Springer Science+Business Media, LLC 2009

Co-transmission of two classical transmitters has now also been demonstrated in both invertebrate and vertebrate preparations (Kupfermann 1991; Marder et al. 1995; Seal and Edwards 2006). In this chapter we will summarize the evidence that mammalian sympathetic neurons coexpress the classical transmitters acetylcholine (ACh) and norepinephrine (NE) and that, under certain circumstances, co-release both of these transmitters in a regulated manner.

Co-transmission is biologically more costly than utilizing a single transmitter since multiple molecular regulatory systems are necessary to generate different transmitter molecules, package them in vesicles and recycle or break them down in the synaptic cleft. Since it would seem more parsimonious to utilize a single transmitter, the question arises as to why neurons would go through the trouble of using multiple neurotransmitters. It seems that there must be an evolutionary advantage to co-transmission. One possibility is that it limits crosstalk between nearby neurons. Transmission between closely apposed pairs of connected neurons might interact if all cells used the same transmitter-receptor system. Similarly unwanted interactions may occur if closely apposed neurons received common input from a single afferent source. However, co-release of multiple transmitters coupled with distinct transmitter receptor expression patterns in target cells would ensure exclusivity of signaling and remove a potential restriction on the structural complexity of neural circuits.

Another advantage of co-transmission is that multiple transmitter systems allow more complex signaling. For example, if one transmitter were contained in a more readily releasable vesicle pool then a weaker stimulus would release only one transmitter while a stronger stimulus might cause release of both transmitters. Therefore different patterns or levels of activity could have different physiological effects. An example of this is seen in sympathetic neurons controlling vasoconstriction in the rat tail where the release of norepinephrine, adenosine triphosphate, and neuropeptide Y is differentially affected by stimulus burst duration (Bradley et al. 2003).

Co-transmission may also allow for local modulation of signaling. Thus, the release of retrograde signals from innervated tissue could be an important determinant of relative release of coexpressed neurotransmitters. In this way target cells could directly regulate afferent input through the regulation of co-transmission. As will be discussed later, this type of signaling appears to play an important role in co-transmission of ACh and NE in the sympathetic nervous system.

4.2 Developmental Regulation of Neurotransmitter Expression in Sympathetic Neurons by Target-Derived Signals

Sympathetic neurons project from ganglia located in the neck and on either side of the spinal column to innervate somatic organs and tissues. The sympathetic nervous system regulates metabolic, contractile and secretory functions that are essential for physiological homeostasis. Developmentally, sympathetic neurons derive from neural crest precursors and express noradrenergic markers (i.e., enzymes necessary to produce the neurotransmitter norepinephrine) early in development (Ernsberger and Rohrer 1996; Francis and Landis 1999; Ernsberger 2000). Cholinergic markers are coexpressed somewhat later in development in a subpopulation of neurons (Schafer et al. 1997; Ernsberger and Rohrer 1999). These expression patterns occur relatively early in development before neurons contact target tissue, suggesting that developing sympathetic neurons have an autonomous capacity to express both cholinergic and noradrenergic markers.

The expression of neurotransmitter phenotype in postnatal sympathetic neurons is developmentally influenced by interactions with target tissues. Most adult sympathetic neurons utilize norepinephrine as their principal neurotransmitter at synaptic contacts with body tissues (Elfvin et al. 1993). However, some tissues, such as sweat glands, periosteum and some vascular beds are cholinergically innervated by sympathetic nerves (Schotzinger and Landis 1988; Elfvin et al. 1993; Francis and Landis 1999; Asmus et al. 2000). Sympathetic innervation of these tissues is initially noradrenergic and gradually switches to cholinergic over the first one to two weeks after birth (Francis and Landis 1999).

The final transmitter profile of noradrenergic and cholinergic sympathetic neurons is dependent upon the target of innervation. In rodents, the glabrous skin of the footpad contains sweat glands that receive cholinergic sympathetic innervation, whereas hairy skin is noradrenergically innervated. Several studies have demonstrated that interactions between the neurons and target tissue are necessary to cause the cholinergic neurotransmitter switch (Schotzinger and Landis 1988; Schotzinger and Landis 1990; Asmus et al. 2000). Schotzinger and Landis 1988 removed a patch of hairy skin from newborn rats and transplanted a section of glabrous skin from a donor animal. They were able to observe the initial innervation of the transplanted skin since rat sympathetic fibers do not enter the dermis until several days after birth. Sympathetic fibers that would have innervated hairy skin and remained noradrenergic switched to a cholinergic phenotype when they terminated on transplanted glabrous skin. Catecholaminergic histofluorescence in the sympathetic fibers innervating sweat glands, but not surrounding hairy skin or blood vessels, decreased sharply. Conversely, immunostaining for acetylcholinesterase and the activity of choline acetyltransferase (markers of cholinergic transmission) gradually increased in fibers innervating the transplanted glabrous skin. Together, these experiments demonstrated that target interactions play an instructive role in setting the neurotransmitter phenotype of innervating sympathetic neurons.

In an analogous set of experiments, the parotid salivary gland, which is innervated by noradrenergic fibers, was transplanted under glabrous skin in the footpad of newborn rats. Sympathetic fibers innervating the transplanted salivary gland retained noradrenergic markers, while fibers innervating surrounding sweat glands switched to a cholinergic phenotype (Schotzinger and Landis 1990). Finally, transplantation of periosteum, the connective tissue surrounding bone that is cholinergically innervated in the adult, to hairy skin in newborn rats promoted a switch to a cholinergic phenotype in sympathetic fibers innervating the transplanted tissue (Asmus et al. 2000). These studies strongly suggest that target-derived signaling plays a crucial developmental role in the determination of sympathetic neurotransmitter phenotype.

The switch from noradrenergic to cholinergic neurotransmission that occurs in sympathetic neurons innervating the sweat glands is dependent on a soluble factor released by the target tissue (Rao and Landis 1990; Rao et al. 1992; Rohrer 1992). Whole protein extracts purified from rat glabrous skin induced a dose-dependent cholinergic shift in cultured sympathetic neurons, while cultures treated with liver, salivary gland or hairy skin extracts, or with extracts from mutant mice that lack sweat glands remained noradrenergic (Rao and Landis 1990; Habecker et al. 1995).

Several growth factors have been identified that induce cholinergic and reduce noradrenergic marker expression in cultured sympathetic neurons. These include ciliary neurotrophic factor (CNTF), leukemia inhibitory factor (LIF), cardiotrophin -1 (CT-1), and neurotrophin-3 (NT-3) (Saadat et al. 1989; Yamamori et al. 1989; Habecker et al. 1995; Brodski et al. 2000). It was therefore initially proposed that one or more of these secreted proteins might be the cholinergic differentiating factor released by sweat gland tissue (Rao et al. 1992; Rohrer 1992). However, several lines of evidence suggest that the in vivo cholinergic differentiating factor is not one of these proteins and that the relevant activity has not yet been identified. Pre-treating glabrous skin extracts with antisera to CNTF resulted in a 55 to 80 % decrease in cholinergic differentiation in cultured sympathetic neurons as measured by acetylcholinesterase activity (Rao et al. 1992; Rohrer 1992), However, Rao et al. (1992) found no evidence for expression of CNTF protein or mRNA in sweat glands. This led to the conclusion that CNTF in glabrous skin extracts was likely derived from sensory neurons innervating the dermis and that it probably played no physiological role in the determination of sympathetic neurotransmitter phenotype at sweat glands. Further, cholinergic sympathetic innervation of sweat glands developed normally in mice lacking either or both the CNTF and LIF genes (Francis et al. 1997), supporting the hypothesis that CNTF is not the cholinergic differentiation factor. Similarly, antisera directed against LIF or CT-1 did not prevent cholinergic differentiation of cultured sympathetic neurons induced by glabrous skin extract (Rao et al. 1992; Habecker et al. 1995). Thus, none of the factors that have been identified as a cholinergic differentiation factor has been shown to play that role during the cholinergic switch of sympathetic neurons in vivo.

Whatever the identity of the released cholinergic differentiation factor turns out to be it does seem to act through the same receptor and signaling systems as CNTF, LIF and CT-1. These three growth factors show structural similarities and signal through LIFRβ and gp130 receptors. The cholinergic differentiation activity of glabrous skin extract was blocked by pharmacological blockade of LIFRβ signaling and the extract induced gp130 specific signaling cascades in cell culture (Bazan 1991; Habecker et al. 1997). Furthermore, deletion of the gp130 gene in mice led to a failure of sympathetic neurons innervating sweat glands to switch to a cholinergic transmitter phenotype in vivo (Stanke et al. 2006). Thus, the physiologically relevant cholinergic differentiation factor is likely to be related to CNTF, LIF, and CT-1 and may be a new member of this cytokine family.

The release of the cholinergic differentiation factor by sweat glands appears to be dependent upon release of NE from sympathetic neurons upon initial synaptic contact during postnatal development. Extracts of glabrous skin selectively deprived of noradrenergic sympathetic input by treatment with the neurotoxin 6-hydroxydopamine were unable to promote a cholinergic transmitter switch in cultured sympathetic neurons (Habecker and Landis 1994). Sweat glands cultured alone or with sensory neurons (which do not produce NE) do not produce cholinergic differentiation factor as assayed by the ability of conditioned media to induce the expression of cholinergic markers in sympathetic neuronal cultures (Habecker et al. 1995). However, sweat gland tissue produced a cholinergic differentiation factor when cultured with sympathetic neurons. Production of this factor was blocked by pharmacological inhibition of NE transmission (Habecker and Landis 1994; Habecker et al. 1995). These results demonstrate that initial noradrenergic synaptic contact is necessary to induce release of the cholinergic differentiation factor and affect a transmitter phenotype switch in the sympathetic neurons.

The neurotrophins are another family of target-derived growth factors that play a key role in regulating physiology in both developing and mature sympathetic neurons (Bibel and Barde 2000). Neurotrophin signaling may also play an important role in determining mature sympathetic neurotransmitter phenotypes. The expression pattern of the high affinity receptor for neurotrophin-3 (NT-3), tropomyosin-related kinase receptor C (TrkC) overlaps with the expression pattern of cholinergic markers in sympathetic ganglia (Brodski et al. 2000), suggesting a possible role for this pathway in setting the cholinergic phenotype. Consistent with this idea, culturing sympathetic chain explants in the presence of NT-3 increased the expression of cholinergic markers while decreasing the expression of noradrenergic markers (Brodski et al. 2000).

Interestingly, the effects of NT-3 could be reversed and were antagonized by another neurotrophin, nerve growth factor (NGF) (Brodski et al. 2000). NGF is a sympathetic neuron survival factor that has been demonstrated to promote noradrenergic properties in the neurons (Chun and Patterson 1977; Bibel and Barde 2000). Expression of noradrenergic markers in the sympathetic chain overlap with the expression of the NGF high affinity tropomyosin-related kinase receptor A (TrkA) (Brodski et al. 2002), consistent with earlier studies showing that NGF increased the expression of noradrenergic markers in cultured sympathetic neurons (Chun and Patterson 1977). Additionally, NGF rapidly increases the activity of tyrosine hydroxylase, an enzyme necessary for the generation of NE, in rat PC12 cell line cultures (Greene et al. 1984). These results suggest that target-derived signaling plays a crucial, yet complex role in determining transmitter expression in sympathetic neurons and raises the possibility of acute changes in neurotransmitter properties as a result of interaction with target-derived factors.

This section has summarized studies that demonstrate that sympathetic neurons in vivo have the ability to express and release two classical transmitters: ACh and NE. The determination of transmitter phenotype appears to be dependent on bidirectional signaling between the sympathetic neurons and their targets via soluble molecules. One question that remains from these studies is whether individual sympathetic neurons are capable of co-releasing these two different classical transmitters. The next section will present evidence from cell culture studies that show that individual sympathetic neurons indeed co-release both ACh and NE and that this co-release is regulated by target-derived soluble signaling molecules.

4.3 Plasticity of Sympathetic Neurotransmitter Phenotype: Cell Culture Studies

A switch in neurotransmitter status from noradrenergic to cholinergic was first observed in sympathetic neurons grown in culture (Bunge et al. 1974; O'Lague et al. 1974). A series of papers published in 1978 described a careful electrophysiological and pharmacological analysis of this phenomenon in cultured sympathetic neurons (O'Lague et al. 1978; O'Lague et al. 1978; O'Lague et al. 1978). These studies used neurons taken from the superior cervical sympathetic ganglion of neonatal rats, which are almost exclusively noradrenergic in vivo (Elfvin et al. 1993). Surprisingly, culturing neurons either with cardiac myocytes or under conditions that allowed proliferation of non-neuronal ganglionic cells (including Schwann cells and fibroblasts) resulted in the formation of functional cholinergic synapses between the sympathetic neurons. When neurons were cultured in media that inhibited non-neuronal cell growth they remained noradrenergic (O'Lague et al. 1978; O'Lague et al. 1978; O'Lague et al. 1978).

Several lines of evidence supported the conclusion that cholinergic transmission occurred in these cultures. Electrophysiological recordings in pairs of neurons demonstrated that stimulation of action potentials in one neuron resulted in synaptic responses in a second neuron (O'Lague et al. 1978; O'Lague et al. 1978; O'Lague et al. 1978). These responses were blocked by decreasing extracellular calcium concentration, which blocks synaptic transmission, and by cholinergic antagonists. Iontophoresis of ACh onto small regions of individual neurons caused depolarizations similar to the synaptic events elicited by electrical stimulation, further supporting the idea that endogenous ACh release could account for the observed synaptic events. In contrast, application of noradrenergic antagonists had no effect on these cholinergic synaptic events, although iontophoretic application of catecholaminergic agonists caused membrane hyperpolarization. This showed that the neurons expressed functional noradrenergic receptors that did not contribute to the cholinergic transmission.

The cholinergic effects of non-neuronal cells appeared to be mediated by a released soluble factor since media conditioned with these cells could also elicit a cholinergic shift in the cultured neurons (Patterson and Chun 1977). Thus, these studies established the ability of sympathetic neurons to form functional cholinergic connections and demonstrated that these cholinergic properties were regulated by extrinsic factors in vitro.

While non-neuronal cells promoted the formation of cholinergic synaptic connections, NE continued to be produced in these cultures. This meant that culture with non-neuronal cells altered the balance of ACh and NE expression in sympathetic neuron co-cultures. Noradrenergic sympathetic neurons grown in the absence of non-neuronal cells produced only small amounts of ACh as assayed by incorporation of radioactive precursors (Patterson and Chun 1974). Addition of increasing numbers of non-neuronal cells increased the amount of ACh produced while decreasing NE levels (Patterson and Chun 1974). While these results suggested that the cultures produced both ACh and NE, it was not clear whether different subpopulations of neurons were responsible for production of each transmitter or whether individual neurons could produce variable amounts of both transmitters.

ACh and NE expression needed to be unambiguously determined in individual neurons to establish if sympathetic neurons were capable of simultaneously expressing both transmitters. The use of microisland cultures containing single sympathetic neurons together with target cells made it possible to address this issue (Furshpan et al. 1976; Furshpan et al. 1986; Potter et al. 1986). The target cells used in these cultures were neonatal rat cardiac myocytes. Sympathetic neurons innervate the heart in vivo to provide excitatory regulation of heart beat rate and cardiac function via noradrenergic transmission (Elghozi and Julien 2007). Interestingly, although sympathetic innervation of the heart is noradrenergic, cardiac myocytes produce cholinergic differentiation factors, and promote the development of cholinergic properties in co-cultured sympathetic neurons (Fukada 1980; Marvin et al. 1984; Furshpan et al. 1986). Thus, co-culture with cardiac myocytes provided a system in which both noradrenergic transmission and the development of cholinergic properties could be examined.

Cardiac myocytes were cultured together with sympathetic neurons in small (300 to 500 μm) microislands on the surface of a culture dish (Furshpan et al. 1976; Furshpan et al. 1986; Potter et al. 1986). In these cultures it was possible to identify individual microislands that contained only a single neuron that grew over the cardiac myocytes, but did not grow past the edge of the microisland (Furshpan et al. 1976; Furshpan et al. 1986). Neurons grown in this way formed functional synapses on themselves and onto the myocytes (Furshpan et al. 1976; Furshpan et al. 1986). This preparation allowed the determination of the identity of released transmitter(s) from individual sympathetic neurons.

One way to determine the identity of the released transmitters in neuron-myocyte microislands was to examine the postsynaptic response of the myocytes to neuronal activity. Myocytes cultured in microislands form electrical junctions with one another and contract, or beat, in a coordinated fashion across the extent of the microisland. These contractions are modulated by neuronal activity via synaptic transmission in a manner similar to the autonomic regulation of heart rate in vivo (Furshpan et al. 1976; Furshpan et al. 1986; Elghozi and Julien 2007). Myocytes have an excitatory response to NE released by sympathetic neurons that is mediated via activation of β–adrenergic receptors. Conversely, myocyte beat rate is inhibited by ACh via activation of muscarinic ACh receptors (Furshpan et al. 1976; Furshpan et al. 1986). Using electrophysiological recording techniques it is possible to monitor the myocyte response to stimulation of the neuron by recording the myocyte action potentials, which correspond to beat rate. Additionally, NE depolarizes and ACh hyperpolarizes myocytes, which can also be monitored by electrophysiological techniques. Therefore by stimulating neurons to release transmitter and recording the myocyte response it is possible to determine whether ACh or NE is being released by a neuron.

In addition to measuring the postsynaptic responses of the myocytes, information about the neurotransmitter identity of the neurons could be gained by examining neuronal postsynaptic responses to synaptic transmission taking place at autapses. While sympathetic neurons grown in microislands released NE, no evidence was found for a noradrenergic neuronal synaptic response. The neurons were excited, however, by synaptically released ACh through activation of nicotinic receptors (Furshpan et al. 1976; Furshpan et al. 1986). Thus, in these microislands, the neurons showed responses to cholinergic transmission, while the myocytes responded differentially to ACh and NE release by the neurons. Thus, recording from individual neurons and myocytes grown in microisland cultures provided an approach to investigate whether individual sympathetic neurons utilized exclusively ACh or NE, or could co-release both transmitters (Furshpan et al. 1976; Furshpan et al. 1986; Potter et al. 1986).

Sympathetic neurons in microislands were found to vary greatly in release of transmitter with some cells expressing almost exclusively ACh or NE while other cells expressed both transmitters in varying proportions. In some neurons, stimulation resulted in excitatory synaptic events recorded in the neuron itself and an inhibition of the myocyte, seen as a membrane hyperpolarization and decrease in spiking and contraction rate. Nicotinic antagonists such as curare and hexamethonium blocked the neuronal synaptic events, while the muscarinic antagonist, atropine blocked the myocyte inhibition. These data demonstrated that the synaptic events in these neurons were due to cholinergic transmission (Furshpan et al. 1976; Furshpan et al. 1986; Potter et al. 1986).

In some neurons stimulation resulted in no synaptic events in the neuron, but caused an increase in spiking and contraction rate of the myocyte. This type of response was found to be sensitive to β–adrenergic antagonists such as propranolol, confirming a noradrenergic phenotype for those neurons. In the majority of cells however, the myocyte response was mixed, showing slower, noradrenergic-mediated excitation preceded by a more rapid cholinergic-mediated inhibition of the myocytes. These effects could be pharmacologically separated using cholinergic or adrenergic antagonists. These data provided evidence for co-release of NE and ACh in individual neurons.

Further evidence for dual neurotransmitter coexpression in sympathetic neurons came from ultrastructural examination. Electron microscopic analysis of microisland neurons revealed the existence of small granular and small clear vesicles, presumably representing noradrenergic and cholinergic vesicles, respectively. In elegant experiments in which physiological responses were linked to ultrastructure in individual neurons, the authors observed that neurons with mixed physiological responses showed both types of vesicles at synaptic endings, while single-function neurons showed mostly one or the other type of vesicle. These studies unambiguously demonstrated that cultured sympathetic neurons have the capacity to co-release both ACh and NE (Furshpan et al. 1976; Furshpan et al. 1986; Potter et al. 1986).

As described in the previous section, initially noradrenergic sympathetic neurons that innervate sweat glands demonstrate a developmental switch to cholinergic transmission over a period of time following the formation of synaptic contacts with target. It is thus possible that co-transmission in sympathetic neurons grown in microislands represents a transitory state that occurs while neurons are switching from a noradrenergic to a cholinergic phenotype in response to soluble signals from non-neuronal cells. Indeed, by recording multiple times on successive days from individual neurons it was found that transmitter phenotype tended to shift from noradrenergic to cholinergic over time. However, many neurons retained dual transmitter status for up to several months in culture, suggesting that dual transmitter expression could be a stable condition (Potter et al. 1986).

Sympathetic neurons contacting sweat glands switch transmitter phenotype in the first one to two postnatal weeks in vivo, raising the question of whether neurotransmitter plasticity is a transient developmental phenomenon. The microisland recordings described above were performed using neonatal sympathetic neurons, which are still in the process of forming final target contacts and hence may be more plastic in respect to transmitter phenotype expression (Francis and Landis 1999). However, adult superior cervical sympathetic neurons grown on microislands also demonstrated neurotransmitter plasticity, albeit in a smaller proportion of cells and with more cells tending to remain noradrenergic (Potter et al. 1986). These results demonstrate that transmitter plasticity, while more pronounced in younger neurons, is not lost in the adult.

4.4 Neurotrophins Induce a Rapid Switch in Neurotransmitter Status of Sympathetic Neurons

Thus far, the neurotransmitter plasticity displayed by sympathetic neurons has been described as a developmental process that takes place over a period of days or weeks. Recent studies however, show that the regulation of neurotransmitter properties can also take place over a rapid time scale. In addition to their role in the developmental regulation of neurotransmitter phenotype (described above), neurotrophins are able to rapidly regulate the co-release properties of sympathetic neurons.

As will be described in detail below, neurotrophins have been shown to cause rapid differential regulation of neurotransmitter release in sympathetic neurons by acting through different receptors. Therefore it will be helpful to detail the types of neurotrophin receptors expressed by sympathetic neurons. Neurotrophins act through two types of receptors; the pan-neurotrophin receptor (p75), which binds all family members with similar affinity, and the tropomyosin-related kinases (Trks), which show specific binding affinities to various neurotrophin family members (Reichardt 2006). Sympathetic neurons express p75 and two of the three tropomyosin-related kinases; TrkA, the NGF receptor and TrkC, which specifically binds NT-3. These neurons do not express the TrkB receptor, the specific receptor for brain-derived neurotrophic factor (BDNF) and neurotrophin 4/5 (Dixon and McKinnon 1994; Wyatt and Davies 1995; Bamji et al. 1998; Reichardt 2006). Trk receptors are thought to underlie many described neuronal responses to neurotrophins including neuronal differentiation, growth, survival and synaptic modulation (Bibel and Barde 2000; Reichardt 2006). However, in recent years a number of distinct and sometimes opposing functions have been described for the p75 receptor. These include promoting apoptotic neuronal death, modulation of growth dynamics, and the regulation of long-term depression (Chao and Hempstead 1995; Lu et al. 2005; Woo et al. 2005). For neurons expressing both Trk and p75 receptors, including sympathetic neurons, responses to neurotrophins may reflect the output of complex interactions between the p75 and Trk signaling pathways (Chao and Hempstead 1995; Huber and Chao 1995; Lu et al. 2005; Woo et al. 2005).

Evidence for differential regulation of co-transmission by p75 and Trk receptors has been provided by analysis of synaptic transmission in neuron-myocyte co-cultures. Superior cervical sympathetic neurons form functional synapses onto the co-cultured cardiac myocytes and these connections are initially almost entirely noradrenergic. Electrophysiological stimulation of neurons in young, noradrenergic neuron-myocyte co-cultures results in an increase in beat rate of myocytes that are connected to the stimulated neuron. This increase can be observed visually and quantified by counting contractions (Conforti et al. 1991; Lockhart et al. 1997). Treatment of the cultures with NGF for ten minutes caused a rapid and reversible potentiation of synaptic transmission that was observed as an increase in myocyte beat rate in response to neuronal stimulation (Lockhart et al. 1997).

The effect of NGF on excitatory transmission was likely to be due to a presynaptic action of NGF on neuronally expressed TrkA receptors. Cardiac myocytes do not express appreciable levels of TrkA and the myocyte response to pressure ejection application of NE was not altered in the presence of NGF (Lockhart et al. 1997). The potentiation of neuronal regulation of myocyte beat rate was blocked by application of the Trk antagonist K252a suggesting that the effect was mediated via TrkA and not p75 activation (Lockhart et al. 1997). This study provided evidence that NGF, acting through TrkA, promotes noradrenergic neurotransmission in sympathetic neurons.

Further investigation into the actions of neurotrophins on sympathetic presynaptic properties revealed the surprising result that, in addition to potentiating the release of NE, neurotrophin signaling regulated co-release of ACh. Treatment of young, noradrenergic neuron-myocyte co-cultures with BDNF rapidly and reversibly resulted in the inhibition of myocyte beat rate during neuronal stimulation (Yang et al. 2002). This result was in contrast to the excitation of myocyte beat rate seen during neuronal stimulation in the presence of NGF, a related member of the neurotrophin family. In these neuron-myocyte co-cultures a fifteen minute application of BDNF was sufficient to switch the myocyte beat rate response during neuron stimulation from excitation to inhibition. The effect of BDNF was reversible and was blocked by the muscarinic cholinergic antagonist, atropine. Similar to the finding with NGF, the effect of BDNF was likely to be presynaptic since myocyte responses to pressure ejection application of muscarine, a muscarinic ACh receptor agonist, or NE were not altered in BDNF. BDNF did not induce expression of cholinergic markers over the time frame of the experiment and, unlike classic cholinergic differentiation factors such as CNTF, did not promote the expression of cholinergic markers even over a three-day culture period (Slonimsky et al. 2003). These experiments suggest that BDNF acts to rapidly promote ACh release through a mechanism independent of the cholinergic differentiation factors described in the previous section.

Evidence suggests that the switch from excitatory to inhibitory neurotransmission in neuron-myocyte co-cultures is mediated through p75 signaling. BDNF is likely to be a selective ligand for p75 in sympathetic neurons as TrkB is not expressed and BDNF does not activate the TrkA or TrkC receptors (Dixon and McKinnon 1994; Wyatt and Davies 1995; Bamji et al. 1998; Yang et al. 2002; Reichardt 2006). In contrast to the finding for potentiation of noradrenergic transmission by NGF, the BDNF-dependent induction of inhibitory synaptic function was not blocked by application of K252a, which blocks the function of Trk family receptors. This finding suggested that BDNF was not acting through a Trk receptor (Yang et al. 2002). Application of C_2-ceramide, a second messenger produced in response to p75 activation mimicked the BDNF response however, suggesting that BDNF acted to modulate cholinergic transmission via p75 receptor signaling (Yang et al. 2002). Further evidence implicating p75 came from overexpression of human p75 in neuron-myocyte co-cultures which led to inhibitory transmission even in the absence of exogenous BDNF (Yang et al. 2002). Finally, the effect of BDNF was not seen in neuron-myocyte co-cultures derived from mice carrying a targeted deletion of the p75 receptor gene (Lee et al. 1992; Yang et al. 2002). Together, these data suggest that even in apparently noradrenergic sympathetic neurons there exists a releasable cholinergic vesicle pool that is mobilized via cell signaling through the p75 receptor.

Pharmacological evidence suggests that the relative level Trk to p75 signaling determines the ratio of ACh and NE released. Sympathetic neurons in myocyte co-cultures co-released ACh and NE to varying degrees after being cultured for three days in different growth factors (Yang et al. 2002). Again, this was assessed by stimulating neurons and observing changes in beat rate of connected myocytes. Growth in control media, containing 5 ng/ml NGF, or media containing 50 ng/ml NGF, resulted in only excitatory responses. Application of the β–adrenergic antagonist propranolol blocked the excitatory responses and revealed a small inhibitory response suggesting that under these conditions neurons were predominately noradrenergic, but did release some ACh. Conversely, growth in BDNF resulted in only inhibitory myocyte responses when neurons were stimulated. This effect was blocked by the muscarinic ACh antagonist, atropine that revealed a small excitatory response presumably due to co-release of NE. These experiments demonstrated the co-release of ACh and NE and the modulation of that release by neurotrophic signaling.

The rapid effects of BDNF are specific to neurotrophin signaling pathways. Growth in CNTF required three weeks before neurons showed a significant functional cholinergic shift in transmission, suggesting that CNTF and BDNF work through different pathways to regulate cholinergic transmission (Yang et al. 2002; Slonimsky et al. 2003).

In addition to a requirement for p75 signaling, probably via the ceramide pathway, the effects of BDNF on promoting cholinergic transmission appear to be dependent on activation of calcium/calmodulin-dependent protein kinase II (CamKII) (Slonimsky et al. 2006). Inhibition of CamKII in the presynaptic neuron blocked the induction of cholinergic neurotransmission by BDNF, whereas transfection of constitutively active CamKII into presynaptic neurons induced a cholinergic shift in the absence of exogenous BDNF (Slonimsky et al. 2006). However an interaction between the p75 receptor and CamKII appears to be necessary since transfection of constitutively active CamKII into p75 knockout animals did not cause a cholinergic shift in neurotransmission (Slonimsky et al. 2006).

This section has described studies that show that neurotrophins can rapidly modulate the amount of ACh and NE being co-released at sympathetic terminals by signaling through two receptors (Fig. 4.1). Increased activation of the TrkA receptor enhances noradrenergic transmission and increased activation of the p75 receptor enhances cholinergic transmission. This provides a potential mechanism for the regulation of the function of sympathetic innervation via target-derived signaling. Sympathetically innervated targets express distinct profiles of neurotrophin species in vivo (Bierl et al. 2005; Randolph et al. 2007) and sympathetic neurons express both p75 and Trk receptors. Increasing the ratio of Trk to p75 activation would be predicted to increase relative release of NE, and conversely increasing the p75 activation compared to Trk activation would increase ACh release.

While the mechanisms underlying the changes in neurotransmitter release properties are not known, firing pattern has been found to influence co-release

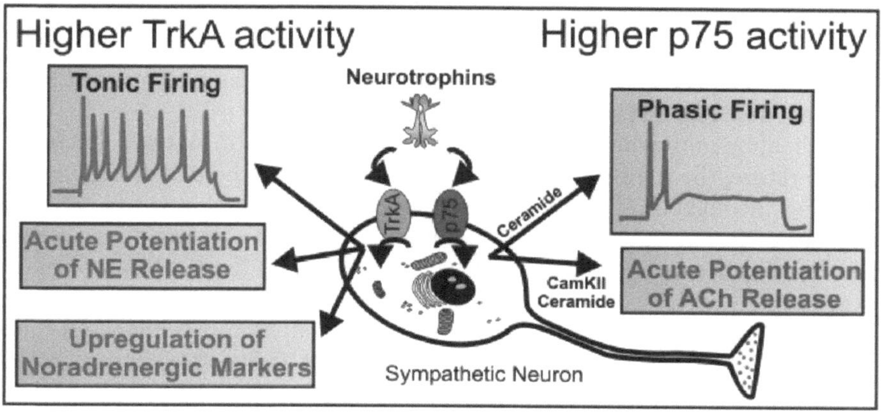

Fig. 4.1 Target-derived neurotrophins rapidly regulate functional neuronal properties differentially through p75 and TrkA signaling pathways. Sympathetic neurons express both TrkA and p75. When TrkA activation is higher than p75 activation the neurons tend to release norepinephrine and fire in a tonic pattern (left). When p75 activity predominates cholinergic transmission is potentiated and cells fire in a phasic pattern. Although the signaling pathways involved are incompletely understood, evidence shows that the second messenger molecule, ceramide, is involved in promotion of both phasic firing and potentiation of acetylcholine release. Calcium/calmodulin-dependent protein kinase II (CamKII) also has been implicated in promoting cholinergic transmission. (*See* Color Plate 2)

of classical transmitters and neuromodulators in different cell types (Bradley et al. 2003; Fulop et al. 2005). Recent evidence shows that in addition to their effects on transmitter choice, neurotrophins also influence sympathetic neuronal firing patterns. In the next section we will explore the possibility that changes in neuronal firing pattern are important determinants of transmitter release.

4.5 Neurotrophins Regulate the Firing Properties of Sympathetic Neurons via Differential Activation of Trk and p75 Receptors

Neurotrophin signaling also plays a role in regulating the membrane electrical properties of sympathetic neurons. Sympathetic neurons fire in one of two characteristic patterns in response to electrophysiological stimulation and the proportion of cells firing in each of these patterns varies among sympathetic ganglia projecting to different targets (Cassell et al. 1986; Wang and McKinnon 1995; Jobling and Gibbins 1999; Anderson et al. 2001). Neurons are found to fire either phasically, that is, they fire once and then go silent, or fire tonically throughout the duration of the stimulus (Cassell et al. 1986). When firing patterns were surveyed over a population of cultured neonatal superior cervical neurons it was found that most cells fired tonically with firing rates being distributed evenly around a mean (Luther and Birren 2006). Application of NGF to these cultures resulted in a shift of firing pattern to a bimodal distribution: increasing the number of phasic cells and increasing the number of high-frequency tonic cells (Luther and Birren 2006).

As mentioned previously, NGF acts through both TrkA and p75 receptors, and therefore, the bimodal shift in firing patterns could be due to differential signaling through those two receptor types in different populations of cells (Fig. 4.1). Indeed, it was found that application of C_2-ceramide, a second messenger generated subsequent to p75 activation, induced phasic firing in most neurons (reported in abstract form, Society for Neuroscience Abstracts, 32, 2006). Application of NGF in cultures prepared from p75 knockout mice resulted in most cells firing tonically with higher firing rates compared to control, suggesting that TrkA signaling increased tonic firing (unpublished observations). These results demonstrate that, in addition to their role in determining neurotransmitter phenotype, soluble, target-derived signaling molecules can also influence repetitive firing in sympathetic neurons.

Firing pattern has been shown to play a role in co-release of transmitters and neuromodulators; however, this has not previously been shown for multiple classical transmitters. For example, adrenal chromaffin cells release catecholamines, but when stimulated more strongly they also release neuropeptides (Fulop et al. 2005). Co-release of NE, adenosine triphosphate and neuropeptide Y from sympathetic neurons are differentially modulated by firing frequency and spike number per burst (Bradley et al. 2003). It is tempting to speculate that regulation of the neurotransmitter phenotype and firing pattern by neurotrophins are related. Perhaps p75 mediated changes in firing pattern lead to conditions that favor release of ACh containing vesicles, and conversely TrkA mediated changes in firing pattern favor release of NE containing vesicles. However, it is unknown whether transmitter status and firing pattern correspond in cultured sympathetic neurons. Future studies will be needed to determine if firing pattern is an important determinant of differential release of ACh and NE in sympathetic neurons.

4.6 Future Directions

Sympathetic neurons have the capacity to switch between cholinergic and noradrenergic phenotypes in vivo and to co-release both transmitters in cell culture. Neurotransmitter status in sympathetic neurons appears to depend on soluble, target-derived signaling molecules. This raises the possibility that sympathetic neuronal properties (i.e., neurotransmitter phenotype and firing patterns) are fine-tuned to the physiological needs of the specific target tissue through local signaling interactions. Target-derived signaling varies with physiological state, age and in response to tissue damage. For example, evidence suggests that a heart disease condition such as congestive heart failure can result in abnormalities of neurotrophin release from cardiac tissue including release of BDNF, which is normally not produced in heart tissue (Cai et al. 2006; Kreusser et al. 2007). Neurotrophin production also changes with age in sympathetically innervated tissues (Cai et al. 2006; Bierl and Isaacson 2007). These physiological changes could bring about adaptive (or maladaptive) shifts in neurotransmitter co-release and firing patterns in sympathetic neurons. Indeed, changes in sympathetic nerve activity occur in both hypertension and heart disease (Esler et al. 2001; Watson et al. 2006). A deeper understanding of how target-derived signaling regulates sympathetic neuronal properties could lead to clinical treatments for these diseases and to a better general understanding of how functional neuronal properties are regulated.

Sympathetic neurons co-release ACh and NE in culture conditions, and even apparently noradrenergically committed cells can rapidly be induced to release ACh via p75 signaling. It is less clear however, whether or not this actually occurs in vivo in mature animals. Interestingly, in adult humans, but not rodents, parasympathetic neurons innervating the heart and sympathetic neurons innervating sweat glands and vascular specializations in the hands and feet, called Hoyer-Grosser organs, were found to express all necessary proteins to co-release ACh and NE (Weihe et al. 2005). Data suggestive of potential co-transmission properties of sympathetic neurons in vivo is sparse however, and experiments that directly address co-transmission in physiologically realistic contexts are technically difficult. However, determining the role of co-transmission in the sympathetic system is an important step for understanding both diseases related to the sympathetic nervous system and neurotransmission in general.

References

Anderson RL, Jobling P and Gibbins IL (2001) Development of electrophysiological and morphological diversity in autonomic neurons. J Neurophysiol 86(3):1237–1251

Asmus SE, Parsons S and Landis SC (2000) Developmental changes in the transmitter properties of sympathetic neurons that innervate the periosteum. J Neurosci 20(4): 1495–1504

Bamji SX, Majdan M, Pozniak CD, Belliveau DJ, Aloyz R, Kohn J, Causing CG and Miller FD (1998) The p75 neurotrophin receptor mediates neuronal apoptosis and is essential for naturally occurring sympathetic neuron death. J Cell Biol 140(4):911–923

Bazan JF (1991) Neuropoietic cytokines in the hematopoietic fold. Neuron 7(2):197–208

Bibel M and Barde YA (2000) Neurotrophins: key regulators of cell fate and cell shape in the vertebrate nervous system. Genes Dev 14(23):2919–2937

Bierl MA, Jones EE, Crutcher KA and Isaacson LG (2005) "Mature" nerve growth factor is a minor species in most peripheral tissues. Neurosci Lett 380(1-2):133–137

Bierl MA and Isaacson LG (2007) Increased NGF proforms in aged sympathetic neurons and their targets. Neurobiol Aging 28(1):122–134

Bradley E, Law A, Bell D and Johnson CD (2003) Effects of varying impulse number on cotransmitter contributions to sympathetic vasoconstriction in rat tail artery. Am J Physiol Heart Circ Physiol 284(6):H2007–14

Brodski C, Schnurch H and Dechant G (2000) Neurotrophin-3 promotes the cholinergic differentiation of sympathetic neurons. Proc Natl Acad Sci U S A 97(17):9683–9688

Brodski C, Schaubmar A and Dechant G (2002) Opposing functions of GDNF and NGF in the development of cholinergic and noradrenergic sympathetic neurons. Mol Cell Neurosci 19(4):528–538

Bunge RP, Rees R, Wood P, Burton H and Ko C (1974) Anatomical and physiological observations on synapses formed on isolated autonomic neurons in tissue culture. Brain Research 66:401–412

Cai D, Holm JM, Duignan IJ, Zheng J, Xaymardan M, Chin A, Ballard VL, Bella JN and Edelberg JM (2006) BDNF-mediated enhancement of inflammation and injury in the aging heart. Physiol Genomics 24(3):191–197

Cassell JF, Clark AL and McLachlan EM (1986) Characteristics of phasic and tonic sympathetic ganglion cells of the guinea-pig. J Physiol 372:457–483

Chao MV and Hempstead BL (1995) p75 and Trk: a two-receptor system. Trends Neurosci 18(7):321–326

Chun LL and Patterson PH (1977) Role of nerve growth factor in the development of rat sympathetic neurons in vitro. I. Survival, growth, and differentiation of catecholamine production. J Cell Biol 75(3):694–704

Conforti L, Tohse N and Sperelakis N (1991) Influence of sympathetic innervation on the membrane electrical properties of neonatal rat cardiomyocytes in culture. J Dev Physiol 15(4):237–246

Dale H (1935) Pharmacology and nerve-endings. Proc. R. Soc. Med. 28:319–332

Dixon JE and McKinnon D (1994) Expression of the trk gene family of neurotrophin receptors in prevertebral sympathetic ganglia. Brain Res Dev Brain Res 77(2):177–182

Eccles J (1976) From electrical to chemical transmission in the central nervous system. Notes Rec R Soc Lond 30(2):219–230

Elfvin LG, Lindh B and Hokfelt T (1993) The chemical neuroanatomy of sympathetic ganglia. Annu Rev Neurosci 16:471–507

Elghozi JL and Julien C (2007) Sympathetic control of short-term heart rate variability and its pharmacological modulation. Fundam Clin Pharmacol 21(4):337–347

Ernsberger U and Rohrer H (1996) The development of the noradrenergic transmitter phenotype in postganglionic sympathetic neurons. Neurochem Res 21(7):823–829

Ernsberger U and Rohrer H (1999) Development of the cholinergic neurotransmitter phenotype in postganglionic sympathetic neurons. Cell Tissue Res 297(3):339–361

Ernsberger U (2000) Evidence for an evolutionary conserved role of bone morphogenetic protein growth factors and phox2 transcription factors during noradrenergic differentiation of sympathetic neurons. Induction of a putative synexpression group of neurotransmitter-synthesizing enzymes. Eur J Biochem 267(24):6976–6981

Esler M, Rumantir M, Kaye D, Jennings G, Hastings J, Socratous F and Lambert G (2001) Sympathetic nerve biology in essential hypertension. Clin Exp Pharmacol Physiol 28(12):986–989

Francis NJ, Asmus SE and Landis SC (1997) CNTF and LIF are not required for the target-directed acquisition of cholinergic and peptidergic properties by sympathetic neurons in vivo. Dev Biol 182(1):76–87

Francis NJ and Landis SC (1999) Cellular and molecular determinants of sympathetic neuron development. Annu Rev Neurosci 22:541–566

Fukada K (1980) Hormonal control of neurotransmitter choice in sympathetic neurone cultures. Nature 287(5782):553–555

Fulop T, Radabaugh S and Smith C (2005) Activity-dependent differential transmitter release in mouse adrenal chromaffin cells. J Neurosci 25(32): 7324–7332

Furshpan EJ, MacLeish PR, O'Lague PH and Potter DD (1976) Chemical transmission between rat sympathetic neurons and cardiac myocytes developing in microcultures: evidence for cholinergic, adrenergic, and dual-function neurons. Proc Natl Acad Sci U S A 73(11):4225–4259

Furshpan EJ, Landis SC, Matsumoto SG and Potter DD (1986) Synaptic functions in rat sympathetic neurons in microcultures. I. Secretion of norepinephrine and acetylcholine. J Neurosci 6(4):1061–1079
Greene LA, Seeley PJ, Rukenstein A, DiPiazza M and A Howard (1984) Rapid activation of tyrosine hydroxylase in response to nerve growth factor. J Neurochem 42(6):1728–1734
Habecker BA and Landis SC (1994) Noradrenergic regulation of cholinergic differentiation. Science 264(5165):1602–1604
Habecker BA, Pennica D and Landis SC (1995) Cardiotrophin-1 is not the sweat gland-derived differentiation factor. Neuroreport 7(1):41–44
Habecker BA, Tresser SJ, Rao MS and Landis SC (1995) Production of sweat gland cholinergic differentiation factor depends on innervation. Dev Biol 167(1):307–316
Habecker BA, Symes AJ, Stahl N, Francis NJ, Economides A, Fink JS, Yancopoulos GD and Landis SC (1997) A sweat gland-derived differentiation activity acts through known cytokine signaling pathways. J Biol Chem 272(48):30421–30428
Huber LJ and Chao MV (1995) A potential interaction of p75 and trkA NGF receptors revealed by affinity crosslinking and immunoprecipitation. J Neurosci Res 40(4):557–563
Jobling P and Gibbins IL (1999) Electrophysiological and morphological diversity of mouse sympathetic neurons. J Neurophysiol 82(5):2747–2764
Kreusser MM, Buss SJ, Krebs J, Kinscherf R, Metz J, Katus HA, Haass M and Backs J (2007) Differential expression of cardiac neurotrophic factors and sympathetic nerve ending abnormalities within the failing heart. J Mol Cell Cardiol
Kupfermann I (1991) Functional studies of cotransmission. Physiol Rev 71(3):683–732
Lee KF, Li E, Huber LJ, Landis SC, Sharpe AH, Chao MV and Jaenisch R (1992) Targeted mutation of the gene encoding the low affinity NGF receptor p75 leads to deficits in the peripheral sensory nervous system. Cell 69(5):737–749
Lockhart ST, Turrigiano GG and Birren SJ (1997) Nerve growth factor modulates synaptic transmission between sympathetic neurons and cardiac myocytes. J Neurosci 17(24): 9573–9582
Lu B, Pang PT and Woo NH (2005) The yin and yang of neurotrophin action. Nat Rev Neurosci 6(8):603–614
Luther JA and Birren SJ (2006) Nerve growth factor decreases potassium currents and alters repetitive firing in rat sympathetic neurons. J Neurophysiol 96(2):946–958
Marder E, Christie AE and Kilman VL (1995) Functional organization of cotransmission systems: lessons from small nervous systems. Invert Neurosci 1(2):105–112
Marvin WJ, Jr., Atkins DL, Chittick VL, Lund DD and Hermsmeyer K (1984) In vitro adrenergic and cholinergic innervation of the developing rat myocyte. Circ Res 55(1):49–58
O'Lague PH, Obata K, Claude P, Furshpan EJ and Potter DD (1974) Evidence for cholinergic synapses between dissociated rat sympathetic neurons in cell culture. Proc Natl Acad Sci U S A 71(9):3602–3606
O'Lague PH, Furshpan EJ and Potter DD (1978) Studies on rat sympathetic neurons developing in cell culture. II. Synaptic mechanisms. Dev Biol 67(2):404–423
O'Lague PH, Potter DD and Furshpan EJ (1978) Studies on rat sympathetic neurons developing in cell culture. I. Growth characteristics and electrophysiological properties. Dev Biol 67(2):384–403
O'Lague PH, Potter DD and Furshpan EJ (1978) Studies on rat sympathetic neurons developing in cell culture. III. Cholinergic transmission. Dev Biol 67 (2):424–443
Patterson PH and Chun LL (1974) The influence of non-neuronal cells on catecholamine and acetylcholine synthesis and accumulation in cultures of dissociated sympathetic neurons. Proc Natl Acad Sci U S A 71 (9):3607–3610
Patterson PH and Chun LL (1977) The induction of acetylcholine synthesis in primary cultures of dissociated rat sympathetic neurons. I. Effects of conditioned medium. Dev Biol 56 (2):263–280

Potter DD, Landis SC, Matsumoto SG and Furshpan EJ (1986) Synaptic functions in rat sympathetic neurons in microcultures. II. Adrenergic/cholinergic dual status and plasticity. J Neurosci 6 (4):1080–1098

Potter DD, Matsumoto SG, Landis SC, Sah DW and Furshpan EJ (1986) Transmitter status in cultured sympathetic principal neurons: plasticity, graded expression and diversity. Prog Brain Res 68:103–120

Randolph CL, Bierl MA and Isaacson LG (2007) Regulation of NGF and NT-3 protein expression in peripheral targets by sympathetic input. Brain Res 1144:59–69

Ro, MS and Landis SC (1990) Characterization of a target-derived neuronal cholinergic differentiation factor. Neuron 5 (6):899–910

Rao MS, Patterson PH and Landis SC (1992) Multiple cholinergic differentiation factors are present in footpad extracts: comparison with known cholinergic factors. Development 116 (3):731–744

Reichardt LF (2006) Neurotrophin-regulated signalling pathways. Philos Trans R Soc Lond B Biol Sci 361 (1473):1545–1564

Rohrer H (1992) Cholinergic neuronal differentiation factors: evidence for the presence of both CNTF-like and non-CNTF-like factors in developing rat footpad. Development 114 (3):689–698

Saadat S, Sendtner M and Rohrer H (1989) Ciliary neurotrophic factor induces cholinergic differentiation of rat sympathetic neurons in culture. J Cell Biol 108 (5):1807–1816

Schafer MK, Schutz B, Weihe E and Eiden LE (1997) Target-independent cholinergic differentiation in the rat sympathetic nervous system. Proc Natl Acad Sci U S A 94 (8):4149–4154

Schotzinger RJ and Landis SC (1988) Cholinergic phenotype developed by noradrenergic sympathetic neurons after innervation of a novel cholinergic target in vivo. Nature 335 (6191):637–639

Schotzinger RJ and Landis SC (1990) Acquisition of cholinergic and peptidergic properties by sympathetic innervation of rat sweat glands requires interaction with normal target. Neuron 5 (1):91–100

Seal RP and Edwards RH (2006) Functional implications of neurotransmitter co-release: glutamate and GABA share the load. Curr Opin Pharmacol 6 (1):114–119

Slonimsky JD, Yang B, Hinterneder JM, Nokes EB and Birren SJ (2003) BDNF and CNTF regulate cholinergic properties of sympathetic neurons through independent mechanisms. Mol Cell Neurosci 23 (4):648–660

Slonimsky JD, Mattaliano MD, Moon JI, Griffith LC and Birren SJ (2006) Role for calcium/calmodulin-dependent protein kinase II in the p75-mediated regulation of sympathetic cholinergic transmission. Proc Natl Acad Sci U S A 103 (8):2915–2919

Stanke M, Duong CV, Pape M, Geissen M, Burbach G, Deller T, Gascan H, Otto C, Parlato R, Schutz G and Rohrer H (2006) Target-dependent specification of the neurotransmitter phenotype: cholinergic differentiation of sympathetic neurons is mediated in vivo by gp 130 signaling. Development 133 (1):141–150

Wang HS and McKinnon D (1995) Potassium currents in rat prevertebral and paravertebral sympathetic neurones: control of firing properties. J Physiol 485 (Pt 2):319–335

Watson AM, Hood SG and May CN (2006) Mechanisms of sympathetic activation in heart failure. Clin Exp Pharmacol Physiol 33 (12):1269–1274

Weihe E, Schutz B, Hartschuh W, Anlauf M, Schafer MK and Eiden LE (2005) Coexpression of cholinergic and noradrenergic phenotypes in human and nonhuman autonomic nervous system. J Comp Neurol 492 (3):370–379

Woo NH, Teng HK, Siao CJ, Chiaruttini C, Pang PT, Milner TA, Hempstead BL and Lu B (2005) Activation of p75NTR by proBDNF facilitates hippocampal long-term depression. Nat Neurosci 8 (8):1069–1077

Wyatt S and Davies AM (1995) Regulation of nerve growth factor receptor gene expression in sympathetic neurons during development. J Cell Biol 130 (6):1435–1446

Yamamori T, Fukada K, Aebersold R, Korsching S, Fann MJ and Patterson PH (1989) The cholinergic neuronal differentiation factor from heart cells is identical to leukemia inhibitory factor. Science 246 (4936):1412–1416

Yang B, Slonimsky JD and Birren SJ (2002) A rapid switch in sympathetic neurotransmitter release properties mediated by the p75 receptor. Nat Neurosci 5 (6):539–545

Chapter 5
GABA, Glycine, and Glutamate Co-Release at Developing Inhibitory Synapses

Deda C. Gillespie and Karl Kandler

Abstract Neurobiologists have long classified synaptic phenotype by a single neurotransmitter released at that synapse. Research over the past two decades has made it clear, however, that the classification of neurons and synapses as purely GABAergic, or even as purely inhibitory or excitatory, is no longer valid. In this chapter we review evidence showing that inhibitory synapses co-release multiple inhibitory neurotransmitters, and that some classical inhibitory synapses also release excitatory neurotransmitters. As multiple transmitter release is particularly prevalent at immature synapses, we pay special attention to developmental plasticity in considering possible mechanisms and functions for release of these seemingly antagonistic neurotransmitters.

List of Abbreviations

ACh	acetylcholine
AMPAR	amino-3-hydroxy-5-methyl-4-isoxazolepropionic acid receptor
ATP	adenosine triphosphate
CN	cochlear nucleus
GABA	gamma-aminobutyric acid
$GABA_AR$	GABA (A) receptor
$GABA_BR$	GABA (B) receptor
GAD	glutamic acid decarboxylase
GlyR	glycine receptor
GLYT2	glycine transporter 2
IPSC	inhibitory postsynaptic current
LSO	lateral superior olive
mIPSC	miniature inhibitory postsynaptic current
MNTB	medial nucleus of the trapezoid body

D.C. Gillespie (✉)
Department of Psychology, Neuroscience & Behaviour, McMaster University, Hamilton, ON L8S 4K1, Canada
e-mail: dgillespie@mcmaster.ca

R. Gutierrez (ed.), *Co-Existence and Co-Release of Classical Neurotransmitters*,
DOI 10.1007/978-0-387-09622-3_5,

mPSC	miniature postsynaptic current
MSO	medial superior olive
NMDAR	N-methyl D-aspartic acid receptor
P*n*	postnatal day *n*
SPN	superior paraolivary nucleus
VGAT	vesicular GABA transporter
VGLUT2	vesicular glutamate transporter 2
VGLUT3	vesicular glutamate transporter 3
VIAAT	vesicular inhibitory amino acid transporter

5.1 Introduction

Classically, many neuroscientists have used Dale's principle as a first-order descriptor of neuronal phenotype. Although referred to as the idea that the same neurotransmitter is released from all terminals of a neuron (Eccles 1964), this principle has been reduced and simplified into the widely accepted dogma: "one neuron-one neurotransmitter." In fact, many students still learn this version of Dale's principle, despite a significant and growing number of studies that have provided convincing counterexamples to invalidate this overly simplified view of neuron and synapse. In this chapter we will focus on a subset of the evidence supporting a more nuanced view of the synapse: release of multiple transmitters at inhibitory synapses. We will first consider co-release of the classic small amino acid neurotransmitters GABA and glycine and we will then consider release of GABA or glycine with other neurotransmitters, in particular glutamate. We will pay special attention to a synapse in the auditory brainstem where the three major fast neurotransmitters of the brain—GABA, glycine and glutamate—are all released during a developmentally significant period, and we will consider potential hypotheses for the function of multiple transmitter release.

5.1.1 Co-release of GABA and Glycine

Early evidence that fast inhibitory synapses might use multiple neurotransmitters came from immunohistochemical studies at the light and electron microsopic levels, which showed colocalization of markers for GABA and glycine in cerebellum, spinal cord, auditory brainstem, dorsal cochlear nucleus, and oculomotor nucleus, among others (Triller et al. 1987; Ottersen et al. 1988; Todd and Sullivan 1990; Helfert et al. 1992; Kolston et al. 1992; Wentzel et al. 1993; Juiz et al. 1996).

5.1.1.1 Spinal Cord and Brainstem

More recent studies have presented physiological evidence for concomitant release of GABA and glycine. Using whole-cell patch clamp recordings in acute slices of neonatal spinal cord, Jonas et al. (1998) recorded simultaneously from interneurons and postsynaptic presumed motor neurons. By stimulating the presynaptic interneuron to induce monosynaptic unitary inhibitory postsynaptic currents (IPSCs), and applying antagonists of $GABA_A$ or glycine receptors ($GABA_ARs$ or GlyRs), the authors showed that the unitary IPSC comprises two components, a strychnine-sensitive component with fast kinetics characteristic of GlyR-mediated currents and a bicuculline-sensitive component with slower kinetics characteristic of $GABA_AR$-mediated currents (Fig. 5.1). Physiologically, the most convincing evidence for co-release of GABA and glycine from single vesicles comes from

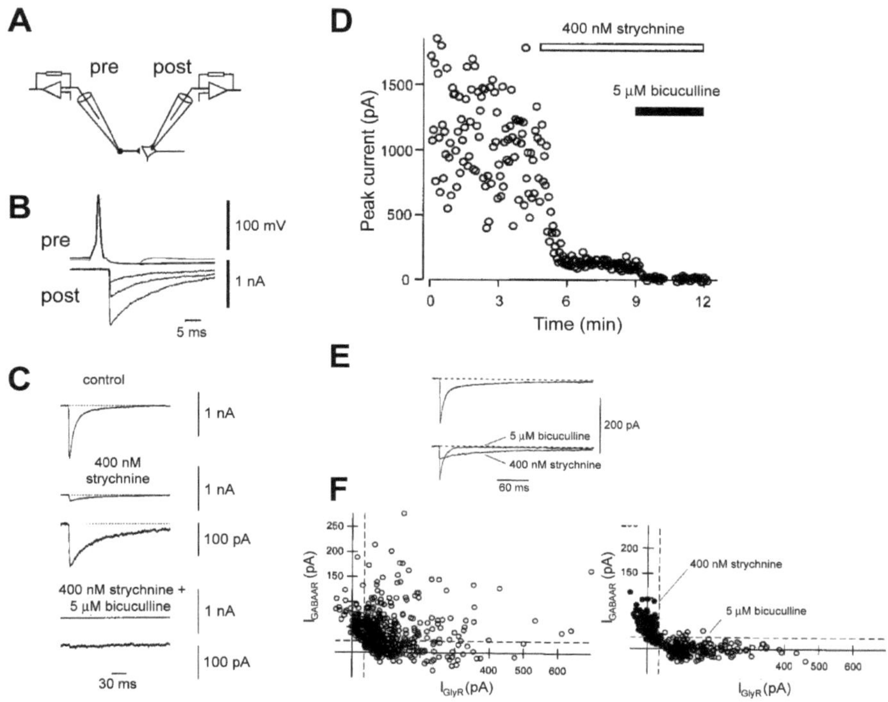

Fig. 5.1 GABA-glycine co-release at immature synapses in the spinal cord. **(a)** Schematic illustration of simultaneous whole-cell recordings from a presynaptic interneuron and a postsynaptic motoneuron. **(b)** Examples of three individual inhibitory postsynaptic currents in the motoneuron (post) evoked by three presynaptic action potentials in the interneuron (pre) (traces are overlaid). **(c, d)** Unitary postsynaptic currents are mediated by both glycine and GABA receptors. The GlyR antagonist strychnine strongly reduces postsynaptic responses but leaves a component that is blocked by the $GABA_AR$ antagonist bicuculline. C) Illustrates the average of 3–10 single sweeps. In D) the peak amplitudes of single responses are plotted against time before and during antagonist application. **(e)** Glycine and GABA components of miniature IPSCs can be distinguished by their decay times. Flunitrazepam, which prolongs specifically $GABA_AR$-mediated currents, increases the decay times of $GABA_AR$- mediated currents. Average miniature IPSCs in control conditions (upper traces) and in the presence of bicuculline or strychnine (lower traces). **(f)** Scatter plots of the amplitudes of the $GABA_AR$-mediated component against amplitudes of the GlyR-mediated component of mIPSCs without (left plot) and in the presence of antagonists (right plot). Points falling outside the dashed lines indicate individual mIPSCs with dual GABA and glycine components. *Adapted with permission from Jonas et al.* 1998.

analysis of miniature IPSCs (mIPSCs), as each mIPSC is generally considered to be a single quantal event resulting from the release of the contents of a single synaptic vesicle. Miniature IPSCs recorded at the spinal interneuron-motor neuron synapse exhibit mixed components, distinguished by differing receptor pharmacology and kinetics (Fig. 5.1e-f). A subset of mIPSCs is mediated purely by $GABA_ARs$, and a larger subset purely by GlyRs, whereas nearly half of all mIPSCs are mediated by both $GABA_ARs$ and GlyRs (points above and to right of dotted lines in Fig. 5.1f, left panel). The group of mixed mIPSCs comprising both GABAergic and glycinergic components indicates that spinal interneurons' terminals contain synaptic vesicles that include both GABA and glycine, and confirms that GABA and glycine are packaged together in individual synaptic vesicles from which they are co-released. Co-release of GABA and glycine is likely to be a common property of developing inhibitory synapses, as other groups subsequently have revealed mixed GABA and glycine release onto functional GlyRs and $GABA_ARs$ at the sympathetic preganglionic neurons of spinal cord lamina X, dorsal horn laminae I-II, and abducens and hypoglossal motoneuron synapses (O'Brien and Berger 1999; Keller et al. 2001; Russier et al. 2002; Seddik et al. 2007).

5.1.1.2 Auditory Brainstem

The lateral superior olive (LSO), a binaural nucleus in the auditory brainstem that computes interaural level differences (Boudreau and Tsuchitani 1968) (Fig. 2a), receives a prominent inhibitory input from the medial nucleus of the trapezoid body (MNTB; Moore and Caspary 1983, Caspary and Finlayson 1991). As in the spinal cord, early electron microscopic studies of immunoreactivity for the amino acid neurotransmitters in the LSO showed label for both GABA and glycine in the same synaptic terminals, and pointed to the possibility that GABA and glycine might both be released from single synapses onto principal neurons of the LSO (Helfert et al. 1992) (Fig. 5.2b).

Fig. 5.2 (continued) **(c)** MNTB-LSO synapses switch from being mainly GABAergic in newborn animals to being mainly glycinergic around hearing onset. Examples illustrate MNTB-elicited postsynaptic currents obtained in whole-cell recordings of LSO neurons in slices from 4-day-old (P4) and 14-day- old (P14) gerbils. At P4, postsynaptic currents are strongly reduced by the $GABA_AR$ antagonist bicuculline (BIC) but only slightly affected by the GlyR antagonist strychnine. At P14, bicuculline has little effect whereas strychnine almost completely abolishes the responses. **(d)** Developmental down-regulation of $GABA_ARs$ and up-regulation of GlyRs in the LSO. Immunoreactivity for the β2, 3 subunits of the $GABA_AR$ decreases in the gerbil from P4 to P14 while immunoreactivity for the α 1 subunit increases in the rat LSO from P4 to P10. **(e)** Co-release of GABA and glycine from single synaptic vesicles in isolated LSO neurons. Traces show examples of miniature PSCs that are glycinergic (fast decay), GABAergic (slow decay), and mixed GABA/glycine (fast and slow decay components). The plot shows changes in the proportion of the three types of mPSCs during development. In newborn rats (P1–2) most mPSCs are GABAergic whereas at P16–17 most mPSCs are glycinergic. Mixed GABA/glycine mPSCS are encountered at all ages investigated. **(f)** Immunoreactivity for glycine increases from P5 to P18 but immunoreactivity for glutamic acid decarboxylase (GAD) decreases, indicating a developmental decrease in the presynaptic release of GABA. *Adapted with permission from: B) Helfert et al.* 1992; *C, D) Kotak et al.* 1998; *D) Friauf et al. 1997; E,F) Nabekura et al.* 2004.

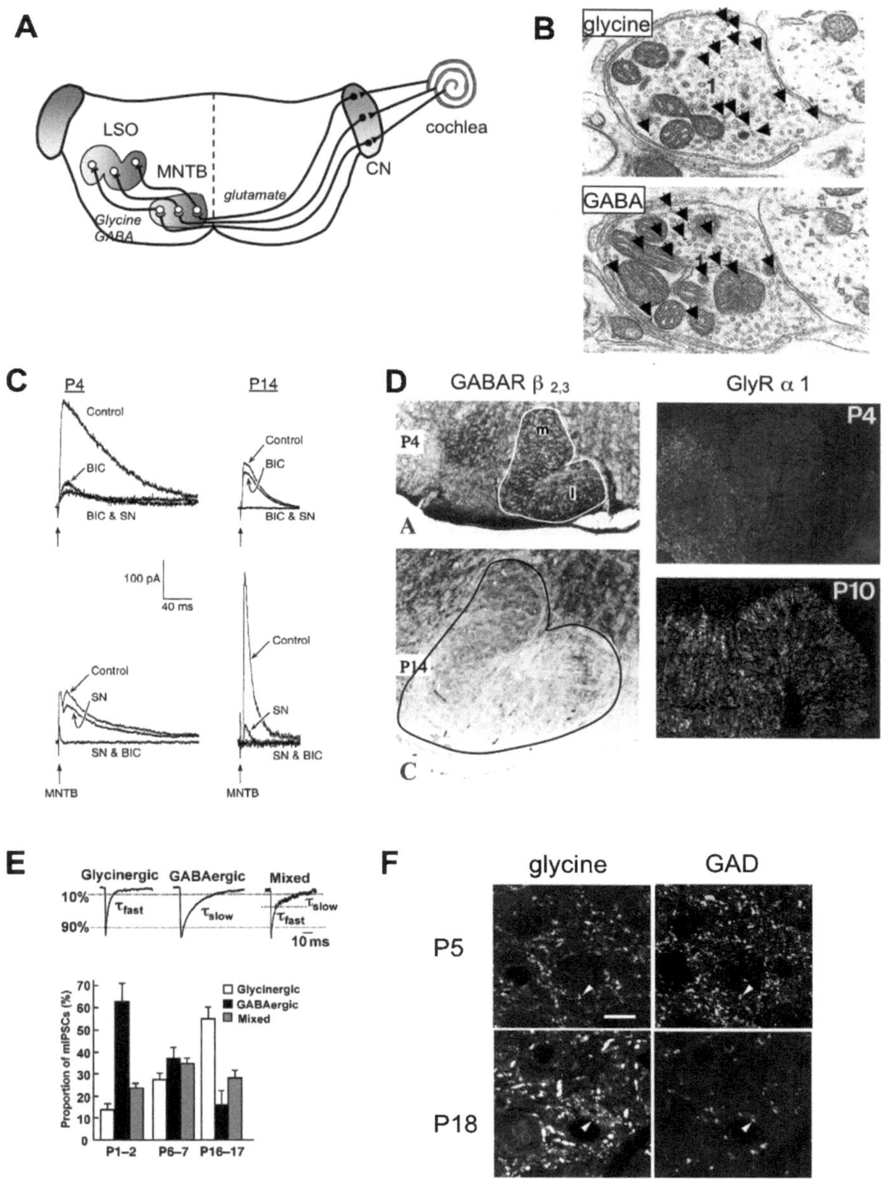

Fig. 5.2 GABA-glycine co-release at developing auditory synapses. **(a)** Schematic illustration of the inhibitory MNTB-LSO pathway in the auditory brainstem. Neurons in the medial nucleus of the trapezoid body (MNTB) receive glutamatergic inputs from the contralateral cochlear nucleus (CN). MNTB neurons give rise to a tonotopically organized inhibitory pathway to the lateral superior olive (LSO). This pathway is glycinergic in mature animals but during development is primarily GABAergic. **(b)** Electron micrographs of an individual terminal from serial sections immunolabeled for glycine and GABA. The terminal labels positively for both glycine and GABA. Gold particles that tag immunopositive sites.

Further physiological and histological studies not only corroborated coincident GABA and glycine release, but also showed that the relative balance of GABAergic and glycinergic components in the MNTB-LSO response shifts during early development. This has been shown in acute brain slices, using whole-cell recordings from LSO principal neurons to measure their physiological response to electrical stimulation of the MNTB. Using the receptor antagonists strychnine and bicuculline to separate the GlyR and $GABA_AR$ components of the response to electrical stimulation of the MNTB shows that in early neonatal (postnatal day 4; P4) gerbil, a majority of the synaptic current is mediated by $GABA_ARs$, whereas by P14 the majority of the current passes through GlyRs (Kotak et al. 1998) (Fig. 5.2c). This developmental shift from a primarily GABAergic to primarily glycinergic response is accompanied by a shift in the population of receptors expressed postsynaptically in the dendrites of LSO neurons (Fig. 5.2d). Immunoreactivity for $GABA_AR$ subunits is relatively high in the postsynaptic membrane of LSO neurons at P4 but decreases over the next 10 days. The inverse pattern is seen for markers of GlyR, as immunoreactivity for the GlyR-associated protein gephyrin increases between P4 and P14 (Korada and Schwartz 1999). Because transcripts for the GlyR subunit $\alpha 2$ are present only at low levels in the LSO throughout postnatal development, whereas transcripts for the GlyR subunit $\alpha 1$ increase over the first few weeks (Piechotta et al. 2001), this increased expression of GlyRs is probably determined largely by addition of the GlyR subunit $\alpha 1$. The decrease in $GABA_AR$ expression, concomitant with an increase in GlyR expression, causes a striking decrease in the $GABA_AR$/GlyR ratio between P4 and P14.

The developmental progression from primarily GABAergic to primarily glycinergic transmission at the developing MNTB-LSO synapse, which has now been confirmed in several studies (Henkel and Brunso-Bechtold 1998; Kullmann and Kandler 2001; Kullmann et al. 2002), results not only from a shift in receptor expression, but also from a shift in neurotransmitter release. In order to isolate and closely examine spontaneous mIPSCs, Nabekura et al. (2004) mechanically dissociated LSO principal cells along with adherent presynaptic (MNTB) terminals and then distinguished the GABAergic and glycinergic components of spontaneous mIPSCs using receptor pharmacology and kinetics. They found a mixed population of mIPSCs: at birth the majority of mIPSCs were purely GABAergic, with the remainder split between purely glycinergic and mixed gly/GABA mIPSCs; at one week the three populations were roughly equal in proportion, and at two weeks the majority of the mIPSCs were purely glycinergic. Although mixed gly/GABA mIPSCs (and hence mixed gly/GABA vesicles) were present at all ages, a clear developmental trend was seen, shifting from predominantly GABA release toward predominantly glycine release (Fig. 5.2e). Over the same period, immunolabeling for glycine increases in presynaptic terminals, while immunolabeling for the GABA marker glutamic acid decarboxylase (GAD) decreases (Fig. 5.2f). Thus, the physiological shift from GABAergic to glycinergic transmission (Kotak et al. 1998) is due to both a shift in postsynaptic receptor expression (Korada and Schwartz 1999) and a shift in vesicle content (Nabekura et al. 2004). Whether the downregulation of $GABA_AR$ expression and the decrease in GABA content of synaptic vesicles occur simultaneously, or whether one event leads—or even induces—the other, is not known. Additionally, the mechanism that accounts for the decrease in GABA-containing synaptic vesicles is an open question, as indeed is the mechanism that determines whether synaptic vesicles contain GABA, glycine, or both.

The progression from GABAergic to glycinergic phenotype at the MNTB-LSO synapse is mirrored at other synapses in the auditory brainstem. For example, in the nearby medial superior olive (MSO), synapses in the inhibitory MNTB-MSO pathway, which are nearly exclusively glycinergic in the adult, also exhibit a prominent GABAergic component during the first postnatal week (Smith et al. 2000). Inhibitory synapses within the MNTB also show a mixed glycinergic and GABAergic phenotype during early development, switching to exclusively glycinergic by P25 (Awatramani et al. 2005). Finally, the switch from GABAergic to glycinergic function is not limited to information transfer in the feed-forward direction and the shift in receptor expression is not limited to the postsynaptic membrane. At the well-known Calyx of Held synapse in the MNTB, activation of presynaptic GlyRs normally causes increased transmitter release (Turecek and Trussell 2001). Before approximately P11, however, this GlyR modulation of glutamate release is largely absent (Turecek and Trussell 2002), and glutamate release is enhanced instead by activation of presynaptic $GABA_ARs$.

5.1.2 *VGAT and Co-release of GABA/Glycine*

The molecular basis for the inhibitory phenotype of MNTB-LSO synapses, as at all synapses that release inhibitory amino acids, is expression of the vesicular GABA transporter (VGAT; also known as vesicular inhibitory amino acid transporter, VIAAT). VGAT, which is localized to synaptic vesicles of glycinergic and GABAergic neurons, was first identified as a proton-coupled high-affinity GABA transporter that also transports glycine, though with lower affinity (McIntire et al. 1997; Sagne et al. 1997; Chaudhry et al. 1998). The only known vesicular transporter for inhibitory amino acids, VGAT underlies co-release of GABA and glycine (Wojcik et al. 2006). Because GABA and glycine share the same vesicular transporter, the mechanism that specifies whether vesicles are GABA- or glycinergic is unknown, though one possibility is that the relative abundance of glycine and GABA in synaptic vesicles is determined by the availability of glycine in the presynaptic terminal. At least *in vitro?* a glycinergic phenotype can be achieved in cell lines by coexpression of VGAT with GLYT2, a membrane-bound, high-affinity, Na^+-coupled, glycine uptake transporter expressed in glycinergic neuronal terminals (Liu et al. 1993; Zafra et al. 1995; Spike et al. 1997; Aubrey et al. 2007), and this mechanism may also regulate GABA/glycine vesicular content in the MNTB terminals. In the LSO, GLYT2 is expressed in presumed presynaptic terminals and GLYT2 expression levels increase during the first two postnatal weeks (Friauf et al. 1999), the same period during which the glycinergic component of MNTB-LSO synapses increases. Because, however, GLYT2 in the LSO is already present prenatally (Friauf et al. 1999)—when MNTB-LSO synapses are predominantly GABAergic—GLYT2 expression alone cannot account for the switch from predominantly GABA- to predominantly glycine-containing vesicles.

5.1.3 Functional Role for Co-release of GABA and Glycine in Developing Auditory Brainstem

Although the progression from release of mixed GABA and glycine to release of glycine alone is common in several areas during development, it is currently not known whether this developmental change is primarily a non-functional epiphenomenon reflecting other developmental processes (such as the maturation of glycine transporters) or whether early GABAergic signaling is important in establishing glycinergic networks. A number of reasons have been proposed for why early GABAergic transmission might be developmentally significant.

5.1.3.1 Trophic Actions of GABA

GABAergic neurotransmission appears to have trophic effects on several early developmental processes including synaptogenesis (for reviews, see Owens and Kriegstein 2000; Represa and Ben-Ari 2005). Although the possible trophic role of GABA is controversial, and may primarily be due to the depolarizing effect GABA exerts during early development, GABAergic neurotransmission is a common feature at many nominally non-GABAergic synapses during development (Ben-Ari et al. 1997, rev; Overstreet-Wadiche et al. 2005). If depolarization is the critical feature of early putative trophic effects of GABA, then is glycine, which also induces depolarization during early development, able to accomplish the same task? Glycine might be sufficiently depolarizing for this scenario, but if longer depolarizations were required, GABAergic transmission would be more effective, due to the slower kinetics of $GABA_A$Rs. Results from a VGAT-knockout mouse (Wojcik et al. 2006), however, showing that synaptogenesis and postsynaptic receptor clustering can occur in the absence of vesicular GABA release, argue against a critical trophic role for GABA in synapse formation.

5.1.3.2 Receptor Kinetics

At excitatory synapses in which activity-dependent mechanisms strengthen or weaken synapses, the timing of inputs relative to postsynaptic membrane depolarizations can determine both the direction and the amplitude of the plasticity (Bi and Poo 1998). This dependence on timing means that the width of the depolarization window, which is itself determined by the kinetics of response to neurotransmitter, can influence synaptic plasticity. For example, decay times for NMDAR-mediated currents in many systems are long early in development and decrease with age (Hestrin 1992; Carmignoto and Vicini 1992), following a timecourse that corresponds to a decrease in developmental plasticity (Crair and Malenka 1995). The early expression of subunits that confer slower kinetics may lengthen the postsynaptic membrane depolarization, increasing the window for coincidence detection and allowing developing circuits access to mechanisms of synaptic plasticity during a period of long synaptic delays and low conduction velocities. Our understanding of plasticity at developing inhibitory synapses is more rudimentary (Gaiarsa et al. 2002, rev; Woodin et al. 2003; Haas et al. 2006), but if glycinergic synapses do undergo analogous timing-dependent plasticity, it is possible that the slower kinetics of $GABA_A$Rs might be better suited to mediating plasticity at the relatively slow speeds of synaptic transmission and action potential conduction of developing circuits. In this receptor kinetics hypothesis, synapses could be established using the slower kinetics of $GABA_A$Rs; the subsequent replacement of $GABA_A$Rs with GlyRs over time would cause the maturing synapse to switch to a predominantly fast glycinergic phenotype.

5.1.3.3 Receptor Clustering

An alternate scenario posits that $GABA_A$Rs are required to establish initial receptor clusters at developing synapses, and that with maturity $GABA_A$Rs are replaced within the synapse by GlyRs. Although much about the development of inhibitory synapses in general is still unknown, the GlyR- and $GABA_A$R- associated protein gephyrin is understood to be critical for clustering GlyRs at functional synapses, as gephyrin-deficient mice lack postsynaptic GlyRs clusters (Feng et al. 1998). Although loss of gephyrin also results in reduced $GABA_A$R clusters (Kneussel et al. 1999), $GABA_A$Rs that include certain receptor subunits can cluster independently of gephyrin (Kneussel et al. 2001). The additional finding that $GABA_A$R clusters can induce associated gephyrin clustering in cultured hippocampal cells over a period of several hours (Levi et al. 2004) correlates with a clustering function for early GABAergic transmission. The clustering hypothesis would assume that for some reason (as might occur, for example, with developmental regulation of splice variants for gephyrin or the gephyrin-binding GlyR β subunit; Paarmann et al. 2006; Oertel et al. 2007), gephyrin is not available to mediate the clustering and maturation of GlyRs during early developmental stages. In the absence of gephyrin, the postsynaptic components of developing inhibitory synapses could nevertheless be established by $GABA_A$R clustering. $GABA_A$R clusters in these nascent synapses would then induce gephyrin clustering, and the gephyrin clusters would in turn seed and organize GlyR clusters. Together with a subsequent loss of $GABA_A$Rs, this increase in functional GlyRs would effect the switch from GABAergic to glycinergic postsynaptic phenotype.

5.1.4 Function of Co-release at Mature Synapses

Although synapses that are mixed GABA- and glycin- ergic during development become primarily glycinergic in the adult, a GABAergic component remains in some cases (Chery and De Koninck 2000; Russier et al. 2002). In the LSO, the large-scale shift in proportion of GABA to glycine release and the upregulation of GlyRs are consistent with a developmental role for GABA. The presence of a lingering GABAergic component (Helfert et al. 1992; Nabekura et al. 2004), however, is consistent with an additional role of GABA in the physiology of the mature synapse. For example, the presence of two transmitters with differing kinetics could allow the synapse to use a greater range of IPSC shapes that might be informationally relevant (Russier et al. 2002). Alternatively, co-packaging of two transmitters could allow a single vesicle released from a presynaptic terminal to activate receptors in distinct neuronal subpopulations. For example, Golgi cells in the cerebellar granular layer release both glycine and GABA, but onto distinct postsynaptic targets: glycine acts on GlyRs on unipolar brush cells, whereas GABA acts on granule cells (Dugue et al. 2005). At present, two such separate targets are unknown in the LSO, as is also the relative location of GlyRs and $GABA_A$Rs. More generally, it has been suggested that synapses using vesicles with varying proportions of two neurotransmitters could achieve a more finely graded range of information transfer than that generally achieved with single-neurotransmitter quantal release (Somogyi 2006). Because GABA and glycine share the same vesicular transporter, however, at GABA/glycinergic synapses this scenario would require involvement of other mechanisms, such as GLYT2 (Aubrey et al. 2007), to regulate glycine concentrations in presynaptic terminals and to control the GABA:glycine ratio in synaptic vesicles. Finally, it is possible that GABA co-released in the adult LSO does not reach postsynaptic receptors, but acts only on presynaptic $GABA_B$Rs. Precedence for this idea exists in the response to glycine and GABA co-release in the adult spinal cord, where release fails to activate postsynaptic $GABA_A$Rs, but does activate postsynaptic GlyRs and presynaptic $GABA_B$Rs (Chery and De Koninck 2000). Although the expression of presynaptic $GABA_B$Rs at the MNTB-LSO synapse is not well understood, $GABA_B$Rs have been shown to modulate both glutamate and glycine release at other synapses in the auditory brainstem (Isaacson 1998, Lim et al. 2000; though note also presynaptic GlyRs and $GABA_A$Rs in developing MNTB (Turecek and Trussell 2001).

5.2 Dual Release of GABA or Glycine and Other Neurotransmitters

The discovery that GABA and glycine can be co-released at many synapses has necessitated a change in the stereotypical model of the inhibitory synapse, but because GABA and glycine are both fast inhibitory neurotransmitters, it has not forced a complete rethinking of the inhibitory synapse. More surprising have been studies suggesting that GABA, and/or glycine, is released with quite different neurotransmitters at several synapses. The following represent a subset of a growing list of such examples.

5.2.1 Release of Multiple Transmitters in the Retina

GABA or glycine co-release with other neurotransmitters appears to be a common theme in amacrine cells of the retina. One population of glycine-immunoreactive amacrine cells is immunoreactive for the vesicular glutamate transporter VGLUT3 (Johnson et al. 2004; Haverkamp and Wassle 2004), and is thought to release glutamate vesicularly. It is unclear whether glycine functions primarily as a neurotransmitter in these synapses, though the lack of VGAT in these cells suggests that any synaptic release is likely via a membrane transporter; regardless, as glycine and glutamate are not both vesicularly released they are unlikely to be co-released from single vesicles.

A second retinal population, the starburst amacrine cells, release both ACh and GABA. Although at these synapses GABA may be released either vesicularly or via reversal of a membrane transporter, it is unlikely to be released together with ACh from single vesicles (Vaney et al. 1988; O'Malley and Masland 1989; O'Malley et al. 1992; Zheng et al. 2004). The role of cholinergic amacrine cells in generating spontaneous retinal waves that drive early visual plasticity (Meister et al. 1991; Feller et al. 1996; Hooks and Chen 2006; Huberman et al. 2006), suggests that release of GABA in the cholinergic amacrine network may affect developmental refinement by shaping spontaneous retinal activity (Wang et al. 2007).

Yet a third retinal amacrine synapse comprises both dopaminergic and GABAergic elements. At synapses between the dopaminergic amacrine cell and the AII amacrine cell, immunoreactivity for GABA colocalizes with immunoreactivity for the dopamine marker tyrosine hydroxylase (Wulle and Wagner 1990). In addition, VGAT and the dopamine-associated vesicular monoamine transporter 2 are both expressed presynaptically, while $GABA_A$Rs are expressed postsynaptically (Contini and Raviola 2003), suggesting that at these synapses both dopamine and GABA are released as functional neurotransmitters.

5.2.2 Release of Multiple Transmitters in Other Brain Areas

Retinal amacrine cells are not the only neurons that may share cholinergic and GABAergic phenotypes, as subpopulations of neurons in the dorsal horn and basal forebrain likely release both GABA and ACh (Todd 1991, Tkatch et al. 1998). GlyRs and nAChRs are found together in the same postsynaptic membranes in chick ciliary ganglion, though glycine and ACh are not truly co-released at this synapse, as ACh undergoes vesicular release and glycine is released by reversal of the glycine transporter GLYT1 (Tsen et al. 2000). In culture, a majority of dorsal horn laminae I-III neurons co-release GABA and ATP, which probably acts on presynaptic receptors to modulate transmitter release (Jo and Schlichter 1999; Hugel and Schlichter 2003). Moving up the neuraxis, brainstem medullary raphe neurons exhibit markers correlating with release of serotonin, glutamate and GABA (Stornetta et al. 2005). Even more surprising is a recent study suggesting that the developing neuromuscular junction can express an array of phenotypes ranging from cholinergic, to glutamatergic, to glycinergic or GABAergic (Borodinsky and Spitzer 2007, see chapter 3 in this volume).

Immuno studies at the light and electron microscopic level showing GABA-immunoreactivity in glutamatergic mossy fiber terminals of hippocampus (Ottersen and Storm-Mathisen 1984; Sandler and Smith 1991) initially appeared to pose a paradox, as they suggested the possibility of co-release of excitatory and inhibitory neurotransmitters. More recently, these findings have been validated and expanded on by physiological studies showing GABA release from glutamatergic hippocampal mossy fibers in young brains (Walker et al. 2001) or after epileptic activity (Gutierrez 2000), a phenomenon discussed in a separate chapter of this volume (Gutierrez, Chapter 10).

5.2.3 Release of GABA, Glycine, and Glutamate in Auditory Brainstem

A surprise in the co-release field was that developing synapses in the MNTB-LSO projection, already shown to co-release GABA and glycine (Nabekura et al. 2004), also release glutamate as a third, and seemingly opposing, neurotransmitter (Gillespie et al. 2005) (Fig. 5.3b). The authors used whole-cell voltage-clamp recordings in acute slices of auditory brainstem to demonstrate dual release of GABA/glycine and glutamate in the LSO in response to photo-uncaging of glutamate in MNTB (to focally activate MNTB cell bodies) or to single-fiber electrical stimulation of MNTB fibers (Fig. 5.3c). Additional evidence came from immunocytochemistry showing that markers for GABAergic (VGAT) and glutamatergic (VGLUT3) transmission colocalize within the LSO in synaptic terminals of the MNTB (Fig. 5.3d, e). Just as GABA release at MNTB-LSO synapses predominates initially and declines during postnatal life, so too glutamate release is highest during the first postnatal week and declines thereafter. In the LSO, VGLUT3 expression is highest during the first two weeks, after which it declines rapidly (Gillespie et al. 2004; Blaesse et al. 2005). Glutamatergic transmission at these synapses is mediated by ionotropic glutamate receptors, largely by NMDARs. These data provide the first physiological evidence for glutamate release from GABA/glycine terminals in the mammalian brain and offer physiological support for earlier anatomical studies suggesting that glutamate and glycine might both be released at MNTB-LSO synapses (Glendenning et al. 1991; Helfert et al. 1992) (Fig. 5.3a).

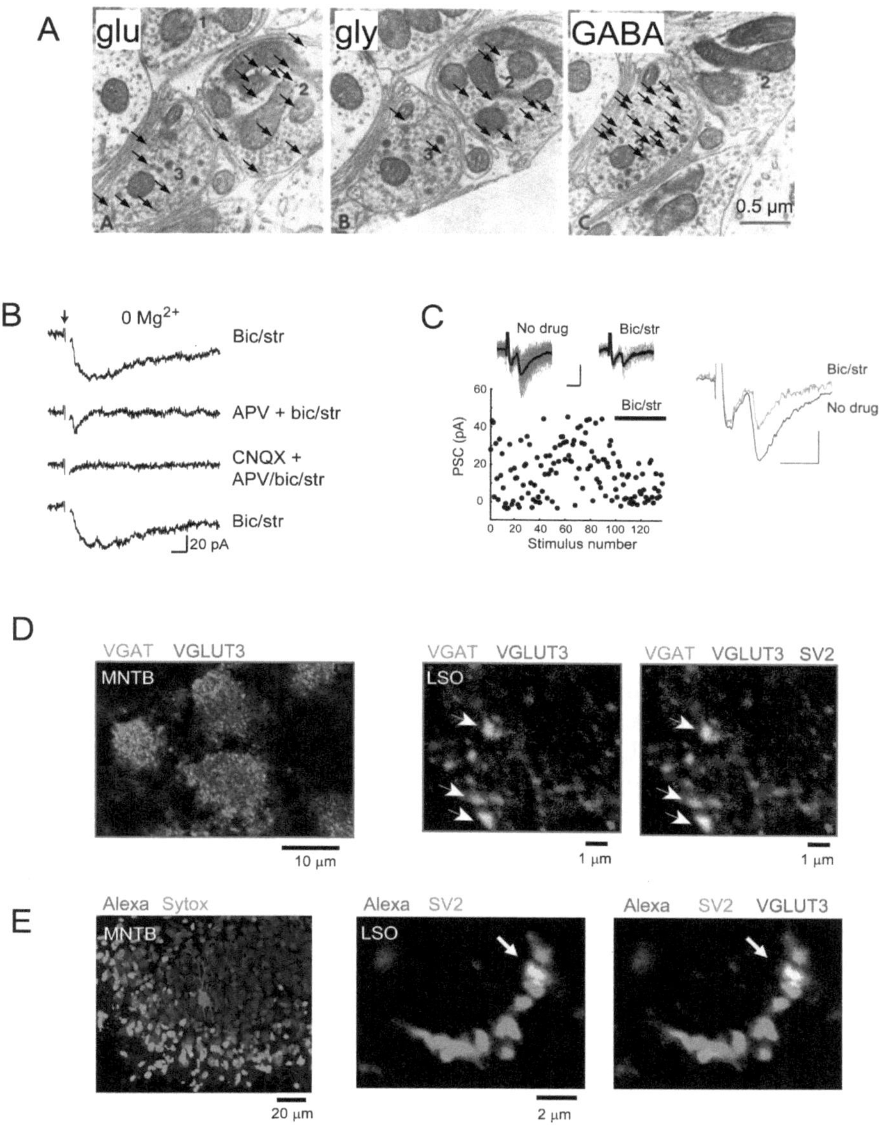

Fig. 5.3 Glutamate co-release at developing GABA/glycine auditory synapses. **(a)** Electron micrographs of an individual terminal from serial sections immunolabeled for glutamate, glycine, and GABA. Terminal 3 is immunoreactive for all three neurotransmitters, terminal 2 for glutamate and glycine. Arrows point to gold particles that tag immunopositive sites. **(b)** Blocking $GABA_A$Rs and GlyRs with bicuculline and strychnine uncovers a MNTB-elicited glutamatergic response in LSO neurons. In these recordings, magnesium was excluded from the bath to unblock NMDA receptors. Glutamatergic responses are mediated by NMDA and AMPA receptors, as they are partially blocked by the NMDAR antagonist APV and completely blocked by addition of the AMPAR antagonist CNQX. **(c)** Single

A second auditory brainstem nucleus, the superior paraolivary nucleus (SPN) also receives a prominent inhibitory input from the MNTB (Banks and Smith 1992). Early expression of VGLUT3 in the SPN is at least as striking as that in the LSO, and immunohistochemistry in the SPN shows a decrease in VGLUT3 expression that parallels the decrease in the LSO (Gillespie et al. 2004; Boulland et al. 2004; Blaesse et al. 2005). Although the synaptic circuitry of the SPN, and even the role of this nucleus in auditory processing, is still poorly understood and controversial (Dehmel et al. 2002; Behrend et al. 2002; Kulesza et al. 2003), the high neonatal expression levels of VGLUT3 and the subsequent developmental decline of VGLUT3 expression suggest that, similar to what has been proposed for the LSO, VGLUT3 also plays an important role in the development of SPN circuitry.

5.2.4 VGLUT3 and Co-release of Glutamate

The basis for glutamate release from MNTB terminals is almost certainly expression of the vesicular glutamate transporter 3 (VGLUT3), as immunohistochemistry for vesicular transporters has revealed high levels of VGLUT3 expression in MTNB cell bodies and in MNTB synaptic terminals within the LSO (Gillespie et al. 2005). The colocalization of immunofluorescence for VGLUT3 with that for VGAT in identified synaptic terminals in the LSO is consistent with the idea that individual terminals release both glutamate and GABA/glycine. VGLUT3 is a relatively rare vesicular glutamate transporter whose function has been a puzzle since its first description, when it was found to be expressed at many non-glutamatergic synapses (Fremeau et al. 2002, Gras et al. 2002, Schafer et al. 2002, Takamori et al. 2002, Seal and Edwards 2006). The temporal correlation of glutamatergic synaptic transmission with high levels of VGLUT3 expression in the LSO constituted the first experimental corroboration for the hypothesis that VGLUT3 in fact underlies vesicular glutamate release at nominally non-glutamatergic synapses.

Fig. 5.3 (continued) GABA/glycinergic MNTB axons can co-release glutamate. Activation of single MNTB axons by minimal stimulation elicits postsynaptic currents in LSO neurons (black trace is average of grey, overlaid responses) that are only partially blocked by bicuculline and strychnine. **(d)** Expression of the vesicular glutamate transporter 3 (VGLUT3) in the MNTB-LSO pathway. MNTB neurons are immunopositive for both the vesicular GABA transporter (VGAT) and VGLUT3. In the LSO, VGAT co-labels with VGLUT3 in small clusters, which also label for the synaptic vesicle protein 2 (SV2). Arrows point to presumed presynaptic endings (SV2-positive) that also label with both VGLUT3 and VGAT. **(e)** Identified MNTB terminals in the LSO express VGLUT3. A single MNTB neuron was filled with the dye Alexa 568 (red). In the LSO an Alexa-filled terminal of this neuron is identified by expression of SV2 (yellow). This terminal also expresses VGLUT3 (white). *Adapted with permission from: A) Helfert et al.* 1992; *B-E) Gillespie et al.* 2005. (*See* Color Plate 3)

5.2.4.1 Possible Glutamate Co-release at GABA/Glycinergic or Other Synapses in Other Brain Areas

The expression of VGLUT3 in many nominally non-glutamatergic synapses suggests that glutamate co-release may be a widespread phenomenon in the mammalian brain (Fremeau et al. 2002; Gras et al. 2002; Schafer et al. 2002; Takamori et al. 2002; Herzog et al. 2004; Somogyi et al. 2004; Gabellec et al. 2007). VGLUT3 expression correlates with markers for non-glutamatergic synapses within certain restricted neuronal populations, and in other cases within a restricted temporal window. For example, VGLUT3 mRNA and protein are found at high levels in the developing cerebellar nuclei, where VGAT and VGLUT3 colocalize in synaptic terminals of presumed Purkinje cells (Boulland et al. 2004; Gras et al. 2005), and where VGLUT3 is down-regulated during the first postnatal weeks. Like the MNTB-LSO synapses, these cerebellar synapses are inhibitory in the adult; unlike the MNTB-LSO synapses, glutamate release at these developing synapses has not (yet?) been demonstrated. Nevertheless, an attractive hypothesis is that VGLUT3 supports an early glutamate release that is important for establishing these inhibitory synapses.

VGLUT3 is not the only vesicular glutamate transporter found in close proximity to GABA release machinery, as both mRNA and protein for both vesicular glutamate transporter 2 (VGLUT2) and glutamic acid decarboxylase (GAD) appear to colocalize in neurons of the anteroventral periventricular nucleus of the preoptic area (Ottem et al. 2004). The expression of both VGLUT2 and VGAT in the same cells further suggests that these neurons may release both GABA and glutamate, though at present no physiological evidence supports this prediction.

5.2.5 Functional Role for Coincident Release of Glutamate, GABA, and Glycine

Glutamate release at GABA/glycinergic MNTB-LSO terminals is a new finding and the functional role of this triple release is unknown. Understanding the role of the inhibitory MNTB-LSO synapse in auditory processing, and what is known about synaptic refinement in the LSO, may cast light on this question.

Sound localization and binaural detection of signal in noise depend on the precise tonotopic alignment of inputs to the principal cells of the LSO. Neurons of the ipsilateral cochlear nucleus (CN) project directly to the LSO where they form glutamatergic synapses (Cant and Casseday 1986, Wu and Kelly 1992). Neurons from the contralateral CN make glutamatergic synapses onto principal cells of the MNTB (Smith et al. 1991), which in turn make inhibitory synapses onto LSO neurons (Moore and Caspary 1983, Caspary and Finlayson 1991). Both the excitatory and the inhibitory projections are tonotopic, but in order to establish an adult LSO in which individual principal neurons receive inhibitory and excitatory inputs responding to the same frequency of sound, each projection also must achieve a precise tonotopic match with the complementary projection.

Why might developing GABA/glycinergic MNTB-LSO synapses release glutamate? Several converging strands of evidence lead to the hypothesis that glutamate co-release plays a critical role in developmental plasticity and refinement of the MNTB-LSO pathway. During LSO development, glutamatergic transmission at MNTB-LSO synapses is developmentally regulated, as is also the expression level of VGLUT3, the protein that presumably supports this glutamate release. In addition, the period when glutamatergic transmission is most prominent corresponds to the period of major functional refinement in the MNTB-LSO projection by synapse elimination (Kim and Kandler 2003). Additionally, decreased glutamatergic transmission and VGLUT3 expression persist for a short period after hearing onset, during a time of increased sharpness in the frequency tuning and alignment of excitatory and inhibitory responses of LSO neurons (Sanes and Rubel 1988). Furthermore, glutamate released at MNTB-LSO synapses activates postsynaptic NMDA receptors (NMDARs), the subtype of ionotropic glutamate receptor closely linked to induction of synaptic plasticity in a variety of excitatory and inhibitory synapses (Kullmann et al. 2000, rev). Finally, the peak of the period of glutamatergic transmission corresponds to the peak period when GABA and glycine are still depolarizing in the LSO (Kandler and Friauf 1995). The depolarizing action of GABA and glycine may be functionally relevant, as it could provide the critical step necessary for NMDAR-dependent plasticity: it could relieve the voltage-sensitive magnesium block of NMDARs, thus allowing co-released glutamate to activate NMDARs.

The temporal correlation of these three transient developmental periods: (1) major functional refinement, (2) glutamate release onto NMDARs, and (3) depolarizing action of GABA and glycine—supports the hypothesis that glutamate release from MNTB terminals onto NMDAR-containing LSO dendrites participates in synaptic refinement in this system. This could occur at one (or both) of two stages: either (a) functional refinement of the MNTB-LSO pathway that occurs before hearing onset and is thought to be directed by spontaneous patterned activity from the cochlea (Lippe 1994; Kros et al. 1998; Beutner and Moser 2001), or (b) subsequent fine-tuning of CN-LSO and MNTB-LSO pathway alignment that may be guided by auditory experience after hearing onset (Sanes and Rubel 1988; Echteler et al. 1989).

During the first stage of synaptic refinement in the MNTB-LSO pathway, local GABA/glycinergic depolarization could relieve the Mg^{++} block of NMDARs (Leinekugel et al. 1997), inducing Ca-influx through NMDA receptors at active MNTB-LSO synapses. In many developing systems, NMDAR-mediated calcium influx is essential for excitatory synaptic plasticity, such as elimination of glutamatergic synapses (Rabacchi et al. 1992, Kakizawa et al. 2000) or insertion of AMPA receptors at "silent" synapses (Isaac et al. 1997, Liao et al. 1999). In the LSO, NMDAR-mediated Ca-influx could play a role in clustering or insertion of GABA and glycine receptors (Kano et al. 1992, Otis et al. 1994; Charpier et al. 1995; Kirsch and Betz 1998; Moss and Smart 2001). Ca-dependent and/or NMDAR-dependent plasticity at inhibitory synapses has been demonstrated in a number of inhibitory systems (Kano et al. 1992, Komatsu 1994, Oda et al. 1995, McLean et al. 1996, Wang and Stelzer 1996, Caillard et al. 1999, Ouardouz and Sastry 2000), although the route by which NMDARs are activated in the absence of synaptically released glutamate has remained an open question. Glutamate release, accompanied by release of depolarizing GABA/glycine, provides an answer to this question by allowing inhibitory synapses to access NMDARs and their downstream machinery of synaptic plasticity, independent of glutamate from other sources. Although at LSO-MNTB synapses a form of LTD can be induced through non-NMDAR-dependent mechanisms involving $GABA_B$Rs (Kotak et al. 2001, Chang et al. 2003), additional forms of activity-dependent synaptic plasticity may also occur at this synapse, hypothetically induced through glutamate release and activation of NMDARs.

Finer scale synaptic refinement during the initial period after hearing onset presents a different problem. Despite our ever-deepening understanding of synaptic plasticity at individual synapses, it has generally been difficult—with some exceptions (e.g., Lien et al. 2006, Nugent et al. 2007)—to determine how the finely tuned coordinate refinement of inhibitory and excitatory inputs to a single neuron might occur. It is tempting to speculate that the reduced levels of VGLUT3 expression, glutamate release, and NMDAR activation that remain in the auditory brainstem after hearing onset might play a role in stimulus-driven alignment of glutamatergic and GABA/glycinergic inputs. By the time of hearing onset, GABA and glycine are hyperpolarizing in the LSO (Kandler and Friauf 1995; Ehrlich et al. 1999), and the major period of functional refinement in the MNTB-LSO pathway is complete (Kim and Kandler 2003), although anatomical refinement occurs during the first few days after hearing onset (Sanes and Siverls 1991). One specific hypothesis for NMDAR-mediated alignment is that the CN-LSO pathway could signal to MNTB-LSO synapses through back-propagating action potentials—or through sufficiently strong depolarization arising in nearby excitatory (CN-LSO) synapses—that relieve the Mg^{++}-block at NMDARs in MNTB-LSO synapses. Simultaneous relief of Mg^{++} block and release of glutamate from MNTB terminals would activate NMDARs in MNTB-LSO synapses, allowing the postsynaptic neuron to detect coincident input from the excitatory CN-LSO pathway and the inhibitory MNTB-LSO pathway. In this strategy, late finescale refinement mediated by NMDAR activation would likely occur not at the excitatory CN-LSO synapses, but rather at the GABA/glycinergic MNTB-LSO synapses. An alternative, reversed, scenario is that glutamate spillover from MNTB-LSO synapses reaches nearby CN-LSO synapses to allow NMDARs in the CN-LSO synapses to detect coincident inputs. Although we know very little about the locations of developing synaptic inputs on LSO principal cells, this scenario would more strongly depend on parameters such as reuptake and the physical locations of excitatory and inhibitory synapses (Rusakov and Kullmann 1998). These additional constraints on NMDAR activation appear to make this scenario less generally applicable than one dependent on backpropagating spiking activity. Nevertheless, both options offer models for how inputs of opposite sign might signal each other through the postsynaptic neuron to achieve coordinated refinement of excitatory and inhibitory synapses.

5.2.6 Molecular Basis for Release of Glutamate with GABA and Glycine

For the strong version of this glutamate-in-inhibitory-plasticity hypothesis to hold in the most limiting case, we might expect that glutamate and GABA/glycine would be released from the same synapse even when stimulation of the presynaptic terminal resulted in release of only a single vesicle. This would require that GABA/glycine and glutamate be packaged together in individual synaptic vesicles. This limiting case version of the hypothesis predicts the existence of individual synaptic vesicles whose membranes contain both VGLUT3 and VGAT. Immuno-EM methods are unfortunately insufficiently precise to answer this question (see e.g. Bergersen et al. 2003), though paired MNTB-LSO recordings or recordings of spontaneous mIPSCs at the MNTB-LSO synapse could offer insight into this question.

Alternatively, glutamate and GABA/glycine may be packaged in distinct populations of synaptic vesicles. This would not necessarily invalidate the hypothesis that glutamate release plays a central role in activity-dependent plasticity at glycinergic synapses. Separate vesicle populations with distinct spatial distributions could participate differently in transmission and plasticity, or release probabilities for GABA/glycinergic vesicles relative to those for glutamatergic vesicles—perhaps via differential expression of distinct synaptotagmin isoforms (Xu et al. 2007)—could be adjusted to maximize the probability of glutamate release within a certain range of firing rates. Under this scenario, the distribution of mPSCs seen at the MNTB-LSO synapses would include at least some purely glutamatergic and/or some purely GABA/glycinergic mPSCs.

Regardless of whether VGLUT3, glutamate release and NMDAR activation mediate plasticity at MNTB-LSO synapses, it will be of great interest to determine whether VGLUT3 and VGAT are in fact inserted in the membrane of the same synaptic vesicles. This possibility seems unlikely, but it may not be preposterous. Although in *Drosophila* a single vesicular glutamate transporter is sufficient to load a glutamatergic vesicle (Daniels et al. 2006), a given mammalian synaptic vesicle may contain approximately 10 transporters (Takamori et al. 2006). Furthermore, two distinct vesicular transporter types, VGLUT1 and VGLUT2, have been found coexpressed in the same synaptic vesicles in the developing hippocampus (Herzog et al. 2006). It may be possible in the developing LSO, by extension, that VGLUT3 and VGAT are transiently co-expressed in the same synaptic vesicles and that single synaptic vesicles contain both excitatory and inhibitory classical fast neurotransmitters.

5.3 Summary

Evidence gleaned over the past decade has forced us to dramatically change our picture of inhibitory synapses. In the first place, nominally inhibitory synapses do not always inhibit their postsynaptic neurons; at early periods, GABA and glycinergic synapses in many parts of the nervous system are depolarizing and even excitatory. Inhibitory information transfer is not strictly unidirectional; presynaptic GABARs and glyRs can modulate release of GABA and glycine. Glycinergic synapses do not always release (much) glycine; immature glycinergic synapses in brainstem and spinal cord in fact release primarily GABA. And finally, at the synapse formerly known as glycinergic, the GABA/glycinergic MNTB-LSO synapse of the auditory brainstem, "inhibitory" synapses do not release solely inhibitory neurotransmitters; nascent glycinergic synapses release the inhibitory neurotransmitters GABA and glycine and the excitatory neurotransmitter glutamate. These findings have forced us to begin to see the inhibitory synapse as a much more complex and exciting unit than it previously appeared.

These findings also force us to consider what the function of multiple transmitter release might be. Of particular interest is the triple release of glutamate, GABA and glycine in the developing MNTB-LSO pathway. This pathway exhibits a rich repertoire of developmental changes, including synapse elimination and strengthening, during the period corresponding to glutamate release and NMDAR activation. The MNTB-LSO pathway has been seen as an elegant model system for understanding the mechanisms by which inhibitory circuits are assembled and refined. With its precisely converging tonotopic projections from excitatory and inhibitory pathways, the LSO offers an exceptionally well-organized model for delving into questions of inhibitory circuit development and of the more complex coordinated refinement of inhibitory and excitatory inputs. The unexpected discovery that the "purely inhibitory" pathway is not so pure has forced us to redraw our model system, but it has also opened up new and exciting research directions.

References

Aubrey KR, Rossi FM, Ruivo R, Alboni S, Bellenchi GC, Le Goff A, Gasnier B, Supplisson S (2007) The transporters GlyT2 and VIAAT cooperate to determine the vesicular glycinergic phenotype. J Neurosci 27:6273–6281

Awatramani GB, Turecek R, Trussell LO (2005) Staggered development of GABAergic and glycinergic transmission in the MNTB. J Neurophysiol 93:819–828

Banks MI, Smith PH (1992) Intracellular recordings from neurobiotin-labeled cells in brain slices of the rat medial nucleus of the trapezoid body. J Neurosci 12:2819–2837

Behrend O, Brand A, Kapfer C, Grothe B (2002) Auditory response properties in the superior paraolivary nucleus of the gerbil. J Neurophysiol 87:2915–2928

Ben-Ari Y, Khazipov R, Leinekugel X, Caillard O, Gaiarsa JL (1997) GABAA, NMDA and AMPA receptors: a developmentally regulated 'ménage à trois'. Trends Neurosci 20:523–529

Bergersen L, Ruiz A, Bjaalie JG, Kullmann DM, Gundersen V (2003) GABA and GABAA receptors at hippocampal mossy fibre synapses. Eur J Neurosci 18:931–941
Beutner D, Moser T (2001) The presynaptic function of mouse cochlear inner hair cells during development of hearing. J Neurosci 21:4593–4599
Bi GQ, Poo MM (1998) Synaptic modifications in cultured hippocampal neurons: dependence on spike timing, synaptic strength, and postsynaptic cell type. J Neurosci 18:10464–10472
Blaesse P, Ehrhardt S, Friauf E, Nothwang HG (2005) Developmental pattern of three vesicular glutamate transporters in the rat superior olivary complex. Cell Tissue Res 320:33–50
Borodinsky LN, Spitzer NC (2007) Activity-dependent neurotransmitter-receptor matching at the neuromuscular junction. Proc Natl Acad Sci USA. 104:335–340
Boudreau JC, Tsuchitani C (1968) Binaural interaction in the cat superior olive S segment. J Neurophysiol 31:442–454
Boulland JL, Qureshi T, Seal RP, Rafiki A, Gundersen V, Bergersen LH, Fremeau RT Jr, Edwards RH, Storm-Mathisen J, Chaudhry FA (2004) Expression of the vesicular glutamate transporters during development indicates the widespread corelease of multiple neurotransmitters. J Comp Neurol 480:264–280
Caillard O, Ben-Ari Y, Gaiarsa JL (1999) Mechanisms of induction and expression of long-term depression at GABAergic synapses in the neonatal rat hippocampus. J Neurosci 19:7568–7577
Cant NB, Casseday JH (1986) Projections from the anteroventral cochlear nucleus to the lateral and medial superior olivary nuclei. J Comp Neurol 247:457–476
Carmignoto G, Vicini S (1992) Activity-dependent decrease in NMDA receptor responses during development of the visual cortex. Science 258:1007–1011
Caspary DM, Finlayson PG (1991) Superior olivary complex: functional neuropharmacology of the principal cell types. In Neurobiology of hearing: the central auditory system (ed. RA Altschuler et al), 141–161
Chang EH, Kotak VC, Sanes DH (2003) Long-term depression of synaptic inhibition is expressed postsynaptically in the developing auditory system. J Neurophysiol 90:1479–1488
Charpier S, Behrends JC, Triller A, Faber DS, Korn H l (1995) "Latent" inhibitory connections become functional during activity-dependent plasticity. Proc Natl Acad Sci 92:117–120
Chaudhry FA, Reimer RJ, Bellocchio EE, Danbolt NC, Osen KK, Edwards RH, Storm-Mathisen J (1998) The vesicular GABA transporter, VGAT, localizes to synaptic vesicles in sets of glycinergic as well as GABAergic neurons. J Neurosci 18:9733–9750
Chery N, De Koninck Y (2000) GABA(B) receptors are the first target of released GABA at lamina I inhibitory synapses in the adult rat spinal cord. J Neurophysiol 84:1006–1011
Contini M, Raviola E (2003) GABAergic synapses made by a retinal dopaminergic neuron. Proc Natl Acad Sci USA. 100:1358–1363
Crair MC, Malenka RC (1995) A critical period for long-term potentiation at thalamocortical synapses. Nature 375:325–328
Daniels RW, Collins CA, Chen K, Gelfand MV, Featherstone DE, DiAntonio A (2006) A single vesicular glutamate transporter is sufficient to fill a synaptic vesicle. Neuron 49:11–16
Dehmel S, Kopp-Scheinpflug C, Dorrscheidt GJ, Rubsamen R (2002) Electrophysiological characterization of the superior paraolivary nucleus in the Mongolian gerbil. Hear Res 172:18–36
Dugue GP, Dumoulin A, Triller A, Dieudonne S (2005) Target-dependent use of co-released inhibitory transmitters at central synapses. J Neurosci 25:6490–6498
Eccles JC (1964) The physiology of synapses. Springer, Berlin
Echteler, SM, Arjmand, E, Dallos, P (1989) Developmental alterations in the frequency map of the mammalian cochlea. Nature 341:147–149
Ehrlich I, Lohrke S, Friauf E (1999) Shift from depolarizing to hyperpolarizing glycine action in rat auditory neurones is due to age-dependent Cl- regulation. J Physiol 520:121–137

Feller MB, Wellis DP, Stellwagen D, Werblin FS, Shatz CJ (1996) Requirement for cholinergic synaptic transmission in the propagation of spontaneous retinal waves. Science 272:1182–1187

Feng G, Tintrup H, Kirsch J, Nichol MC, Kuhse J, Betz H, Sanes JR (1998) Dual requirement for gephyrin in glycine receptor clustering and molybdoenzyme activity. Science 282:1321–1324

Fremeau RT Jr, Burman J, Qureshi T, Tran CH, Proctor J, Johnson J, Zhang H, Sulzer D, Copenhagen DR, Storm-Mathisen J, Reimer RJ, Chaudhry FA, Edwards RH (2002) The identification of vesicular glutamate transporter 3 suggests novel modes of signaling by glutamate. Proc Natl Acad Sci USA. 99:14488–14493

Friauf E, Aragon C, Lohrke S, Westenfelder B, Zafra F (1999) Developmental expression of the glycine transporter GLYT2 in the auditory system of rats suggests involvement in synapse maturation. J Comp Neurol 412:17–37

Gabellec MM, Panzanelli P, Sassoe-Pognetto M, Lledo PM (2007) Synapse-specific localization of vesicular glutamate transporters in the rat olfactory bulb. Eur J Neurosci 25:1373–1383

Gaiarsa JL, Caillard O, Ben-Ari Y (2002) Long-term plasticity at GABAergic and glycinergic synapses: mechanisms and functional significance. Tr Neurosci 25: 564–570

Gillespie DC, Cihil K, Kandler K (2004) Developmental expression patterns of the vesicular glutamate transporters VGLUT1–3 in the auditory brainstem. Soc Nsci Abstr 947.12

Gillespie DC, Kim G, Kandler K (2005) Inhibitory synapses in the developing auditory system are glutamatergic. Nat Neurosci 8:332–338

Glendenning KK, Masterton RB, Baker BN, Wenthold RJ (1991) Acoustic chiasm. III: Nature, distribution, and sources of afferents to the lateral superior olive in the cat. J Comp Neurol 310:377–400

Gras C, Herzog E, Bellenchi GC, Bernard V, Ravassard P, Pohl M, Gasnier B, Giros B, El Mestikawy S (2002) A third vesicular glutamate transporter expressed by cholinergic and serotoninergic neurons. J Neurosci 22:5442–5451

Gras C, Vinatier J, Amilhon B, Guerci A, Christov C, Ravassard P, Giros B, El Mestikawy S (2005) Developmentally regulated expression of VGLUT3 during early post-natal life. Neuropharmacology 49:901–911

Gutierrez R (2000) Seizures induce simultaneous GABAergic and glutamatergic transmission in the dentate gyrus-CA3 system. J Neurophysiol 84:3088–3090

Haas JS, Nowotny T, Abarbanel HD (2006) Spike-timing-dependent plasticity of inhibitory synapses in the entorhinal cortex. J Neurophysiol 96:3305–3313

Haverkamp S, Wassle H (2004) Characterization of an amacrine cell type of the mammalian retina immunoreactive for vesicular glutamate transporter 3. J Comp Neurol 468:251–263

Helfert RH, Juiz JM, Bledsoe SC Jr, Bonneau JM, Wenthold RJ, Altschuler RA (1992) Patterns of glutamate, glycine, and GABA immunolabeling in four synaptic terminal classes in the lateral superior olive of the guinea pig. J Comp Neurol 323:305–325

Henkel CK, Brunso-Bechtold JK (1998) Calcium-binding proteins and GABA reveal spatial segregation of cell types within the developing lateral superior olivary nucleus of the ferret. Microsc Res Tech 41:234–245

Herzog E, Gilchrist J, Gras C, Muzerelle A, Ravassard P, Giros B, Gaspar P, El Mestikawy S (2004) Localization of VGLUT3, the vesicular glutamate transporter type 3, in the rat brain. Neuroscience 123:983–1002

Herzog E, Takamori S, Jahn R, Brose N, Wojcik SM (2006) Synaptic and vesicular co-localization of the glutamate transporters VGLUT1 and VGLUT2 in the mouse hippocampus. J Neurochem 99:1011–1018

Hestrin S (1992) Developmental regulation of NMDA receptor-mediated synaptic currents at a central synapse. Nature 357:686–689

Hooks BM, Chen C (2006) Distinct roles for spontaneous and visual activity in remodeling of the retinogeniculate synapse. Neuron 52:281–291

Huberman AD, Speer CM, Chapman B (2006) Spontaneous retinal activity mediates development of ocular dominance columns and binocular receptive fields in v1. Neuron 52:247–254
Hugel S, Schlichter R (2003) Convergent control of synaptic GABA release from rat dorsal horn neurones by adenosine and GABA autoreceptors. J Physiol 551:479–489
Isaac JT, Crair MC, Nicoll RA, Malenka RC (1997) Silent synapses during development of thalamocortical inputs. Neuron 18:269–280
Isaacson JS (1998) GABAB receptor-mediated modulation of presynaptic currents and excitatory transmission at a fast central synapse. J Neurophysiol 80:1571–1576
Jo YH, Schlichter R (1999) Synaptic corelease of ATP and GABA in cultured spinal neurons. Nat Neurosc. 2:241–245
Johnson J, Sherry DM, Liu X, Fremeau RT Jr, Seal RP, Edwards RH, Copenhagen DR (2004) Vesicular glutamate transporter 3 expression identifies glutamatergic amacrine cells in the rodent retina. J Comp Neurol 477:386–398
Jonas P, Bischofberger J, Sandkuhler J. (1998) Corelease of two fast neurotransmitters at a central synapse. Science 281:419–424
Juiz JM, Helfert RH, Bonneau JM, Wenthold RJ, Altschuler RA l (1996) Three classes of inhibitory amino acid terminals in the cochlear nucleus of the guinea pig. J Comp Neurol 373:11–26
Kakizawa S, Yamasaki M, Watanabe M, Kano M (2000) Critical period for activity-dependent synapse elimination in developing cerebellum. J Neurosci 20:4954–4961
Kandler K, Friauf E (1995) Development of glycinergic and glutamatergic synaptic transmission in the auditory brainstem of perinatal rats. J Neurosci 15:6890–6894
Kano M, Rexhausen U, Dresse, J, Konnerth A (1992) Synaptic excitation produces a long-lasting rebound potentiation of inhibitory synaptic signals in cerebellar Purkinje cells. Nature 356:601–604
Keller AF, Coull JA, Chery N, Poisbeau P, De Koninck Y (2001) Region-specific developmental specialization of GABA-glycine cosynapses in laminas I-II of the rat spinal dorsal horn. J Neurosci 21:7871–7880
Kim GS, Kandler K (2003) Elimination and strengthening of glycinergic/GABAergic connections during tonotopic map formation. Nat Neurosci 6:282–290
Kirsch J, Betz H (1998) Glycine-receptor activation is required for receptor clustering in spinal neurons. Nature 392:717–720
Kneussel M, Brandstatter JH, Gasnier B, Feng G, Sanes JR, Betz H (2001) Gephyrin-independent clustering of postsynaptic GABA(A) receptor subtypes. Mol Cell Neurosci 17:973–982
Kneussel M, Brandstatter JH, Laube B, Stahl S, Muller U, Betz H (1999) Loss of postsynaptic GABA(A) receptor clustering in gephyrin-deficient mice. J Neurosci 19:9289–9297
Kolston J, Osen KK, Hackney CM, Ottersen OP, Storm-Mathisen J (1992) An atlas of glycine- and GABA-like immunoreactivity and colocalization in the cochlear nuclear complex of the guinea pig. Anat Embryol (Berl) 186:443–465
Komatsu Y (1994) Age-dependent long-term potentiation of inhibitory synaptic transmission in rat visual cortex. J Neurosci 14:6488–6499
Korada S, Schwartz IR (1999) Development of GABA, glycine, and their receptors in the auditory brainstem of gerbil: a light and electron microscopic study. J Comp Neurol 409:664–681
Kotak VC, DiMattina C, Sanes DH (2001) GABA(B) and Trk receptor signaling mediates long-lasting inhibitory synaptic depression. J Neurophysiol 86:536–540
Kotak VC, Korada S, Schwartz IR, Sanes DH (1998) A developmental shift from GABAergic to glycinergic transmission in the central auditory system. J Neurosci 18:4646–4655
Kros CJ, Ruppersberg JP, Rusch A (1998) Expression of a potassium current in inner hair cells during development of hearing in mice. Nature 394:281–284
Kulesza RJ Jr, Spirou GA, Berrebi AS (2003) Physiological response properties of neurons in the superior paraolivary nucleus of the rat. J Neurophysiol 89:2299–2312.
Kullmann DM, Asztely F, Walker MC (2000) The role of mammalian ionotropic receptors in synaptic plasticity: LTP, LTD and epilepsy. Cell Mol Life Sci 57:1551–1561

Kullmann PH, Kandler K (2001) Glycinergic/GABAergic synapses in the lateral superior olive are excitatory in neonatal C57Bl/6 J mice. Brain Res Dev Brain Res 131:143–147
Kullmann PH, Ene FA, Kandler K (2002) Glycinergic and GABAergic calcium responses in the developing lateral superior olive. Eur J Neurosci 15:1093–1104
Leinekugel X, Medina I, Khalilov I, Ben-Ari Y, Khazipov R (1997) Ca^{2+} oscillations mediated by the synergistic excitatory actions of $GABA_A$ and NMDA receptors in the neonatal hippocampus. Neuron 18:243–255
Levi S, Logan SM, Tovar KR, Craig AM (2004) Gephyrin is critical for glycine receptor clustering but not for the formation of functional GABAergic synapses in hippocampal neurons. J Neurosci 24:207–217
Liao D, Zhang X, O'Brien R, Ehlers MD, Huganir RL (1999) Regulation of morphological postsynaptic silent synapses in developing hippocampal neurons. Nat Neurosci 2:37–43
Lien CC, Mu Y, Vargas-Caballero M, Poo MM (2006) Visual stimuli-induced LTD of GABAergic synapses mediated by presynaptic NMDA receptors. Nat Neurosci 9:372–380
Lim R, Alvarez FJ, Walmsley B (2000) GABA mediates presynaptic inhibition at glycinergic synapses in a rat auditory brainstem nucleus. J Physiol 525:447–459
Lippe WR (1994) Rhythmic spontaneous activity in the developing avian auditory system. J Neurosci 14:1486–1495
Liu QR, Lopez-Corcuera B, Mandiyan S, Nelson H, Nelson N (1993) Cloning and expression of a spinal cord- and brain-specific glycine transporter with novel structural features. J Biol Chem 268:22802–22808
McIntire SL, Reimer RJ, Schuske K, Edwards RH, Jorgensen EM (1997) Identification and characterization of the vesicular GABA transporter. Nature 389:870–876
McLean HA, Caillard O, Ben-Ari Y, Gaiarsa JL (1996) Bidirectional plasticity expressed by GABAergic synapses in the neonatal rat hippocampus. J Physiol 496:471–477
Meister M, Wong RO, Baylor DA, Shatz CJ (1991) Synchronous bursts of action potentials in ganglion cells of the developing mammalian retina. Science 252:939–943
Moore MJ, Caspary DM (1983) Strychnine blocks binaural inhibition in lateral superior olivary neurons. J Neurosci 3:237–242
Moss, SJ, Smart, TG (2001) Constructing inhibitory synapses. Nat Rev Neurosci 2:240–250
Nabekura J, Katsurabayashi S, Kakazu Y, Shibata S, Matsubara A, Jinno S, Mizoguchi Y, Sasaki A, Ishibashi H (2004) Developmental switch from GABA to glycine release in single central synaptic terminals. Nat Neurosci 7:17–23
Nugent FS, Penick EC, Kauer JA (2007) Opioids block long-term potentiation of inhibitory synapses. Nature 446:1086–10890
O'Brien JA, Berger AJ (1999) Cotransmission of GABA and glycine to brain stem motoneurons. J Neurophysiol 82:1638–1641
Oda Y, Charpier S, Murayama Y, et al (1995) Long-term potentiation of glycinergic inhibitory synaptic transmission. J Neurophysiol 74:1056–1074
Oertel J, Villmann C, Kettenmann H, Kirchhoff F, Becker CM (2007) A novel glycine receptor beta subunit splice variant predicts an unorthodox transmembrane topology Assembly into heteromeric receptor complexes. J Biol Chem 282:2798–2807
O'Malley DM, Masland RH (1989) Co-release of acetylcholine and gamma-aminobutyric acid by a retinal neuron. Proc Natl Acad Sci USA. 86:3414–3418
O'Malley DM, Sandell JH, Masland RH (1992) Co-release of acetylcholine and GABA by the starburst amacrine cells. J Neurosci 12:1394–408
Otis TS, De Koninck Y, Mody I (1994) Lasting potentiation of inhibition is associated with an increased number of gamma-aminobutyric acid type A receptors activated during miniature inhibitory postsynaptic currents. Proc Natl Acad Sci USA. 91:7698–7702
Ottem EN, Godwin JG, Krishnan S, Petersen SL (2004) Dual-phenotype GABA/glutamate neurons in adult preoptic area: sexual dimorphism and function. J Neurosci 24:8097–8105

Ottersen OP, Storm-Mathisen J (1984) Glutamate- and GABA-containing neurons in the mouse and rat brain, as demonstrated with a new immunocytochemical technique. J Comp Neurol 229:374–392
Ottersen OP, Storm-Mathisen J, Somogyi P (1988) Colocalization of glycine-like and GABA-like immunoreactivities in Golgi cell terminals in the rat cerebellum: a postembedding light and electron microscopic study. Brain Res 450:342–353
Ouardouz M, Sastry BR (2000) Mechanisms underlying LTP of inhibitory synaptic transmission in the deep cerebellar nuclei. J Neurophysiol 84:1414–1421
Overstreet-Wadiche L, Bromberg DA, Bensen AL, Westbrook GL (2005) GABAergic signaling to newborn neurons in dentate gyrus. J Neurophysiol 94:4528–4532
Owens DF, Kriegstein AR (2002) Is there more to GABA than synaptic inhibition? Nat Rev Neurosci 3:715–727
Paarmann I, Schmitt B, Meyer B, Karas M, Betz H (2006) Mass spectrometric analysis of glycine receptor-associated gephyrin splice variants. J Biol Chem 281:34918–34925
Piechotta K, Weth F, Harvey RJ, Friauf E (2001) Localization of rat glycine receptor alpha1 and alpha2 subunit transcripts in the developing auditory brainstem. J Comp Neurol 438:336–352
Rabacchi S, Bailly Y, Delhaye-Bourchaud N, Mariani J (1992) Involvement of the N-methyl D-aspartate (NMDA) receptor in synapse elimination during cerebellar development. Science 256:1823–1825
Represa A, Ben-Ari Y (2005) Trophic actions of GABA on neuronal development. Trends Neurosci 28:278–283
Rusakov DA, Kullmann DM (1998) Extrasynaptic glutamate diffusion in the hippocampus: ultrastructural constraints, uptake, and receptor activation. J Neurosci 18:3158–3170
Russier M, Kopysova IL, Ankri N, Ferrand N, Debanne D (2002) GABA and glycine co-release optimizes functional inhibition in rat brainstem motoneurons in vitro. J Physiol 541:123–137
Sagne C, El Mestikawy S, Isambert MF, Hamon M, Henry JP, Giros B, Gasnier B (1997) Cloning of a functional vesicular GABA and glycine transporter by screening of genome databases. FEBS Lett 417:177–183
Sandler R, Smith AD (1991) Coexistence of GABA and glutamate in mossy fiber terminals of the primate hippocampus: an ultrastructural study. J Comp Neurol 303:177–192
Sanes DH, Rubel EW (1988) The ontogeny of inhibition and excitation in the gerbil lateral superior olive. J Neurosci 8:682–700
Sanes DH, Siverls V (1991) Development and specificity of inhibitory terminal arborizations in the central nervous system. J Neurobiol 22:837–854
Schafer MK, Varoqui H, Defamie N, Weihe E, Erickson JD (2002) Molecular cloning and functional identification of mouse vesicular glutamate transporter 3 and its expression in subsets of novel excitatory neurons. J Biol Chem 277:50734–50748
Seal RP, Edwards RH. (2006) The diverse roles of vesicular glutamate transporter 3. Handb Exp Pharmacol (175):137–150
Seddik R, Schlichter R, Trouslard J (2007) Corelease of GABA/glycine in lamina-X of the spinal cord of neonatal rats. Neuroreport 18:1025–1029
Smith AJ, Owens S, Forsythe ID (2000) Characterisation of inhibitory and excitatory post-synaptic currents of the rat medial superior olive. J Physiol 529:681–698
Smith PH, Joris PX, Carney LH, Yin TCT (1991) Projections of physiologically characterized globular bushy cell axons from the cochlear nucleus of the cat. J Comp Neurol 304:387–407
Somogyi J (2006) Functional significance of co-localization of GABA and Glu in nerve terminals: a hypothesis. Curr Top Med Chem 6:969–973
Somogyi J, Baude A, Omori Y, Shimizu H, El Mestikawy S, Fukaya M, Shigemoto R, Watanabe M, Somogyi P (2004) GABAergic basket cells expressing cholecystokinin contain vesicular glutamate transporter type 3 (VGLUT3) in their synaptic terminals in hippocampus and isocortex of the rat. Eur J Neurosci 19:552–569

Spike RC, Watt C, Zafra F, Todd AJ (1997) An ultrastructural study of the glycine transporter GLYT2 and its association with glycine in the superficial laminae of the rat spinal dorsal horn. Neuroscience 77:543–551

Stornetta RL, Rosin DL, Simmons JR, McQuiston TJ, Vujovic N, Weston MC, Guyenet PG (2005) Coexpression of vesicular glutamate transporter-3 and gamma-aminobutyric acidergic markers in rat rostral medullary raphe and intermediolateral cell column. J Comp Neurol 492:477–494

Takamori S, Malherbe P, Broger C, Jahn R (2002) Molecular cloning and functional characterization of human vesicular glutamate transporter 3. EMBO Rep 3:798–803

Takamori S, Holt M, Stenius K, Lemke EA, Gronborg M, Riedel D, Urlaub H, Schenck S, Brugger B, Ringler P, Muller SA, Rammner B, Grater F, Hub JS, De Groot BL, Mieskes G, Moriyama Y, Klingauf J, Grubmuller H, Heuser J, Wieland F, Jahn R (2006) Molecular anatomy of a trafficking organelle. Cell 127:831–846

Tkatch T, Baranauskas G, Surmeier DJ (1998) Basal forebrain neurons adjacent to the globus pallidus coexpress GABAergic and cholinergic marker mRNAs. Neuroreport 9:1935–1939

Todd AJ (1991) Immunohistochemical evidence that acetylcholine and glycine exist in different populations of GABAergic neurons in lamina III of rat spinal dorsal horn. Neuroscience 44:741–746

Todd AJ, Sullivan AC (1990) Light microscope study of the coexistence of GABA-like and glycine-like immunoreactivities in the spinal cord of the rat. J Comp Neurol 296:496–505

Triller A, Cluzeaud F, Korn H (1987) gamma-Aminobutyric acid-containing terminals can be apposed to glycine receptors at central synapses. J Cell Biol 104:947–956

Tsen G, Williams B, Allaire P, Zhoru YD, Ikonomov O, Kondova I, Jacob MH (2000) Receptors with opposing functions are in postsynaptic microdomains under one presynaptic terminal. Nat Neurosci 3:126–132

Turecek R, Trussell LO (2001) Presynaptic glycine receptors enhance transmitter release at a mammalian central synapse. Nature 411:587–590

Turecek R, Trussell LO (2002) Reciprocal developmental regulation of presynaptic ionotropic receptors. Proc Natl Acad Sci USA. 99:13884–13889

Vaney DI, Young HM (1988) GABA-like immunoreactivity in cholinergic amacrine cells of the rabbit retina. Brain Res. 438:369–373

Walker MC, Ruiz A, Kullmann DM (2001) Monosynaptic GABAergic signaling from dentate to CA3 with a pharmacological and physiological profile typical of mossy fiber synapses. Neuron 29:703–715

Wang CT, Blankenship AG, Anishchenko A, Elstrott J, Fikhman M, Nakanishi S, Feller MB (2007) GABA(A) receptor-mediated signaling alters the structure of spontaneous activity in the developing retina. J Neurosci 27:9130–9140

Wang, JH, Stelzer, A (1996) Shared calcium signalling pathways in the induction of long-term potentiation and synaptic disinhibition in CA1 pyramidal cell dendrites. J Neurophysiol 75:1687–1702

Wentzel PR, De Zeeuw CI, Holstege JC, Gerrits NM (1993) Colocalization of GABA and glycine in the rabbit oculomotor nucleus. Neurosci Lett 164:25–29

Wojcik SM, Katsurabayashi S, Guillemin I, Friauf E, Rosenmund C, Brose N, Rhee JS (2006) A shared vesicular carrier allows synaptic corelease of GABA and glycine. Neuron 50:575–587

Woodin MA, Ganguly K, Poo MM (2003) Coincident pre- and postsynaptic activity modifies GABAergic synapses by postsynaptic changes in Cl- transporter activity. Neuron 39:807–820

Wu SH, Kelly JB (1992) Synaptic pharmacology of the superior olivary complex studied in mouse brain slice. J Neurosci 12:3084–3097

Wulle I, Wagner HJ. (1990) GABA and tyrosine hydroxylase immunocytochemistry reveal different patterns of colocalization in retinal neurons of various vertebrates. J Comp Neurol 296:173–178

Xu J, Mashimo T, Sudhof TC (2007) Synaptotagmin-1, -2, and -9: Ca(2+) sensors for fast release that specify distinct presynaptic properties in subsets of neurons. Neuron 54:567–581

Zafra F, Gomeza J, Olivares L, Aragon C, Gimenez C (1995) Regional distribution and developmental variation of the glycine transporters GLYT1 and GLYT2 in the rat CNS. Eur J Neurosci 7:1342–1352

Zheng JJ, Lee S, Zhou ZJ (2004) A developmental switch in the excitability and function of the starburst network in the mammalian retina. Neuron 44:851–864

Chapter 6
GABA is the Main Neurotransmitter Released from Mossy Fiber Terminals in the Developing Rat Hippocampus

Victoria F. Safiulina, Majid H. Mohajerani, Sudhir Sivakumaran, and Enrico Cherubini

Abstract Early in postnatal development, correlated activity in the hippocampus is characterized by giant depolarizing potentials (GDPs). GDPs are generated by the interplay between glutamate and GABA, which in the immediate postnatal period is depolarizing and excitatory. Here, we review some recent data obtained in our laboratory concerning neuronal signaling at immature MF connections. MF responses were identified on the basis of their strong paired-pulse facilitation, short-term frequency-dependent facilitation and sensitivity to group III mGluR agonist L-AP4. Unlike adulthood, during the first week of postnatal life minimal stimulation of MF evoked responses that were potentiated by flurazepam and abolished by picrotoxin indicating that they were GABAergic. In addition, using a pairing procedure we found that GDPs and associated calcium transients act as coincident detectors for enhancing synaptic efficacy at poorly developed MF-CA3 and MF-interneurons connections. This may be crucial for synaptogenesis and for establishing the adult neuronal circuit.

Abbreviations

AMPA	α-amino-3-hydroxy-5-methyl-4-isoxazolepropionic acid
CNS	Central Nervous System
CNQX	6-cyano-7-nitroquinoxaline-2,3-dione
D-APV	D-(-)-2-Amino-5-phosphonopentanoic acid
DCG-IV	(2S,2'R,3'R)-2-(2',3'-Dicarboxycyclopropyl)glycine
DNQX	6,7-Dinitroquinoxaline-2,3-dione
GAT-1	a high-affinity GABA plasma membrane transporter
GABA	γ-Amino-butyric acid
GAD	glutamic acid decarboxylase

E. Cherubini (✉)
Neurobiology Sector, International School for Advanced Studies, Via Beirut 2–4, 34014 Trieste, Italy
e-mail: cher@sissa.it

R. Gutierrez (ed.), *Co-Existence and Co-Release of Classical Neurotransmitters*, DOI 10.1007/978-0-387-09622-3_6, © Springer Science+Business Media, LLC 2009

GDPs	Giant Depolarizing Potentials
IPSC	inhibitory postsynaptic current
KCC2	neuronal Potassium-Chloride cotransporter
L-AP4	2-amino-4-phosphonobutyric acid
MF	Mossy fibers
mGluR	metabotropic glutamate receptors
NKCC1	Sodium, Potassium Chloride cotransporter
NMDA	N-methyl-D-aspartate
P	postnatal day
VGAT	Vesicular GABA Transporter

6.1 γ-Aminobutiric Acid (GABA) Plays a Crucial Role in Developmental Networks

GABA is the main inhibitory transmitter in the adult mammalian CNS. It reduces cell excitability by activating $GABA_A$ receptor channels, which are mainly permeable to chloride ions. In addition, GABA inhibits neuronal firing by acting on $GABA_B$ receptors coupled to potassium or calcium channels (Cherubini and Conti, 2001). GABA plays a crucial role in many physiological processes including network synchronization and generation of theta and gamma rhythms, thought to be associated with higher cognitive functions (Buzsaki and Draguhn, 2004). Dysfunction of GABAergic signaling leads to several neurological disorders, including epilepsy which is triggered by the unbalance between excitation and inhibition (Roberts, 1986). Interestingly, in the immediate postnatal period, when glutamatergic synapses are still poorly developed (Hosokawa et al., 1994; Tyzio et al., 1999), GABA depolarizes and excites target cells through an outwardly directed flux of chloride (Cherubini et al., 1991; Ben-Ari et al., 1997; Ben-Ari, 2002; Owens and Kriegstein, 2002; Mohajerani and Cherubini, 2005). The intracellular chloride concentration is under control of two main Cl^- co-transporters the NKCC1 and KCC2 that enhance and lower $[Cl^-]_i$, respectively (Payne et al., 2003). Due to the low expression of the KCC2 extruder at birth, chloride accumulates inside the neuron *via* NKCC1. The progressive increase in the expression of KCC2 is responsible for the developmental shift of GABA from the depolarizing to the hyperpolarizing direction (Rivera et al., 1999). In the immature hippocampus, the depolarizing action of GABA which occurs well before synapses formation (Demarque et al., 2002) enables the induction of synchronized activity, the so called giant depolarizing potentials or GDPs, which consist in recurrent membrane depolarizations with superimposed fast action potentials, separated by quiescent intervals (Ben-Ari et al., 1989). GDPs which have been proposed to be the in vitro counterpart of "sharp waves" recorded in rat pups during immobility periods, sleep and feeding (Leinekugel et al., 2002) can be considered a primordial form of synchrony between neurons, which precedes more organized forms of activity such as the theta and the gamma rhythms (Buzsaki and Draguhn, 2004). Correlated network activity constitutes a hallmark of developmental networks, well preserved during evolution that has been observed not only in the hippocampus but in almost every brain structure, including the retina (Feller et al., 1997), the neocortex (Owens et al., 1996; Dammerman et al., 2000; Maric et al., 2001), the hypothalamus (Chen et al., 1996), the cerebellum (Yuste and Katz, 1991; Eilers et al., 2001) and the spinal cord (Wang et al., 1994; O'Donovan, 1999).

The depolarizing action of GABA during GDPs results in calcium influx through the activation of voltage-dependent calcium channels and N-methyl-D-aspartate (NMDA) receptors (Leinekugel et al., 1997; Garaschuk et al., 1998). GDPs and associated calcium transients lead to the activation of intracellular signaling pathways thought to contribute to several developmental processes including DNA synthesis, cell migration, morphological maturation and synaptogenesis (Owens and Kriegstein, 2002). More recently, GDPs and associated calcium transients have been shown to act as coincident detectors for enhancing synaptic efficacy at poorly developed synapses (Kasyanov et al., 2004; Mohajerani et al., 2007).

How GDPs are generated is still a matter of debate. In the disinhibited hippocampus, population synchrony has been proposed to depend on an active process consisting in a build up period during which synaptic traffic and cell firing exceeds a certain threshold (Menendez de la Prida et al., 2006). Functionally excitatory synaptic interactions would facilitate neuronal synchronization and the initiation of population bursts (Traub and Wong, 1982; Miles and Wong, 1987; Traub and Miles, 1991). A similar process may be involved in the generation of GDPs early in postnatal life (Menendez de la Prida and Sanchez-Andres, 1999; 2000) when synaptic interactions are facilitated by the excitatory action of GABA (Cherubini et al., 1991; Ben-Ari, 2002).

From the above mentioned examples it emerges that GABA is one of the major players in neuronal development.

In this chapter we will review some recent data obtained in our laboratory demonstrating that, during the first week of postnatal life, the axons of *dentate gyrus* granule cells, the mossy fibers, which in adult are glutamatergic, release into CA3 pyramidal cells and GABAergic interneurons mainly GABA. In addition, we will provide evidence that GABA-mediated GDPs act as coincidence detectors for enhancing synaptic efficacy at mossy fiber-CA3 synapses.

6.2 Mossy Fiber Synapses

The axons of granule cells have been originally called "mossy" by Ramon y Cajal because of their particular appearance at the light microscopic level that reminds, as the mossy fibers in the cerebellum, the shape of the moss on trees (Ramon y

Cajal, 1911). Thus, unlike other principal cells, mossy fibers (MF) give rise to large *en passant* swellings and terminal expansions on CA3 principal neurons or mossy cells seen as giant boutons at the electron microscopic level. These presynaptic swellings adapt very well to specialized postsynaptic elements present on proximal dendrites of CA3 principal cells, called *thorny excrescences*. The MF synaptic complex contains multiple active zones (up to 50) associated with postsynaptic densities. In addition MF make synaptic contacts with GABAergic interneurons present in the *hilus* and in the CA3 area and these represent the majority of all MF connections (Frotscher et al., 2006). In a seminal paper, Acsády et al. (1998) demonstrated that, MF connections with interneurons have either the shape of small boutons or filopodial extensions. Differences in morphology between MF terminals at principal cells and interneurons may account for the distinct functional properties of these synapses which appear to be regulated in a target specific way (Nicoll and Schmitz, 2005). Interestingly, at principal cell synapses, giant boutons develop gradually during the first 21 days (Amaral and Dent, 1981). A light and electron microscopic study has shown that, during the first postnatal days, at the time when our study was performed, immature axons terminate in very small, spherical expansions, which establish both symmetric and asymmetric contacts with pyramidal cell dendrites (Fig. 6.1).

These contacts are made several days before the development of *thorny excrescences* (Stirling and Bliss, 1978; Amaral and Dent, 1981). Expansions markedly increase in size by day 9 while maintaining a relatively spherical shape. During this period, pyramidal cell dendrites show a marked lateral growth and fingers which began indenting into MF expansions. This period is also characterized by an increased number and densities of synaptic vesicles.

In adults, the MF input to CA3 is glutamatergic and comprises the second synapse of the classical trisynaptic hippocampal circuit. Glutamate acts mainly

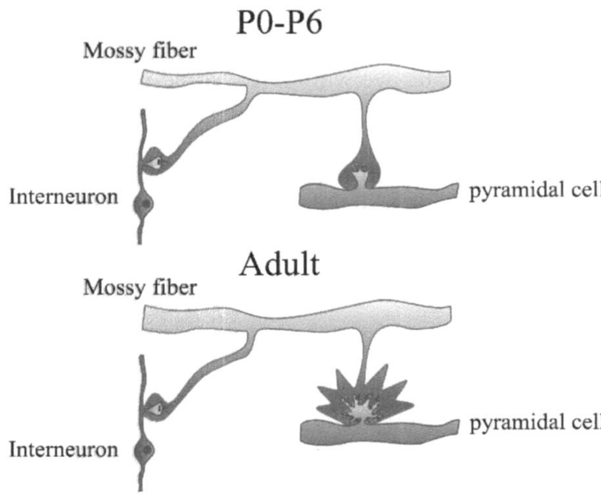

Fig. 6.1 Schematic representation of a mossy fiber (blue) making synaptic contacts with a GABAergic interneuron (red) and a pyramidal cell (green) during the first week of postnatal life (P0–P6) and in adulthood. (Modified from Amaral and Dent, 1981). (*See* Color Plate 4)

on postsynaptic α-amino-3-hydroxy-5-methyl-4-isoxazolepropionic acid (AMPA) and kainate receptors (Henze et al., 2000). It is stored in synaptic vesicles with zinc, which is co-released with glutamate upon nerve stimulation and is known to down regulate both N-methyl-D-aspartate and $GABA_A$ receptors (Westbrook and Mayer, 1987). Besides glutamate, MFs release other substances including dynorphin, known to modulate glutamate release *via* presynaptic receptors (Weisskopf et al., 1993). The latter is stored on large dense-core vesicles, which also contain other peptides such as enkephalins, cholecystokinin and neuropeptide Y (Henze et al., 2000). Moreover, MF are endowed with a variety of different receptors (autoreceptors) whose tonic activation has been shown to depress or enhance transmitter release, respectively. Thus, activation of A1 adenosine receptors (Moore et al., 2003), $GABA_B$ receptors (Hirata et al., 1992) and type II/III metabotropic glutamate receptors (mGluR,) reduces transmitter release (Kamiya et al., 1996; Shigemoto et al., 1997) while activation of kainate receptors facilitates transmitter release (Schmitz et al., 2001).

In pathological conditions, MFs can release GABA in addition to glutamate. Thus, in the hippocampus of epileptic animals (Gutierrez and Heinemann, 2001; Romo-Parra et al., 2003) monosynaptic GABAergic inhibitory postsynaptic potentials (IPSPs) occur in principal cells in response to *dentate gyrus* stimulation. Seizures are associated with a transient upregulation of the GABAergic markers GAD65 and GAD67 (Schwarzer and Sperk, 1995; Sloviter et al., 1996) as well as the mRNA for the vesicular GABA transporter, VGAT (Lamas et al., 2001). Interestingly, both GAD67 and its product GABA appear to be constitutively expressed in MF. Further evidence suggests that MFs can release glutamate and GABA also in physiological conditions. Hence, Walker et al. (2001) and Gutierrez et al. (2003) have demonstrated the presence of both monosynaptic GABAergic and glutamatergic responses following activation of granule cells in the *dentate gyrus* in juvenile guinea pigs and rats. While these pieces of work will be the object of other chapters in this book, here we will focus on the first week of postnatal life when the main neurotransmitter released by the MF has been found to be GABA. It should be stressed that, due to the complexity of *dentate gyrus*-CA3 circuitry (Henze et al., 2000), studying *pure* MF synaptic responses with electrophysiological approaches is quite difficult. This task is, at least partially facilitated in neonatal animals where the small size of neurons and the relatively low extension of dendritic branches allow good space clamp conditions.

6.3 Criteria for Identifying Single Mossy Fiber Responses

The best way for studying synaptic transmission at given synapses is to record simultaneously from interconnected pre and postsynaptic cells. However, in the case of MF synapses this approach is very difficult due to the very low probability of finding interconnected granule cells and CA3 pyramidal neurons. An alternative approach consists in recording single fiber responses (Jonas et al., 1993; Allen and Stevens, 1994). With this technique, a small stimulating electrode is placed in the granule cell layer and the stimulation intensity is decreased until only a single axon is activated. This is achieved when the mean amplitude of the postsynaptic currents and failure probability remain constant over a range of stimulus intensities near threshold for detecting a response. Small movements of the stimulating electrode 20–30 μm away from the initial location, lead to the loss of the evoked response. The example of Fig. 6.2 shows average traces of synaptic currents evoked in a CA3 principal cell in response to stimulation of granule cells in the *dentate gyrus* with different intensities. An abrupt increase in the mean peak amplitude of synaptic currents can be detected by increasing the strength of stimulation. This all-or-none behavior suggests that only a single granule cell is stimulated. When the stimulation intensity is turned down the probability of failures in synaptic transmission is near 1.

In addition, the latency and the shape of individual synaptic responses should remain constant for repeated stimuli. Between P0 and P6, MF-evoked synaptic currents occurred with a latency of 3.8 ± 0.3 ms in principal cells and 3.4 ± 0.4 ms in interneurons. The latencies distribution was unimodal and narrow with an average standard deviation of 0.31 ± 0.03 ms ($n = 10$; Fig. 6.2). Moreover, the latency as well as the rise time of synaptic responses remained constant when the extracellular Ca^{2+}/Mg^{2+} concentration ratio was reduced from 2:1.3 to 1:3 (Fig. 6.2) further supporting the monosynaptic nature of synaptic currents. The 10–90% rise time was 3.5 ± 0.8 ms in principal cells ($n = 12$) and 3.1 ± 0.4 ms in interneurons ($n = 8$; see also Walker et al., 2001).

MF inputs to principal cells or interneurons were identified on the basis of their sensitivity to group III mGluR agonist 2-amino-4-phosphonobutyric acid (L-AP4; Fig. 6.3). In this regard, neonatal rats behave differently from adult animals which are sensitive to: (2S,2'R,3'R)-2-(2',3'-Dicarboxycyclopropyl)-glycine (DCG-IV) but insensitive to L-AP4 (Lanthorn et al., 1984). It should be stressed, however, that both group II and III mGluRs have been found on rat MF terminals: while group III mGluRs are located predominantly in presynaptic active zones, group II are in preterminal rather than terminal portions of axons (Shigemoto et al., 1997). The inability of DCG-IV in modulating MF responses in immature neurons can be attributed to the different expression and/or location of group II/III mGluRs early in postnatal development. One intriguing aspect to be considered is how mGluRs are activated early in postnatal days, when MFs seem to release only GABA. One possibility is that they are constitutively activated by ambient glutamate present in the extracellular medium. However, more work should be done in order to elucidate this point.

Interestingly, monosynaptic inhibitory responses obtained in principal cells by stimulating GABAergic interneurons were found to be insensitive to mGluR agonists (Walker et al., 2001; Doherty et al., 2004; Kasyanov et al., 2004; Gutierrez, 2005; Fig. 6.5). L-AP4-sensitive interneurons have been described in

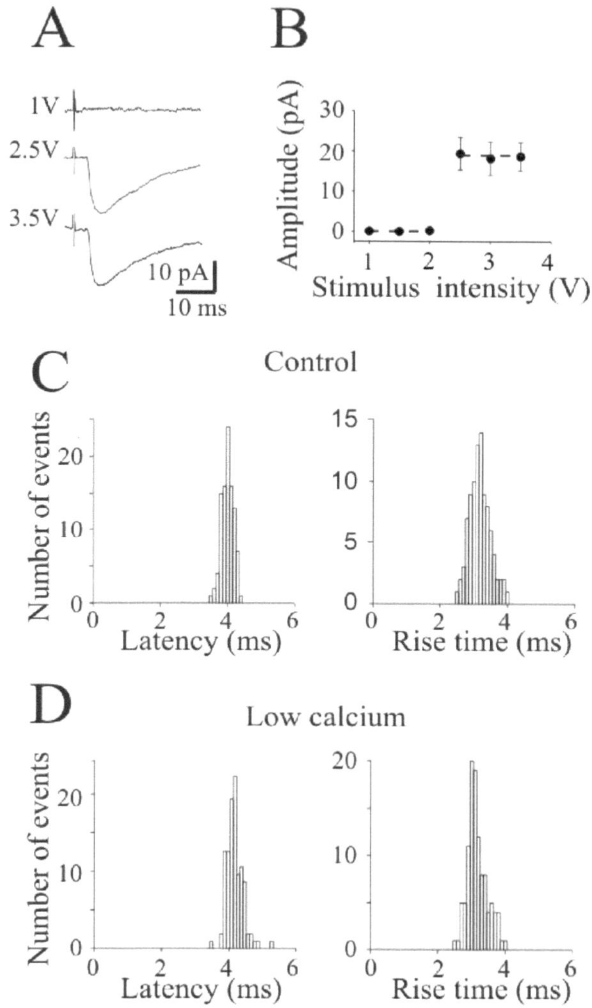

Fig. 6.2 (a) Unitary synaptic currents evoked in a P3 CA3 principal cell by minimal stimulation of granule cells in the *dentate gyrus*. Each trace is the average of 15–20 responses. Holding potential –70 mV. The peak amplitude of synaptic currents represented in A is plotted against different stimulus intensities in **(b)**. Note the all-or-none appearance of synaptic currents with increasing stimulus intensities. Bars are SEM. Dashed lines connect the mean values of individual points within the same group. **(c)**, **(d)** Latency and rise time distributions of individual currents evoked in another CA3 pyramidal cell in the presence of a solution containing a Ca^{2+}/Mg^{2+} ratio of 2:1.3 (C) or 1:3 (D). Note the unimodal distributions of latencies and rise times of individual responses that did not change when the extracellular Ca^{2+}/Mg^{2+} ratio was changed. Modified from Safiulina et al. (2006)

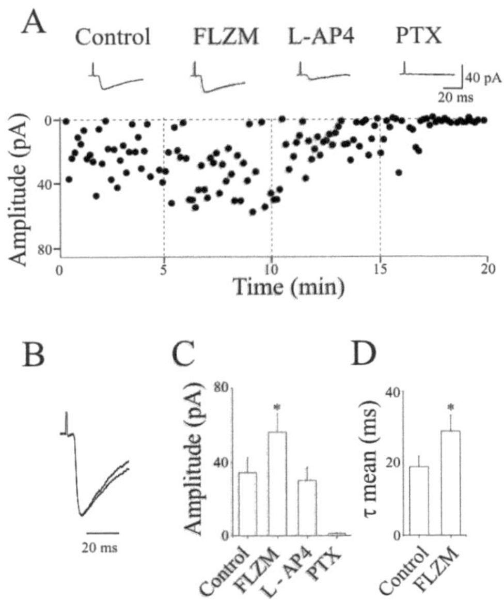

Fig. 6.3 Minimal stimulation of granule cells in the *dentate gyrus* evokes $GABA_A$-mediated monosynaptic responses in CA3 pyramidal cells. **(a)** Amplitude of synaptic responses (dots) evoked by stimulation of granule cells in the *dentate gyrus* are plotted against time in control, during bath application of flurazepam 3 μM (FLZM), L-AP4 (10 μM) and picrotoxin 100 μM (PTX). The inset above show examples of average traces, taken in different experimental conditions (each is the average of at least 20 traces including failures). Note that the currents were enhanced by FLZM, reduced in amplitude by L-AP4 and blocked by PTX. **(b)** Normalized responses obtained in the absence (1) or in the presence (2) of FLZM. Note the slow down of the deactivation kinetics of synaptic currents with FLZM. **(c)** Each column represents the mean amplitude (± SEM) of MF-evoked synaptic currents recorded from 5 CA3 principal cells in control, in the presence of FLZM, L-AP4 and PTX. **(d)** Decay kinetics (τ mean) of synaptic currents recorded in control and in the presence of FLZM ($n = 5$) × $p<0.05$

guinea pig hippocampus (Semyanov and Kullmann, 2000). Here, L-AP4, at the concentration of 50 μM (5 times higher than that used in the present experiments) was able to reduce the amplitude of IPSCs of ~50%. This raises the possibility that different mGluR subtypes are expressed in different animal species.

MF inputs were also identified on the basis of their strong paired pulse facilitation and short-term frequency-dependent facilitation (Salin et al., 1996). Strong paired pulse facilitation was observed particularly at MF-CA3 synapses, which were often "silent" in response to the first stimulus. At MF-interneuron synapses both paired pulse facilitation and depression were observed while at interneuron-CA3 or interneuron-interneuron synapses the most common feature was paired pulse depression (Fig. 6.5).

The degree of frequency-dependent facilitation is another peculiar aspect of MF responses which probably depends on the enhanced probability of neurotransmitter release following the large rise of intraterminal calcium concentration and activation of calcium/calmodulin-dependent kinase II (Salin et al., 1996). Alternatively, synaptic facilitation may occur as the consequence of the progressive and local saturation of calcium by the endogenous fast calcium buffer calbindin, which is highly expressed in MF terminals. This would produce a gradual increase in calcium concentration at releasing sites (Blatow et al., 2003). In our case, frequency facilitation occurred already when the stimulation frequency was shifted from 0.05 to 0.3 Hz. However, the increment in size of synaptic responses was larger at MF-CA3 principal cell connections than at MF-interneuron synapses (see also Toth et al., 2000).

6.4 GABA Is the Main Neurotransmitter Released by MF Early in Postnatal Life

As shown in the illustrative example of Fig. 6.3, minimal stimulation of granule cells in the *dentate gyrus* was able to evoke in CA3 principal cells monosynaptic currents that were completely blocked by picrotoxin or bicuculline, suggesting that they were mediated by $GABA_A$ receptors. As classical MF responses, synaptic currents exhibited strong paired pulse facilitation. Moreover, they were highly sensitive to L-AP4 and underwent short-term frequency-dependent facilitation. As expected for $GABA_A$-mediated responses MF-evoked synaptic currents were potentiated by NO-711, a blocker of the GABA transporter GAT-1 and by flurazepam, an allosteric modulator of $GABA_A$ receptors.

Similar results were found for MF making synaptic contacts with GABAergic interneurons. In this case however, in agreement with previous findings (Toth et al., 2000), changing the stimulation frequency from 0.05 to 0.33 Hz, induced only a moderate facilitation of synaptic responses. This can be attributed to the fact that, in contrast with principal cells, MF contacting interneurons comprise only a single release site (Acsady et al., 1998). In a few cases, a depression of synaptic currents was also observed.

Pressure application of glutamate to granule cells dendrites in *stratum moleculare* (in the presence of the AMPA/kainate receptor antagonist 6,7-Dinitroquinoxaline-2,3-dione (DNQX) to prevent the recruitment of GABAergic interneurons) induced in target cells barrages of L-AP4 sensitive currents that were completely abolished by picrotoxin. It is therefore likely that activation of NMDA receptors localized on granule cells dendrites in *stratum moleculare* causes a membrane depolarization and the release of GABA from MF terminals. This was supported by the observation that, blocking NMDA receptors with D-(-)-2-Amino-5-phosphonopentanoic acid (D-APV) prevented the effects of chemical stimulation of granule cell dendrites on CA3 principal cells and GABAergic interneurons. Moving the pressure pipette few μm away toward the hilus to activate hilar interneurons caused barrage of synaptic currents that were insensitive to L-AP4 but were blocked by picrotoxin, implying that they were mediated by the release of GABA from GABAergic interneurons.

Additional fibers releasing both glutamate and GABA into CA3 principal cells and interneurons could be recruited by increasing the strength of stimulation (Safiulina et al., 2006). In comparison with MF-induced GABAergic currents, glutamatergic responses occurred with a shorter latency. However, these responses involved MF synapses, since they were reversibly depressed by L-AP4 and were abolished when DNQX (10 μM) was added to picrotoxin.

In additional experiments a low chloride intracellular solution (E_{Cl} –90 mV) was used to simultaneously record AMPA- and $GABA_A$-mediated synaptic currents at room temperature (22–24°C) to avoid the activation of polysynaptic pathways and GDPs. Thus, at –50 mV, AMPA-mediated synaptic responses were detected as inward currents while $GABA_A$-mediated responses as outward currents (Fig. 6.4). The Figure shows also that, in comparison with AMPA-mediated

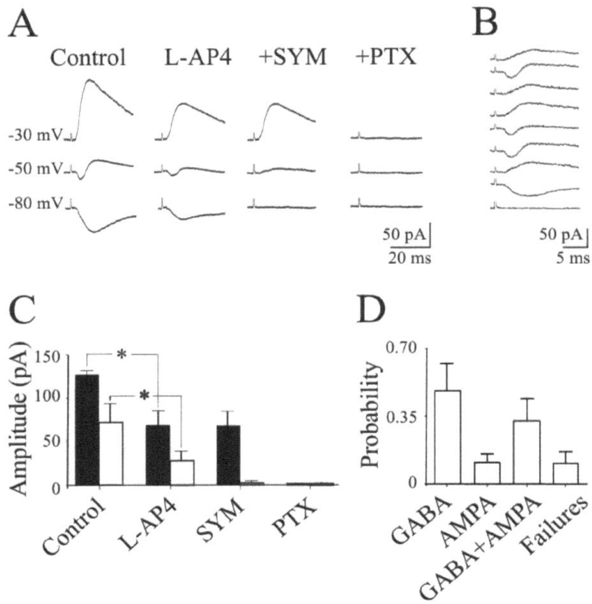

Fig. 6.4 Glutamatergic and GABAergic currents evoked in principal cells by stimulation of the granule cells in the *dentate gyrus* at room temperature (to avoid the activation of GDPs; see Ben-Ari et al. 1989). **(a)** Average responses evoked at three different holding potentials in a CA3 pyramidal cell recorded with a low chloride intrapipette solution (E_{Cl} –90 mV). Note the biphasic currents at –50 mV and the isolated GABAergic and glutamatergic components at –30 and –80 mV, respectively. The two components were reduced in amplitude by L-AP4 and were selectively blocked by the AMPA and $GABA_A$ receptor blockers, SYM-2206 (20 μM) and PTX (100 μM), respectively **(b)** Individual traces from one single cell recorded at –50 mV showing inward, outward, biphasic responses and response failures **(c)** Summary data (n = 5) showing the mean amplitude of GABAergic (black columns) and glutamatergic (white columns) currents in control, in the presence of L-AP4, L-AP4 plus SYM-2206 and L-AP4 plus SYM-2206 and picrotoxin. **(d)** Each column represents the relative frequency of each type of response for 3 cells. Modified from Safiulina et al. (2006)

synaptic currents, those mediated by GABA occurred more frequently. While at –80 mV, close to the chloride reversal potential, pure AMPA-mediated responses could be detected, at –30 mV pure $GABA_A$-mediated synaptic responses. Both components were sensitive to L-AP4 and were blocked by the respective receptor antagonists.

Synaptic currents fluctuated between outward, biphasic and inward and were intermingled with response failures. This suggests that GABA and glutamate can be released independently from the same fiber.

The possibility that the same fiber can release different neurotransmitters has been well documented in several brain structures including the retina (O' Malley and Masland, 1989) and the spinal cord (Jonas et al., 1998). In particular, GABA has been reported to be released from excitatory inputs in CA3 pyramidal cells (Walker et al., 2001; Gutierrez et al., 2003) whereas glutamate from inhibitory terminals in the lateral superior olive in the developing auditory system (Gillepsie et al., 2005).

In contrast with MF responses, synaptic currents mediated by GABA released from GABAergic interneurons were insensitive to L-AP4 and DNQX but were blocked by bicuculline or picrotoxin (Fig. 6.5). These responses were probably generated by interneurons projecting to principal cells or interneurons sending collaterals to the granule cells into the *dentate gyrus*. They occurred with high probability and exhibited a strong paired-pulse depression in response to two closely spaced stimuli. Moreover, in comparison with MF responses they were potentiated by flurazepam in a more pronounced way (Fig. 6.5).

Altogether these experiments are in line with previous reports showing the sequential expression of GABAergic and glutamatergic synapses early in postnatal development (Hosokawa et al., 1994; Tyzio et al., 1999) and clearly demonstrate that GABA is the main neurotransmitter released from MFs during postnatal development.

Further evidence in favor of GABA as a transmitter at MF synapses is the observation that the vesicular GABA transporter VGAT was found in MF terminals (Safiulina et al., 2006; see also Chaudhry et al., 1998). However, in order to generate synaptic responses, GABA should not only be present in synaptic vesicles and released in an activity-dependent manner but should bind to postsynaptic $GABA_A$ receptors. Although evidence for the presence of $GABA_A$ receptors facing immature MF terminals early in development is still lacking, a previous study on adult rats has demonstrated the presence of AMPA and $GABA_A$ receptors co-localized in front of the respective active zones (Bergersen et al., 2003). GABA may also spill out to activate neighboring extrasynaptic $GABA_A$ receptors localized away from the release sites as suggested for juvenile guinea pigs and rats (Walker et al., 2001). However, the relatively fast rise time of GABAergic responses found in the present experiments, similar to that of glutamatergic synaptic currents makes this hypothesis unlikely.

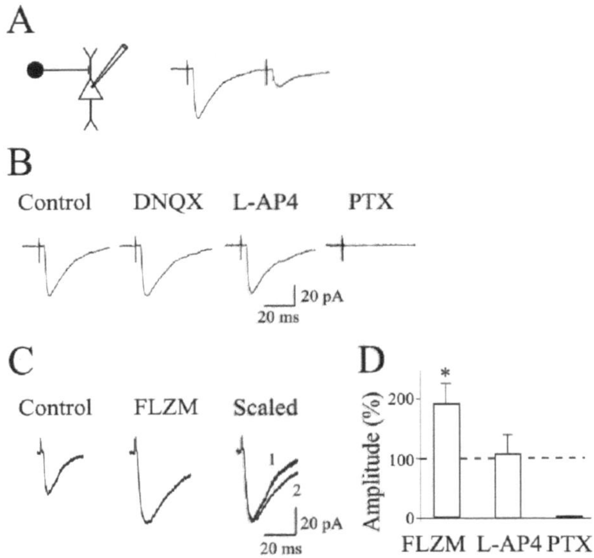

Fig. 6.5 L-AP4-insensitive synaptic currents elicited in principal cells by stimulation of the granule cells in the *dentate gyrus*. **(a)** Left: diagram showing a GABAergic interneurons impinging into a CA3 pyramidal cell. Right: average of 20 individual responses (including failures) to paired stimuli. **(b)** Average of 20 individual traces evoked in control conditions, in the presence of L-AP4 and L-AP4 plus PTX. Note lack of effect of L-AP4. **(c)** Average of 20 individual traces from another pyramidal cell obtained in control or in the presence of FLZM. On the right the two normalized traces are superimposed. **(d)** Summary data showing the amplitude of synaptic currents obtained from 4 pyramidal cells in different experimental conditions normalized to the respective controls (dashed line). Modified from Safiulina et al. (2006)

Overall, these data support the hypothesis that early in development, MF contain two different sets of low and high threshold fibers releasing GABA and GABA plus glutamate respectively. While the first would disappear with maturation the second would persist longer or would reappear in pathological conditions such as in epilepsy (Walker et al., 2001; Gutierrez et al., 2003). Comparable to our results it has been recently shown that minimal stimulation of the granule cell layer in hippocampal slices obtained from P4–P6 mice evoked GABAergic currents that were insensitive to 6-cyano-7-nitroquinoxaline-2,3-dione (CNQX), D-APV but were blocked by picrotoxin (Uchigashima et al., 2007). According to these authors, these currents were due to GABA released from low threshold GABAergic interneurons. It should be stressed however, that unlike putative MF responses described here, which occurred with very low probability (responses were often silent to first pulse) and exhibited strong short-term frequency-dependent facilitation (see Fig. 6.6), those reported by Uchigashima et al. (2007) displayed only minimal facilitation.

Paired recordings from granule cells and postsynaptic neurons would probably allow better understanding how MF establish their contact at early developmental stages. However, we cannot exclude the possibility that, during the

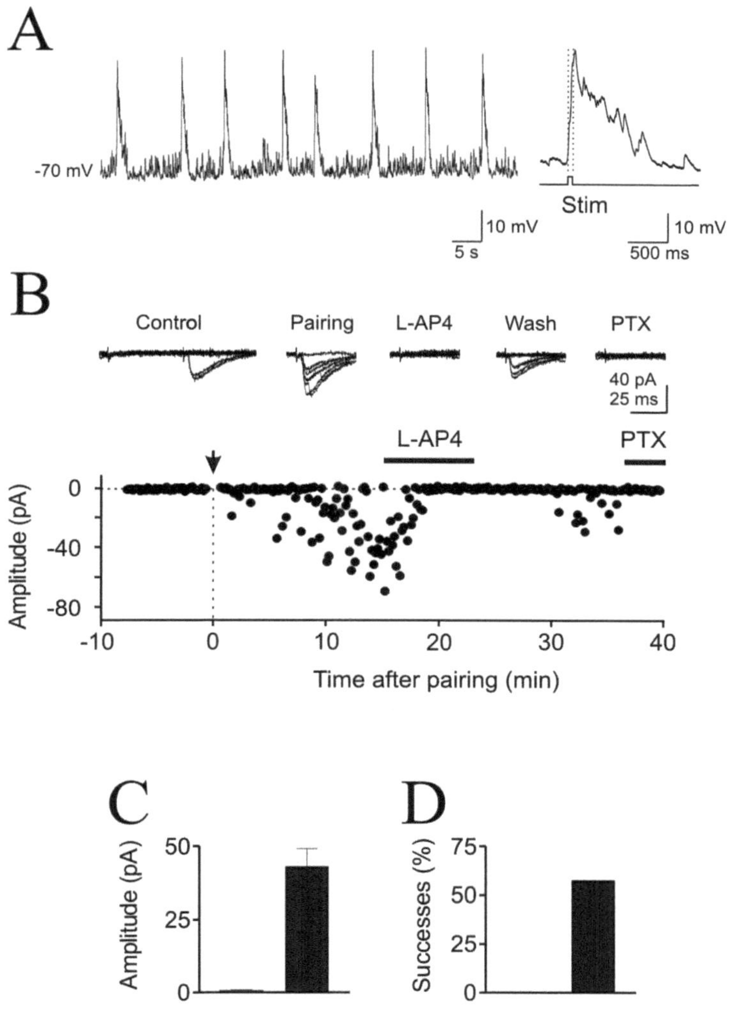

Fig. 6.6 Pairing-induced the appearance of synaptic responses in "presynaptically silent neurones" **(a)** GDPs recorded from a CA3 pyramidal cell in current clamp mode from the hippocampus of a P2 old rat. On the right, a single GDP is shown on an expanded time scale. Note the absence of spikes riding on the top of GDP due to block of the sodium channel with QX 314. The rising phase of GDPs (between the dotted lines) was used to trigger synaptic stimulation (Stim). **(b)** Amplitudes of synaptic responses (dots) evoked by minimal stimulation of MF before and after pairing (arrow at time 0) are plotted against time. The traces above the graph represent individual responses evoked before and after pairing in different experimental conditions as indicated. This synapse was considered "presynaptically" silent because did not exhibit any response to the first stimulus over 48 trials (at 0.1 Hz) but occasional (two) responses to the second one. For clarity after "pairing" only synaptic responses evoked by the first stimulus are shown. C and D. Mean excitatory postsynaptic current (EPSC) amplitude **(c)** and mean percentage of successes **(d)** before and after pairing for the cell shown in B. Modified from Kasyanov et al. (2004)

first week of postnatal life, synaptic currents evoked by minimal stimulation of granule cell in the *dentate gyrus* are generated by GAD or GAD mRNA positive interneurons, which early in development migrate to the upper and middle portions of the granule layer cells (Dupuy and Houser, 1997) while maintaining all the functional properties of later appearing glutamatergic MF responses.

6.5 GDPs as Coincidence Detectors for Enhancing Synaptic Efficacy at Low Probability MF-CA3 Synapses

To assess whether synchronized network activity such as GDPs are able to modify MF connections in an associative type of manner, we have developed a "pairing" procedure consisting of correlating GDPs-associated calcium rise in the postsynaptic cell with stimulation of MF (Kasyanov et al., 2004). To this purpose, after a control period of 5–10 min, the patch was switched from voltage-clamp to current-clamp mode and MF responses were paired for 5 min with GDPs. "Pairing" consisted in triggering MF stimulation with the rising phases of GDPs (Fig. 6.6). After this period the patch was switched back to voltage clamp mode and synaptic currents were recorded as in control. As illustrated in Fig. 6, in the case of presynaptically silent synapses (Gasparini et al., 2000), the pairing protocol caused the appearance of responses to the first stimulus and increased the number of successes to the second one.

In the case of non-silent low probability synapses, the pairing procedure produced a strong and persistent potentiation of MF responses, which was associated with a significant increase in the number of successes and in double pulse experiments, with a significant reduction in the paired-pulse ratio and a significant increase in the inverse squared value of the coefficient of variation. This suggests that an increased probability of transmitter release accounts for the persistent increase in synaptic efficacy.

In the absence of pairing no significant changes in synaptic efficacy could be detected. Moreover, when the interval between GDPs and MF stimulation was progressively increased, the potentiation declined and reached the control level when presynaptic signals were activated 2–3 s after GDPs (Kasyanov et al., 2004). In addition we found that pairing-induced long-lasting increase in synaptic efficacy was prevented when cells were loaded with the calcium chelator BAPTA or when nifedipine was added to the extracellular medium. In contrast, the NMDA receptor antagonist D-APV failed to prevent pairing-induced potentiation, indicating that early in postnatal life calcium rise through voltage-dependent calcium channel activated by the depolarizing action of GABA during GDPs is the common trigger for activity-dependent changes in synaptic efficacy.

In conclusions, during development, coincident detection provided by GDPs can be important for enhancing synaptic transmission at emerging MF-CA3 connections in a Hebbian type of way. How this may contribute to the wiring and proper assembly of adult networks remains to be determined.

Acknowledgments The authors are grateful to Drs. A. Kasyanov, G. Fattorini and F. Conti for participating in some experiments. The original research work was supported by grants from Ministero Istruzione, Universita', Ricerca (MIUR, Italy) and the European Union.

References

Acsady L, Kamondi A, Sik A, Freund T, Buzsaki G (1998) GABAergic cells are the major postsynaptic targets of mossy fibers in the rat hippocampus. J Neurosci 18:3386–3403

Allen C, Stevens CF (1994) An evaluation of causes for unreliability of synaptic transmission. Proc Natl Acad Sci U S A 91:10380–10383

Amaral DG, Dent JA (1981) Development of the mossy fibers of the dentate gyrus: I. A light and electron microscopic study of the mossy fibers and their expansions. J Comp Neurol 195:51–86

Ben-Ari Y (2002) Excitatory actions of GABA during development: the nature of the nurture. Nature Rev Neurosci 3:728–739

Ben-Ari Y, Cherubini E, Corradetti R, Gaiarsa JL (1989) Giant synaptic potentials in immature rat CA3 hippocampal neurones. J Physiol (Lond) 416:303–325

Ben-Ari Y, Khazipov R, Leinekugel X, Caillard O, Gaiarsa JL (1997) GABAA, NMDA and AMPA receptors: a developmentally regulated "menage a trois". Trends Neurosci 20:523–529

Bergersen L, Ruiz A, Bjaalie JG, Kullmann DM, Gundersen V (2003) GABA and GABAA receptors at hippocampal mossy fibre synapses. Eur J Neurosci 18:931–941

Blatow M, Caputi A, Burnashev N, Monyer H, Rozov A (2003) Ca^{2+} buffer saturation underlies paired pulse facilitation in calbindin-D28k-containing terminals. Neuron 38:79–88

Buzsaki G, Draguhn A. (2004) Neuronal oscillations in cortical networks. Science 304:1926–1929

Chaudhry FA, Reimer RJ, Bellocchio EE, Danbolt NC, Osen KK, Edwards RH, Storm-Mathisen J (1998) The vesicular GABA transporter, VGAT, localizes to synaptic vesicles in sets of glycinergic as well as GABAergic neurons. J Neurosci 8:9733–9750

Chen G, Trombley PQ, van den Pol AN (1996) Excitatory actions of GABA in developing rat hypothalamic neurones. J Physiol 494:451–464

Cherubini E, Gaiarsa JL, Ben-Ari Y (1991) GABA: an excitatory transmitter in early post-natal life. Trends Neurosci 14:515–519

Cherubini E and Conti F (2001) Generating diversity at GABAergic synapses. Trends Neurosci 24:155–162

Dammerman RS, Flint AC, Noctor S, Kriegstein AR (2000) An excitatory GABAergic plexus in developing neocortical layer 1. J Neurophysiol 84:428–434

Demarque M, Represa A, Becq H, Khalilov I, Ben-Ari Y and Aniksztejn L (2002) Paracrine intercellular communication by a Ca^{2+} – and SNARE-independent release of GABA and glutamate prior to synapse formation. Neuron 36:1051–1061

Doherty JJ, Alagarsamy S, Bough KJ, Conn PJ, Dingledine R, Mott DD (2004) Metabotropic glutamate receptors modulate feedback inhibition in a developmentally regulated manner in rat dentate gyrus. J Physiol 561:395–401

Dupuy ST, Houser CR (1997) Developmental changes in GABA neurons of the rat dentate gyrus: an in situ hybridization and birthdating study. J Comp Neurol 389:402–418

Eilers J, Plant TD, Marandi N, Konnerth A (2001) GABA-mediated Ca^{2+} signalling in developing rat cerebellar Purkinje neurones. J Physiol 536:429–437

Feller MB, Butts DA, Aaron HL, Rokhsar DS, Shatz CJ (1997) Dynamic processes shape spatiotemporal properties of retinal waves. Neuron 19:293–306

Frotscher M, Jonas P, Sloviter RS (2006) Synapses formed by normal and abnormal hippocampal mossy fibers. Cell Tissue Res 326:361–367

Garaschuk O, Hanse E, Konnerth A (1998) Developmental profile and synaptic origin of early network oscillations in the CA1 region of rat neonatal hippocampus. J Physiol (London) 507:219–236

Gasparini S, Saviane C, Voronin LL, Cherubini E (2000) Silent synapses in the developing hippocampus: lack of functional AMPA receptors or low probability of glutamate release? Proc Natl Acad Sci U S A 97:9741–9746

Gillespie DC, Kim G, Kandler K (2005) Inhibitory synapses in the developing auditory system are glutamatergic. Nat Neurosci 8:332–338

Gutierrez R (2005) The dual glutamatergic-GABAergic phenotype of hippocampal granule cells. Trends Neurosci 28:297–303

Gutierrez R, Heinemann U (2001) Kindling induces transient fast inhibition in the dentate gyrus–CA3 projection. Eur J Neurosci 13:1371–1379

Gutierrez R, Romo-Parra H, Maqueda J, Vivar C, Ramirez M, Morales MA, Lamas M (2003) Plasticity of the GABAergic phenotype of the "glutamatergic" granule cells of the rat dentate gyrus. J Neurosci 23:5594–5598

Henze DA, Urban NN, Barrionuevo G (2000) The multifarious hippocampal mossy fiber pathway: a review. Neuroscience 98:407–427

Hirata K, Sawada S, Yamamoto C (1992) Quantal analysis of suppressing action of baclofen on mossy fiber synapses in guinea pig hippocampus. Brain Res 578:33–40

Hosokawa Y, Sciancalepore M, Stratta F, Martina M and Cherubini E (1994) Developmental changes in spontaneous $GABA_A$-mediated synaptic events in rat hippocampal CA3 neurones. Eur J Neurosci 6:805–813

Jonas P, Major G, Sakmann B (1993) Quantal components of unitary EPSCs at the mossy fibre synapse on CA3 pyramidal cells of rat hippocampus. J Physiol 472:615–663

Jonas P, Bischofberger J, Sandkuhler J (1998) Corelease of two fast neurotransmitters at a central synapse. Science 281:419–424

Kamiya H, Shinozaki H, Yamamoto C (1996) Activation of metabotropic glutamate receptor type 2/3 suppresses transmission at rat hippocampal mossy fibre synapses. J Physiol 493:447–455

Kasyanov AM, Safiulina VF, Voronin LL, Cherubini E (2004) GABA-mediated giant depolarizing potentials as coincidence detectors for enhancing synaptic efficacy in the developing hippocampus. Proc Natl Acad Sci USA 101:3967–3972

Lamas M, Gomez-Lira G, Gutierrez R (2001) Vesicular GABA transporter mRNA expression in the dentate gyrus and in mossy fiber synaptosomes. Brain Res Mol Brain Res 93:209–214

Lanthorn TH, Ganong AH, Cotman CW (1984) 2-Amino-4-phosphonobutyrate selectively blocks mossy fiber-CA3 responses in guinea pig but not rat hippocampus. Brain Res 290:174–178

Leinekugel X, Medina I, Khalilov I, Ben-Ari Y, Khazipov R (1997) Ca^{2+} oscillations mediated by the synergistic excitatory actions of GABA(A) and NMDA receptors in the neonatal hippocampus. Neuron 18:243–255

Leinekugel X, Khazipov R, Cannon R, Hirase H, Ben-Ari Y and Buzsaki G (2002) Correlated bursts of activity in the neonatal hippocampus in vivo. Science 296:2049–2052

Maric D, Liu QY, Maric I, Chaudry S, Chang YH, Smith SV, Sieghart W, Fritschy JM, Barker JL (2001) GABA expression dominates neuronal lineage progression in the embryonic rat neocortex and facilitates neurite outgrowth via GABA(A) autoreceptor/Cl- channels. J Neurosci 21:2343–2360

Menendez de la Prida L, Sanchez-Andres JV (1999) Nonlinear frequency-dependent synchronization in the developing hippocampus. J Neurophysiol 82:202–208

Menendez de la Prida L, Sanchez-Andres JV (2000) Heterogeneous populations of cells mediate spontaneous synchronous bursting in the developing hippocampus through a frequency-dependent mechanism. Neuroscience 97:227–241

Menendez de la Prida LM, Huberfeld G, Cohen I, Miles R (2006) Threshold behavior in the initiation of hippocampal population bursts. Neuron 49:131–142
Miles R, Wong RKS (1987) Latent synaptic pathways revealed after tetanic stimulation in the hippocampus. Nature 329:724–726
Mohajerani MH and Cherubini E (2005) Spontaneous recurrent network activity in organotypic rat hippocampal slices. Eur J Neurosci 22:107–118
Mohajerani MH, Sivakumaran S, Zacchi P, Aguilera P, Cherubini E (2007) Correlated network activity enhances synaptic efficacy via BDNF and the ERK pathway at immature CA3–CA1 connections in the hippocampus. Proc Natl Acad Sci U S A 104:13176–13181
Moore KA, Nicoll RA, Schmitz D (2003) Adenosine gates synaptic plasticity at hippocampal mossy fiber synapses. Proc Natl Acad Sci U S A 100:14397–14402
Nicoll RA, Schmitz D (2005) Synaptic plasticity at hippocampal mossy fibre synapses. Nat Rev Neurosci 6:863–876
O'Donovan MJ (1999) The origin of spontaneous activity in developing networks of the vertebrate nervous system. Curr Opin Neurobiol 9:94–104
O'Malley DM, Masland RH (1989) Co-release of acetylcholine and gamma-aminobutyric acid by a retinal neuron. Proc Natl Acad Sci U S A 86:3414–3418
Owens DF, Boyce LH, Davis MBE, Kriegstein AR (1996) Excitatory GABA responses in embryonic and neonatal cortical slices demonstrated by gramicidin perforated-patch recordings and calcium imaging. J Neurosci 16:6414–6423
Owens DF, Kriegstein AR (2002) Is there more to GABA than synaptic inhibition? Nature Rev Neurosci 3:715–727
Payne JA, Rivera C, Voipio J, Kaila K (2003) Cation-chloride co-transporters in neuronal communication, development and trauma. Trends Neurosci 26:199–206
Ramon y Cajal SR (1911) Histologie du Système Nerveux de l'Homme et des Vertébrés, vol. II. Maloine, Paris
Rivera C, Voipio J, Payne JA, Ruusuvuori E, Lahtinen H, Lamsa K, et al (1999) The K + /Cl- co-transporter KCC2 renders GABA hyperpolarizing during neuronal maturation. Nature 397:251–255
Roberts E (1986) Failure of GABAergic inhibition: a key to local and global seizures. Adv Neurol 44:319–341
Romo-Parra H, Vivar C, Maqueda J, Morales MA, Gutierrez R (2003) Activity-dependent induction of multitransmitter signaling onto pyramidal cells and interneurons of hippocampal area CA3. J Neurophysiol 89:3155–3167
Salin PA, Scanziani M, Malenka RC, Nicoll RA (1996) Distinct short-term plasticity at two excitatory synapses in the hippocampus Proc Natl Acad Sci U S A 93:13304–13309
Safiulina VF, Fattorini G, Conti F, Cherubini E (2006) GABAergic signaling at mossy fiber synapses in neonatal rat hippocampus. J Neurosci 26:597–608
Schmitz D, Mellor J, Nicoll RA (2001) Presynaptic kainate receptor mediation of frequency facilitation at hippocampal mossy fiber synapses. Science. 2001 Mar 9; 291(5510): 1972–1976
Schwarzer C, Sperk G (1995) Hippocampal granule cells express glutamic acid decarboxylase-67 after limbic seizures in the rat. Neuroscience 69:705–709
Semyanov A, Kullmann DM (2000) Modulation of GABAergic signaling among interneurons by metabotropic glutamate receptors. Neuron 25:663–672
Shigemoto R, Kinoshita A, Wada E, Nomura S, Ohishi H, Takada M, Flor PJ, Neki A, Abe T, Nakanishi S, Mizuno N (1997) Differential presynaptic localization of metabotropic glutamate receptor subtypes in the rat hippocampus. J Neurosci 17:7503–7522
Sloviter RS, Dichter MA, Rachinsky TL, Dean E, Goodman JH, Sollas AL, Martin DL (1996) Basal expression and induction of glutamate decarboxylase and GABA in excitatory granule cells of the rat and monkey hippocampal dentate gyrus. J Comp Neurol 373:593–618
Stirling RV, Bliss TV (1978) Hippocampal mossy fiber development at the ultrastructural level. Prog Brain Res 48:191–198

Toth K, Suares G, Lawrence JJ, Philips-Tansey E, McBain CJ (2000) Differential mechanisms of transmission at three types of mossy fiber synapse. J Neurosci 20:8279–8289

Traub RD, Wong RK (1982) Cellular mechanism of neuronal synchronization in epilepsy. Science 216:745–747

Traub RD, Miles R (1991) Multiple modes of neuronal population activity emerge after modifying specific synapses in a model of the CA3 region of the hippocampus. Ann N Y Acad Sci 627:277–290

Tyzio R, Represa A, Jorquera I, Ben-Ari Y, Gozlan H and Aniksztejn L (1999) The establishment of GABAergic and glutamatergic synapses on CA1 pyramidal neurons is sequential and correlates with the development of the apical dendrite. J Neurosci 19:10372–10382

Uchigashima M, Fukaya M, Watanabe M, Kamiya H (2007) Evidence against GABA release from glutamatergic mossy fiber terminals in the developing hippocampus. J Neurosci 27:8088–8100

Walker MC, Ruiz A, Kullmann DM (2001) Monosynaptic GABAergic signaling from dentate to CA3 with a pharmacological and physiological profile typical of mossy fiber synapses. Neuron 29:703–715

Wang J, Reichling DB, Kyrozis A, MacDermott AB (1994) Developmental loss of GABA- and glycine-induced depolarization and Ca^{2+} transients in embryonic rat dorsal horn neurons in culture. Eur J Neurosci 6:1275–1280

Weisskopf MG, Zalutsky RA, Nicoll RA (1993) The opioid peptide dynorphin mediates heterosynaptic depression of hippocampal mossy fibre synapses and modulates long-term potentiation. Nature 362:423–427

Westbrook GL, Mayer ML (1987) Micromolar concentrations of Zn^{2+} antagonize NMDA and GABA responses of hippocampal neurons. Nature 328:640–643

Yuste R, Katz LC (1991) Control of postsynaptic Ca^{2+} influx in developing neocortex by excitatory and inhibitory neurotransmitters. Neuron 6:333–344

Chapter 7
Postsynaptic Determinants of Inhibitory Transmission at Mixed GABAergic/Glycinergic Synapses

Stéphane Dieudonné and Marco Alberto Diana

Abstract In the vertebrate central nervous system, ionotropic inhibition is mediated by two neurotransmitters: GABA and glycine. While inhibitory neurons of the forebrain release mainly GABA, both neurotransmitters coexist in most structures of the hindbrain. More specifically, a majority of hindbrain inhibitory neurons contain both GABA and glycine that accumulate in the same vesicle and are co-released. On the postsynaptic side, GABA and glycine activate separate chloride-permeant ionotropic receptors that display similar biophysical properties. We review here the distribution and organization of inhibitory co-transmission, with an emphasis on the postsynaptic side of the synapse. We show that very different types of functional organization have been adopted by mixed inhibitory circuits. However one rule is always preserved: the GABAergic and glycinergic components of mixed inhibitory synapses display different decay kinetics. $GABA_A$ receptor kinetics are determined by the combination of a rich variety of subunits. In contrast glycinergic receptors are assembled from a small number of subunits and most adult neurons may express the same receptor type. Accumulating evidence suggests that the kinetics of glycine synaptic currents could be determined and modulated in an activity dependant manner, through various mechanisms. In conclusion we propose that tunable glycinergic inhibition timecourse may optimize rate- coding circuits of the hindbrain whereas forebrain coding through oscillations and synchrony may benefit from the rigid yet diverse subunit combination of $GABA_A$ receptors.

7.1 Introduction

Several neuronal types have been known for decades to be responsive to the application of both GABA, and glycine. The hypothesis of the coexistence of distinct inhibitory input types onto individual neurons emerged in the same

S. Dieudonné (✉)
Laboratoire de neurobiologie, CNRS UMR8544, Ecole normale supérieure, 46 rue d'Ulm, 75005 Paris, France
e-mail: dieudon@biologie.ens.fr

R. Gutierrez (ed.), *Co-Existence and Co-Release of Classical Neurotransmitters*,
DOI 10.1007/978-0-387-09622-3_7,

years when GABA and glycine were being identified as neuroactive agents [21]. Early evidence came from studies where goldfish Mauthner cells [44, 55] in-vivo spinal neurons both in in-vivo [182] and in-vitro [11, 142], and cerebellar-brainstem neurons [11] in-vitro were shown to be inhibited by exogenous application of both transmitters. In the following years other reports, mainly based on histological techniques, provided compelling evidence in favor of the coexistence of GABA and glycine in the same cells, and in synapse-like structures in several brain regions [135, 167, 171, 181], also suggesting that the two molecules may be released together. This, amongst other findings, led to the resurgence of the idea that an individual neuron might synthesize and release more than a single transmitter [26].

The first formal demonstration of the coliberation of GABA and glycine from individual neurons was achieved only years later by using paired recordings from synaptically connected neurons in spinal cord slices [80]. A definitive argument in the same direction was finally given by studies concerning the molecular mechanisms responsible for the loading of the two transmitters into synaptic vesicles. After cloning [110, 147], the Vesicular Inhibitory Amino Acids Transporter (VIAAT) was indeed revealed to be able to load GABA and glycine into synaptic vesicles, albeit with different affinity [110], thus confirming previous data suggesting similarities and competition between the 2 vesicle loading systems [25, 33]. At present, the concept of co-release of inhibitory neurotransmitters from single neurons is well established, but its functional consequences are still poorly understood.

In the first section of our review, we will give an overview of the structures in the central nervous system, where co-release of GABA and glycine has either been demonstrated, or suggested. We will then examine some general principles of organization of mixed inhibitory transmission, and some of the possible postsynaptic mechanisms regulating the efficacy of these synapses.

7.2 An Overview of Inhibitory Co-transmission in the Mammalian Brain

7.2.1 Forebrain

The unequivocal identification of glycinergic fibres is a necessary precondition for studying the possible existence of mixed synapses in the rostral part of the brain, given the sparse localization here of glycinergic neurons. In the forebrain, little is indeed known about the distribution of glycinergic synapses [141, 174]. Important progress in their morphological and functional analysis has come from the production (Fig. 7.1) [185] of a bacterial artificial chromosome transgenic mouse, expressing the enhanced green fluorescent protein (EGFP)

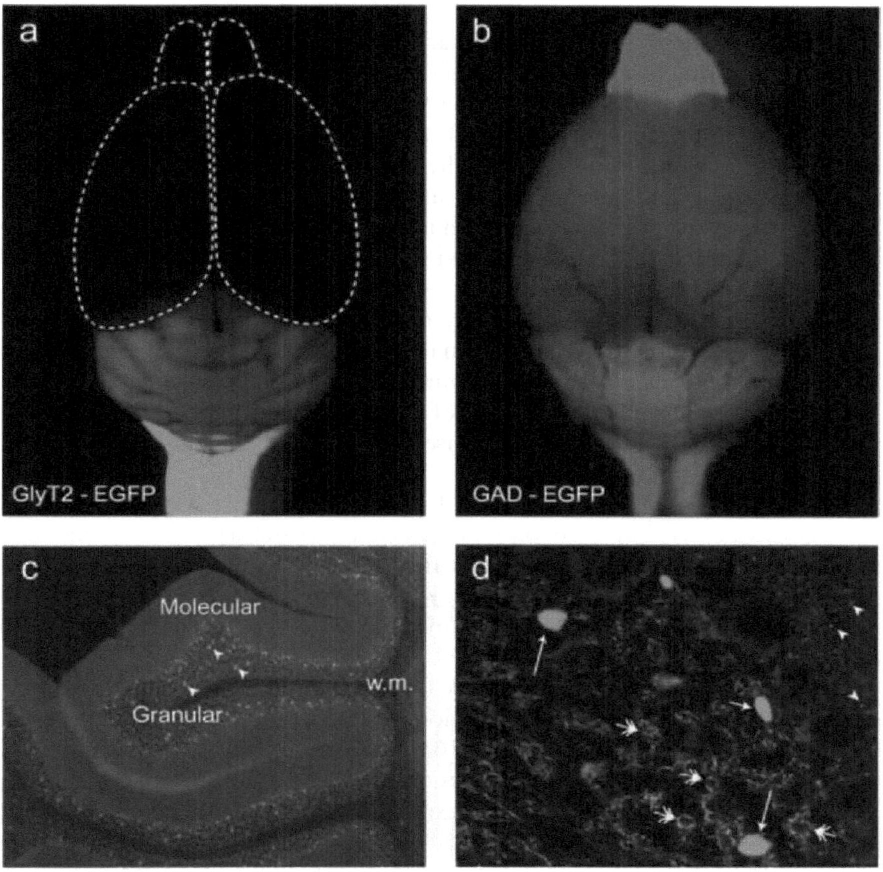

Fig. 7.1 GABAergic and glycinergic transmission in the central nervous system of rats. **(a)** A strong rostro-caudal gradient of glycinergic neuron density is evident in a transgenic mouse expressing GFP under the promoter of the neuronal glycine transporter GlyT2 (kind gift of H.U. Zeilhofer). **(b)** Similar view for a mouse expressing EGFP under the promoter of GAD65 that catalyzes the limiting step for GABA synthesis in GABAergic neurons (kind gift of G. Szabo) (Lopez-Bendito et al., Cereb Cortex, 2004). **(c)** Parasagittal section showing a lobule of the cerebellar cortex of the GlyT2-EGFP mouse. Immunostaining for GABA is shown in red. Note the prevalence of GABA in the molecular layer and the abundance of glycinergic interneurons in the granular layer (arrowheads). Most of the Golgi cell axons in the granular layer appear to co-stain for GABA (yellow background). GFP-positive axons, presumably of granular layer interneurons, are running in the white matter (w.m.). **(d)** Detail of a cerebellar lobule of the GlyT2-EGFP mouse stained for VIAAT, the vesicular cotransporter of GABA and glycine, in red. Golgi cell bodies (long arrows) and Lugaro cell bodies (short arrow) are clearly visible. Golgi cell axonal varicosities delineate the contour of cerebellar glomeruli (double arrows). Note that some of the VIAAT-stained profiles correspond to GFP-negative boutons from GABAergic Golgi cells. Lugaro cell varicosities, that co-release GABA and glycine, are visible in the molecular layer (arrowheads) as a minoritary population. (*See* Color Plate 5)

controlled by the promoter of the membranous glycine transporter GlyT2 gene, which reliably marks glycinergic structures [104, 138, 159]. The analysis of these mice has confirmed that no glycinergic neurons are present in the forebrain, although several anterior areas are innervated by glycinergic fibres, albeit more weakly than in the brainstem and in the spinal cord. Glycinergic innervation is most widespread in the thalamus, the hypothalamus and the preoptic area, the cholinergic basal forebrain nuclei, and the bed nucleus of the stria terminalis. In contrast, the innervation pattern is very weak in cortical and hippocampal areas. Sparse staining appears also in the olfactory bulb, the amygdala and the basal ganglia [141, 185].

Physiological data on glycinergic systems in forebrain regions are dramatically scarce. Functional glycinergic synapses have indeed been revealed only in relay cells of the thalamic ventrobasal nucleus (VBN) of the juvenile rat [68, 69]. In these cells extracellular stimulation of afferent inhibitory fibers does produce mixed GABAergic/glycinergic responses, but GABA and glycine are most likely released by different inputs [69].

In the hypothalamus, no functional glycinergic synapse has been detected and, consequently, no case of GABA/glycine co-release has so far been reported. The only known physiological role of glycine is then represented by the control of oxytocin release by extrasynaptic glycinergic receptors, which can sense the taurine liberated by glial cells in response to hypoosmotic challenges in the supraoptic nucleus [43, 75, 76].

In higher forebrain areas, a veritable discrepancy exists between the extreme sparseness of glycinergic fiber innervation [141, 185], and the moderate levels of detectable glycine receptor subunits [16, 35, 106, 140, 149, 150]. The morphological evidence supporting the expression of GlyRs has been extensively validated by the electrophysiological examination of GlyR agonist-induced currents [28, 29, 36, 40, 48, 52, 77, 78, 94, 107–109, 118, 154, 155, 163, 165, 172, 175, 187]. Nevertheless, it is generally accepted that, if functional, glycinergic synapses do not play a central role in the physiology of higher areas. In contrast, alternative functions have been envisaged for glycinergic receptors, like the control of the release of dopamine and acetylcholine in the striatum and the nucleus accumbens following release of taurine [39, 53]. A development-related role for glycinergic receptors has then been suggested in the cortex [58].

Only a single study reports the presence of morphologically identified mixed GABAergic/glycinergic synapses in the hippocampus [35]. In this report, the authors found that, in the CA3 region, an important percentage of GlyR clusters colocalized with the $GABA_AR$ $\gamma 2$ subunit, around 50% of which were at extrasynaptic sites. These data are evidently at odds with most of the existing literature. Nevertheless, together with the existence of glycinergic innervation in the hippocampus [185], they still leave some space open for questions about the possible role of glycine as neurotransmitter in higher forebrain areas.

7.2.2 Cerebellum and Cerebellar-Like Structures

7.2.2.1 Cerebellum

Many studies have shown the presence of glycinergic structures in the cerebellar cortex, although GABA is the dominant inhibitory transmitter (Fig. 7.1b,c). In early studies glycine immunoreactivity was detected in the granule cell layer [134, 135, 183]. Moreover, the expression of GlyT2 was also revealed in Golgi cell axonal terminals, and in varicosities of the molecular layer [184], whereas glycinergic receptor subunits were localized in cortical areas [106, 140], and functional glycinergic receptors were investigated in Golgi cells [45] and granule cells [84].

Morphological colocalization of GABA and glycine in Golgi cells was also revealed in early reports [134, 135, 183]. At present, the mixed GABAergic/glycinergic synapses of the cerebellum have been precisely identified, both at functional and morphological level [46, 49, 51, 156, 185].

The cerebellum boasts of two GABAergic/glycinergic neuronal types, the Lugaro and the Golgi cells, and two mixed inhibitory synapses, those between Lugaro and Golgi cells and those between Golgi and unipolar brush cells (UBCs). We will examine them separately.

The Lugaro to Golgi Cell Mixed Synapse

Lugaro cells are inhibitory interneurones located in the higher part of the granule cell layer (Fig. 7.1d) [95, 96]. Dumoulin and colleagues [51] studied the synapses between Lugaro cells and Golgi cells by exploiting the selective depolarization of the Lugaro cells by serotonin. Serotonin application triggered a strong inhibitory input in Golgi cells with a clear mixed GABAergic/glycinergic profile. The authors showed morphologically that the vast majority of the axonal varicosities of Lugaro cells were liable to release both GABA and glycine (Fig. 7.1d), and that Golgi cell dendrites showed colocalized clusters of both $GABA_A$Rs and GlyRs, proving that co-transmission occurred at single synaptic sites on Golgi cell dendrites. Interestingly, $GABA_A$R and GlyR clusters remained segregated instead of overlapping at individual postsynaptic sites (Fig. 7.2), hinting at the existence of different mechanisms anchoring the two types of receptors to the synapse.

The analysis of the kinetic properties of spontaneous mixed (from Lugaro cells) and purely GABAergic (from molecular layer basket and stellate cells) IPSCs in Golgi cells offered further details. The authors remarked, in fact, a great variability of the decay of both IPSC populations among postsynaptic cells, with mixed IPSCs typically showing slower kinetics. The slower kinetics could be accounted for entirely by the presence of the glycinergic component. Interestingly, in individual postsynaptic cells, the decays of purely GABAergic and of mixed events were strictly correlated, showing the likely presence of common postsynaptic mechanisms regulating IPSCs duration, independently from their nature and from their presynaptic origin.

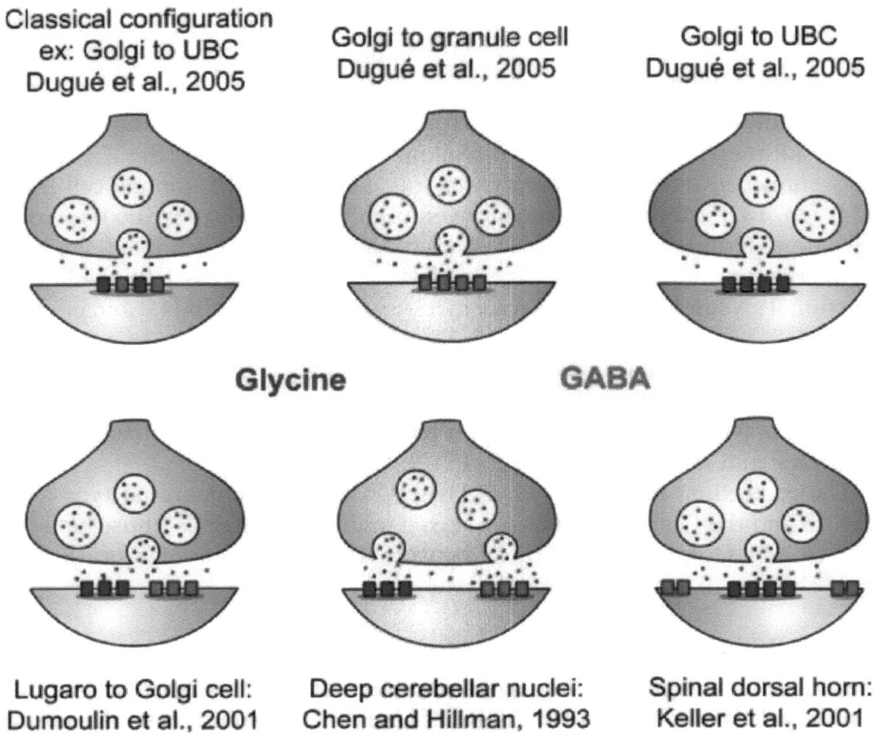

Fig. 7.2 Postsynaptic heterogeneity at synapses co-releasing GABA and glycine. We show here six different cases reported in the literature (see citations in the figure), where distinct postsynaptic configurations are present in front of presynaptic sites co-releasing GABA and glycine. Each instance is described in the text in detail. (*See* Color Plate 6)

The Golgi cell to unipolar brush cell (UBC) mixed synapse

Golgi cells are the most numerous, and important, inhibitory interneurons in the granule cell layer. In all lobules of the cerebellar cortex, Golgi cells form purely GABAergic synapse onto granule cells [23, 71, 144].

Golgi cells are not a homogenous group of cells, although their most widely recognized profile is GABAergic (Fig. 7.1c, d) [67, 156]. In fact, between 70% and 80% of the Golgi cell population is mixed GABAergic/glycinergic, whereas only 15% is purely GABAergic [156].

One postsynaptic target of the glycine released by Golgi cells has been identified only recently in the unipolar brush cells, or UBCs (Fig. 7.2) [49]. UBCs are small glutamatergic interneurons specifically localized in the granule cell layer of the vestibulocerebellar lobules [121]. In paired recordings from synaptically connected Golgi cells and UBCs, Dugué and colleagues [49] found that a presynaptic Golgi cell can co-release GABA and glycine onto an individual UBC. Moreover, the authors found convincing evidence that a Golgi cell can generate mixed IPSCs in

UBCs, and pure GABAergic events in granule cells, thus showing that Golgi cells can perform target-specific GABAergic, glycinergic or mixed transmission (Fig. 7.2). A relevant variability of the nature of inhibition, from purely GABAergic to purely glycinergic, was also observed among UBCs.

The capacity of detecting the release of glycine is dictated by the specific expression of GlyRs in UBCs in contrast to granule cells [49], because morphological analysis has revealed that GABA and glycine co-release is not a specific feature of the vestibular subdivisions, but rather a general feature of Golgi cells in the whole cerebellum [156]. The variability of the inhibitory phenotype of UBCs may thus originate from the postsynaptic side, for example from mechanisms of control of GlyR and $GABA_A$R expression and anchoring to the synapse.

The recently characterized heterogeneity of the Golgi cell population, together with the known heterogeneity of UBCs [127], nonetheless leaves open the possibility that distinct subgroups of UBCs may select their presynaptic partners within specific Golgi cell subgroups. Unpublished data from our laboratory indeed suggest a correlation between afferent inhibitory phenotype and functional UBC subgroups (C. Rousseau, M.A.D. and S.D. unpublished observations).

Finally, further unpublished data (C. Rousseau, M.D. and S.D.) show that, as with the synapses between Lugaro and Golgi cells, spontaneous IPSCs decay kinetics in UBCs display a wide variability, and that, in individual cells, the glycinergic component is always slower than the GABAergic one. The origins of the variability of the nature, and of the kinetics of the inhibition in UBCs are still unknown.

7.2.2.2 Deep Cerebellar Nuclei

The deep cerebellar nuclei (DeCNs), the primary output areas of the cerebellum, are composed of a heterogeneous population of glutamatergic and GABAergic projection neurons, and by smaller local interneurons. Local interneurons have been shown to colocalize GABA and glycine [12, 31]. At ultrastructural level, GABA/glycine colocalization was suggested at single boutons, presynaptically to DeCN neurons and with individual terminals frequently apposed to multiple active zones either showing, or lacking GlyR staining (Fig. 7.2) [31]. Detection of functional glycinergic synapses on DeCN neurons has been arduous. Glycinergic spontaneous IPSCs have been detected only after strong pharmacological stimulation of presynaptic fibers in an age-dependent manner [87, 136]. The presence of mixed GABAergic/glycinergic synapses remains to be demonstrated at the functional level.

7.2.2.3 Ventral and Dorsal Cochlear Nuclei

The ventral (VCN), and the dorsal (DCN) cochlear nuclei are the first central relays of the auditory information from the ears. Anatomically the DCN is layered, and presents a cellular and circuit organization that is similar to that of the cerebellum [131, 132]. These notable similarities notwithstanding, the main inhibitory neurotransmitter in the DCN is glycine, and not GABA.

GABA and glycine are highly colocalized in the DCN [4, 91]. Similar to the cerebellum, the Golgi cells of the DCN likely release GABA and glycine at individual terminals [4, 91], and mixed GABAergic/glycinergic Golgi cells may represent one of the sources of inhibition of granule cells [4, 9]. In striking contrast to the cerebellum, the granule cells of the DCN receive a very strong glycinergic inhibition, although an important contribution from GABAergic synapses is also detected [9]. Nonetheless, as with mixed transmission in the cerebellum, in DCN granule cells glycinergic evoked synaptic currents are slower than GABAergic ones. More data are required to finally establish beyond doubt the mixed nature of the input from Golgi cells.

Finally, GABA is co-released with glycine onto bushy cells in the VCN. In these cells GABA may be relevant for controlling release probability via $GABA_B$ autoreceptors expressed on presynaptic afferences [103].

7.2.3 *Spinal Cord and Brainstem*

The great expression of glycinergic fibres in caudal structures (Fig. 7.1a) exemplifies a dominant role of glycinergic synaptic inhibition in the spinal cord and brainstem [141, 184, 185]. Although less important than in higher regions, GABA nonetheless remains a relevant element of the inhibitory system (Fig.7.1b).

Several early studies showed that spinal and brainstem neurons were responsive to exogenous applications of both GABA, and glycine [11, 142, 182], a fact that evoked the possible existence of mixed synapses.

Over the years, morphological evidence has been accumulating showing colocalization of GABAergic and glycinergic receptors [19, 65, 123, 123], of both GABA and glycine in cells, and at presynaptic boutons [30, 50, 65, 124, 166–168], and pre/postsynaptic juxtaposition of neurotransmitters and receptors [124, 168, 171].

At a functional level, mixed GABAergic/glycinergic transmission was first demonstrated in spinal motoneurones using paired recordings [80], and in hypoglossal neurons [129]. Co-release has been detected and studied in both motoneurons [63] and interneurons [70] of the ventral spinal cord, in the dorsal spinal cord [32, 88, 117], in auditory nuclei of the brainstem, like the lateral superior olive (LSO) [125], and the medial nucleus of the trapezoid body (MNTB) [8], in hypoglossal motoneurons (HMs) [124, 130], and in neurons of the abducens nucleus [146]. All these studies have exploited one invariant feature of mixed inhibitory transmission in lower areas: the notable difference in decay kinetics between the 2 components, with GABA currents being significantly slower than glycinergic currents. By analyzing the decay properties of miniature IPSCs (mIPSCs), this property has allowed authors to associate mIPSCs showing intermediate kinetics with the simultaneous release of GABA and glycine from individual terminals. Analyzing this population of mIPSCs has provided insights on the relative weight of GABA/glycine co-releasing synapses, as well as on its variations during development.

Developmental changes of the phenotype of synaptic inhibition are, indeed, another key feature of lower inhibitory synapses. In general, full morphological and functional maturation of lower structures takes place between two and four weeks after birth, depending on the structures considered. During this variable lapse of time, in the vast majority of the structures where mixed inhibition has been reported, development is associated with a switch in inhibition from either a predominantly GABAergic profile [8, 63, 92, 125], or a profile where both GABA and glycine give notable contributions at early stages [70, 88, 124, 158], to a predominantly glycinergic phenotype at maturation. Evidence in favor of both presynaptic and postsynaptic developmental modifications has been proposed, but a brief overview shows the compelling variability of the strategies deployed to reach the mature state.

In an important example, co-release of GABA and glycine takes place at all stages of development in spinal cord neurons of lamina I and II, but postsynaptic codetection is lost with age, due to either a redistribution of $GABA_A$Rs from intra to extrasynaptic sites, or to a change in affinity of $GABA_A$Rs for the transmitter (Fig. 7.2) [88]. Benzodiazepines, in fact, unmask mixed mIPSCs both in lamina I, where at maturation transmission is otherwise glycinergic, and in lamina II, where mIPSCs are either GABAergic, or glycinergic but never mixed.

Co-release is, instead, a truly transient phenomenon in the developing MNTB [8]. Here, mixed mIPSCs go from 15% to less than 1% of the total, while the overall efficacy of GABAergic transmission, and the sensitivity of neurons to exogenously applied GABA show little maturation. This suggests that, in the MNTB, single presynaptic cells, or individual terminals, might select a single transmitter during development.

In some mature neurons mixed inhibitory synapses do survive development [70, 124, 125]. The percentage of mixed mIPSCs on the total may either remain constant [125], or decrease [70, 124]. This fact notwithstanding, the importance of mixed synapses invariably declines, because of the strong potentiation of the glycinergic part of the mixed IPSCs with time [70, 124, 125].

The fate of co-releasing synapses is inextricably linked to the process of synaptic maturation, which consists of a complex set of structural and functional changes touching both pre and postsynaptic compartments. A full morphological and physiological analysis of the development of inhibition would, thus, probably reveal changes at all levels but, unfortunately, this has rarely been done. An interesting example is represented by hypoglossal neurons [123, 124]. Muller and colleagues reported an age-dependent increase of the number of morphologically identified postsynaptic co-clusters of $GABA_A$ and glycinergic receptors on hypoglossal motoneurons [123]. Expecting an increase in co-release with development, these authors later found that, in contrast, the most relevant changes in transmission were an increase in glycinergic transmission, and an unexpected reduction in mixed synaptic activity [124]. These changes could be explained by the decreases in the presynaptic staining for GAD65 and in the colocalization for GAD65 and GlyT2, and by the simultaneous augmentation of the staining for GlyT2 alone. Therefore, presynaptic neurons lose the capacity of synthesizing and releasing GABA during maturation [124] although, postsynaptically, the level of receptor colocalization increases.

Another example is represented by the LSO. Here, the age-related potentiation of glycinergic transmission has been associated, postsynaptically, with an increase in the staining for the glycinergic receptor-associated protein gephyrin and with a reduction in the staining for the β2-3 $GABA_A$ receptor subunits [92], but also, presynaptically, with an increase in glycine and a decrease in the GABA content of presynaptic terminals, associated with a reduction of GAD65/67 staining [125].

A final point that we should mention concerns the maturation of IPSC time courses. Inhibition in the spinal cord and the brainstem is, indeed, characterized also by an important acceleration of the decay kinetics of both GABAergic and glycinergic currents [8, 63, 70, 88, 124, 157]. The acceleration of glycinergic currents can be explained, at least partially, with the switch from fetal $\alpha 2$ subunit- to adult-like $\alpha 1$ subunit-containing receptors. This switch takes place during the first two–three weeks following birth, and it is a well-known trait of glycinergic receptor maturation [2, 15, 106, 162], consistently linking the structural refinement of glycinergic synapses with the physiological changes observed postnatally.

In a later section we will examine in more detail some of the possible mechanisms setting the kinetic properties of glycinergic receptors at mixed synapses [117].

7.3 Cellular and Molecular Organization of Mixed Inhibitory Circuits

7.3.1 Independence of Presynaptic and Postsynaptic Phenotypes

In the preceding section we have reviewed the organization of inhibitory networks in various regions of the brain. At first glance it is hard to find a general principle of organization because the histochemical evidence in favor of presynaptic GABAergic and glycinergic transmission is not always found to match the postsynaptic expression of GABA and glycinergic receptors [49, 51, 88, 124]. One problem for the interpretation of bulk tissue staining is that each presynaptic neuron, and each postsynaptic target present in the structure may obey specific rules defining the nature of the synaptic contact. Cerebellar Golgi cells and Lugaro cells offer a clear example of this. As described above in more detail, the same presynaptic cell can mediate purely GABAergic inhibition on one of its targets, glycinergic inhibition on a second and mixed inhibition on a third, depending on the postsynaptic receptors present [42, 49, 51]. These examples suggest that inhibitory neurons express the same phenotype at all their varicosities independently of the capacity of their targets to respond to the neurotransmitters released. This type of organization would explain the numerous mismatch between pre and postsynaptic markers observed with immunohistochemical techniques [124]. Future studies should benefit from the recent development of transgenic mice expressing markers in a cell-type specific manner.

7.3.2 Presynaptic Regulations of Receptor Occupancy

During synaptic transmission, postynaptic receptors are activated by extremely brief transients of high agonist concentration which decay in a few hundreds of microseconds [120]. Receptors do not reach equilibrium and their occupancy will define the amplitude of the synaptic currents, that is, the proportion of receptors that will be liganded and will open [120]. At central synapse, the level of occupancy of $GABA_A$ and glycinergic receptors after the release of a single vesicle of neurotransmitter is quite variable [7, 143]. At co-releasing synapses, the number of vesicles released and their content in GABA or glycine will affect the ratio of the GABAergic and glycinergic components. If only one receptor is expressed, the vesicular content of the presynaptic element will still define the occupancy of the postsynaptic receptors and could be subject to presynaptic regulations. It was recently shown that the transporter GlyT2 competes with GABA synthesis and recapture for the loading of vesicles with glycine [6]. Regulatory mechanisms for the transmitter phenotype of inhibitory vesicles have not been identified yet. However biochemical evidence suggest that GlyT2 may interact directly with the synaptic and vesicular machinery [5, 64, 74, 133]. Furthermore GABA loaded in the vesicles may be primarily neosynthetized by GAD from cytoplasmic glutamate [79], implicitly involving an association between GAD and VIAAT. The vesicular content could also vary in an activity-dependent manner since reloading of vesicles after depletion appears to proceed at different rates for GABA and glycine [86].

7.3.3 Molecular Organization of the Postsynaptic Receptors

The mechanisms that govern the clustering of postsynaptic inhibitory receptors have been intensely investigated in the last decade and have been the subject of many excellent reviews [101, 119, 148, 170]. Shortly, both GABA [169] and glycine [17] receptors interact with a major anchoring protein, gephyrin [139]. Gephyrin can form aggregates at the synapse and is thought to stabilize the receptors, which can nevertheless diffuse in and out of the synapse [170]. Many splice variants of gephyrin have been identified [113] and the expression of various cassettes can modify the clustering capacity of gephyrin. Gephyrin variants may play a central role in specifying the nature of the inhibitory synapse but this molecular language is not yet deciphered [111]. A recent report involves certain gephyrin isoforms in the exclusion of glycinergic receptors from GABAergic synapses [112]. At some synapses glycine and GABAergic receptor clusters are apposed, the glycine cluster being the only one associated with gephyrin [51]. In the deep cerebellar nuclei, mixed presynaptic boutons can form two active zones, one of which only expresses glycinergic receptors [31]. Further ultrastructural studies are needed to determine if the presence of two active zones is a hallmark of synaptic articulation harboring separate glycine and GABAergic receptor clusters. At other synapses $GABA_A$ receptors are found perisynaptically [32, 88, 124], with no known mechanism for this localization. GABAC receptors, although generally neglected, may also play a role at mixed inhibitory synapses and explain some of the mismatch seen with classical immunohistochemical markers. Both at the mRNA level [20, 186] and at the immunohistochemical level, rho-containing receptors are abundant in regions where mixed inhibition prevails like the colliculi, brainstem, spinal cord and cerebellum [59, 145]. Moreover a significant proportion of GABAC clusters colocalize with $GABA_A$ gamma2 or glycinergic receptor clusters [59]. Although it is clear that specific GABA and glycinergic receptor isoforms can be accumulated to specific synapses in the same neuron, much remains to be understood in terms of addressing and anchoring mechanisms that could generate the extraordinary functional diversity of mixed inhibitory synapses.

7.4 Functional Correlate of Inhibitory Co-transmission: Tuning the Timecourse of Inhibition at Mixed Synapses

7.4.1 GABA and Glycine Receptors Mediate Different Kinetic Components of the IPSCs

In all known cases of mixed inhibition, the decay kinetics of the GABA and glycine components of IPSCs were found to be significantly different [8, 9, 32, 51, 70, 80, 88, 124, 125, 129, 130, 146]. Glycinergic currents decay faster than GABAergic currents in most structures, except in the cerebellum and in the dorsal cochlear nucleus, where GABAergic inhibition is faster [9, 51]. In cerebellar Unipolar Brush Cells, this is mostly due to the fast kinetics of the GABAergic IPSCs (1–2 ms) [49]. In Golgi cells, in contrast, glycinergic IPSCs display a particularly slow biexponential decay [51]. In most systems the kinetics of both glycine and GABA components accelerate in parallel during development. A change in subunit composition appears to underlie the speeding of IPSCs decay but many different mechanisms, like protein phosphorylation or interaction with synaptic proteins, could participate [1, 161].

Complementary functions have been proposed for the fast and slow decay components of synaptic inhibition. Large and fast synaptic conductances are needed to hyperpolarize neurons toward the reversal potential of chloride ions, because they provide the charges necessary to load the membrane capacitance. The slow component in turn maintains the membrane potential at this hyperpolarized level and determines the overall duration of the IPSP and the shunt of subsequent excitatory inputs [146]. This functional dichotomy may be conceived in a different manner. Whereas the large and fast component of the IPSC will efficiently constrain the timing of spikes, summation of the slow components over time may preferentially tune the gain of the neurons by counteracting the temporal summation of excitatory inputs [116]. A distinctive property of mixed inhibitory synapses may reside in the capacity to tune the kinetics of postsynaptic IPSCs rapidly.

7.4.2 The Decay Kinetics of GABA and Glycine IPSCs Depend on the Temporal Transient of Agonist Concentration in a Different Manner

$GABA_A$ receptors can be assembled from many combinations of subunits that display specific affinities, ranging from hundreds of nanomolar to tens of micromolar. $GABA_A$ receptor subtypes with the highest affinity can be activated by low concentrations of GABA occurring around the release site or by spillover from nearby synapses [10]. The amplitude of these slow IPSCs will depend crucially on the amplitude and decay kinetics of the agonist transient around the synapse. Perisynaptic $GABA_A$ receptors seem to constitute a common feature of mixed inhibitory synapses of the spinal cord and brainstem both from the immunohistochemical and functional points of view [32, 88, 124, 158]. These receptors can be recruited by spillover of GABA during massive stimulation of the inhibitory afferents [32]. Modulation of these receptors by neurosteroids can also unmask a slow GABA component of the IPSCs [89, 137].

In a simple view of postsynaptic receptors activation, the decay of the IPSCs is shaped by deactivation kinetics resulting from the unbinding of the agonist molecules. Agonist timecourse in the synaptic cleft does not affect the decay time course of IPSCs, which proceeds on longer timescales. As expected, deactivation kinetics of $GABA_A$ receptor subtypes are related to the apparent affinity for their agonist [82, 83] (Fig. 7.3). When short (1 ms) applications of GABA were used to mimic synaptic activation, the decay of the evoked GABA currents was not affected by the concentration of the agonist [62]. However $GABA_A$ receptors can also undergo fast desensitization that leads to the appearance of a slow component of decay of the IPSCs [81]. Finally in a modeling study of the variability of mIPSCs at various synapses, Nusser et al. (2001) found that the size of the synaptic transient of agonist concentration did not contribute significantly to the variability of the IPSCs decay except at the smallest connections [128]. All these observations are in striking contrast to the activation behavior of glycinergic receptors, as described below.

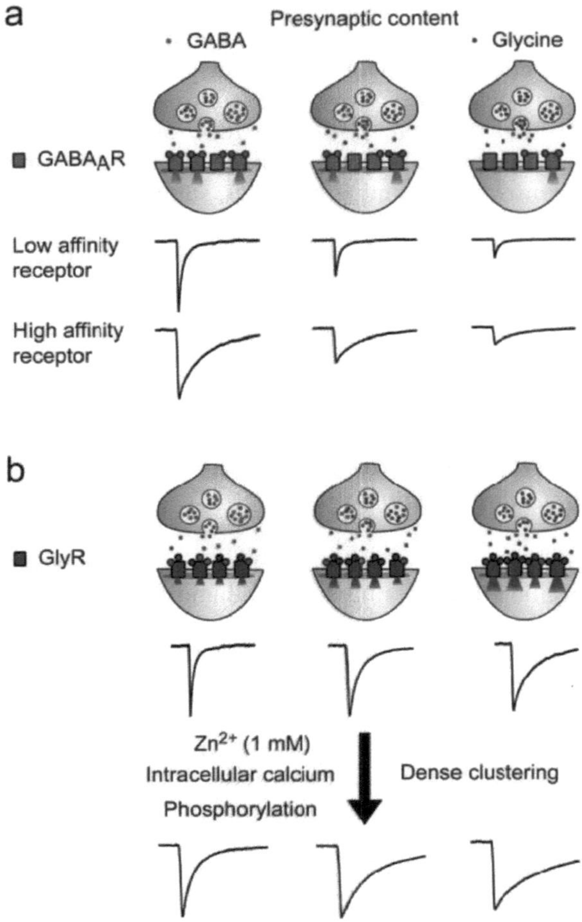

Fig. 7.3 Different mechanisms govern the amplitude and duration of GABAergic and glycinergic IPSCs at mixed synapses. **(a)** $GABA_A$ receptors activation at mixed inhibitory synapses. The amount of GABA released by mixed vesicles determines the amplitude of the GABAergic component of the IPSCs (left to right). The degree of occupancy of agonist binding sites defines the number of fully liganded receptors, which will open. The molecular composition of the $GABA_A$ receptors present at the synapse is the main determinant of the receptors affinity. It fixes the decay kinetics of the IPSCs. **(b)** Glycinergic receptors are allosteric devices with tunable decay kinetics. Partially liganded receptors display a low apparent affinity for their agonists and mediate fast IPSCs. Synaptic current should decay more slowly at glycine-rich synapses (right), where fully liganded receptors show a higher apparent affinity. In GABA-rich synapses, GABA can compete for glycine at the receptor binding sites. The low affinity and fast unbinding rate of GABA result in very brief glycinergic IPSCs. Allosteric transitions governing the gating kinetics of glycinergic receptors are modulated by a variety of environmental effectors that promote high apparent affinity states. (*See* Color Plate 7)

Glycinergic receptors are assembled from only five subunits in rodents and in most regions receptors are composed of alpha2/beta in the young and of alpha1/beta in the adult. Yet the diversity of the decay kinetics of glycinergic IPSCs is astounding, ranging from millisecond at auditory synapses [22] to tens of milliseconds in cerebellar Golgi cells [51]. This heterogeneity cannot be explained easily by differences in the properties of the various subunits but may result from peculiar gating properties. The response of homomeric glycinergic receptors to long agonist applications was tested in recombinant expression systems. Surprisingly the deactivation time was found to vary as the logarithm of the agonist concentration [60]. A similar behavior was found on native channels when decreasing the duration of the agonist pulse from 10 ms to 200 μs [105], suggesting that the level of occupancy of the binding sites on the receptor is the main parameter controlling deactivation. A mechanistic correlate for this mode of gating was sought at the single-channel level. With increasing steady-state concentrations of agonist, channels opened in bursts of increasing duration [13, 14, 60, 173] with characteristic time constants spanning the range 20 μs to 70 ms. This behavior was correctly described by three binding steps of increasing apparent affinity [13, 14, 27, 66], suggesting the occurrence of an allosteric transition in the receptor conformation after the first binding step and before the channel opens [13, 14, 27, 100]. The main consequence of this peculiar gating behavior appears to be the exquisite sensitivity of the IPSC decay on the liganded state of the receptor and therefore on the amplitude and duration of the glycine concentration transient in the cleft. Mixed presynaptic terminals will release less glycine and more GABA [6] than pure glycinergic ones and evoke faster-decaying IPSCs. Co-released GABA will speed decay kinetics in yet another way. GABA is a low affinity agonist of the glycinergic receptors [41, 60, 153] and will compete with glycine for the binding sites on the receptor. Unbinding of GABA will proceed within milliseconds and leave the receptor in a partly liganded state that will deactivate faster [105]. In this elegant study Lu et al. (2008) showed that a vesicular ratio of GABA over glycine of 3 can explain the fast decay kinetics of IPSCs at auditory synapses.

The decay of glycinergic IPSCs may be regulated by yet another allosteric peculiarity of the receptors. When cloned human GlyR alpha1 subunits were first expressed in Xenopus oocyte, it appeared that the apparent affinity increased with the density of receptor expression on the membrane [164]. These results were reproduced for zebra fish receptors and human receptors in HEK cells [41, 60], ruling out perfusion artefacts like rebinding of the agonist or unresolved desensitization. This increase in apparent affinity occurs in parallel for both glycine, taurine and GABA, three agonists with very different affinities [41, 60]. Furthermore it transforms GABA and taurine from partial agonist into full agonists, suggesting that the increase in apparent affinity is linked to an increase in gating efficacy [41]. Interactions between receptors or titration of an interacting protein may explain this behavior. It is tempting to propose that these interactions favor the same allosteric transition leading to higher apparent agonist affinity that was induced by binding of the first agonist molecule in single-channel studies [13, 14, 27]. The occurrence of such modulation during clustering was tested by coexpression of gephyrin and alpha subunits harboring the gephyrin interacting domain of beta subunits [102]. High density expression of the alpha subunit, or gephyrin coexpression, both favored a fast component of desensitization (5 ms) that led to a slow component of deactivation [102]. Interestingly, phosphorylation of alpha1 homomeric receptors was found to produce identical effect on both desensitization and deactivation [66].

Finally a fast and reversible potentiation of glycinergic receptors by cytoplasmic calcium elevations, independent of G protein signaling but sensitive to intracellular dialysis, has been described in recombinant expression systems, spinal cord neurons and brain-stem motoneurons [61, 122]. Not surprisingly, elevated calcium concentrations increased the apparent affinity of the receptor and the deactivation time constant, thus favoring long bursts of single-channel openings and prolonged synaptic IPSCS [61, 122].

In conclusion glycinergic receptors can be considered as delicate molecular machines with a tunable apparent affinity for their agonist. Tunability results from the existence of allosteric transitions that dramatically increase agonist affinity, producing long opening bursts, which shape IPSC decay. All modulatory pathways converge on these transitions, be it intermolecular interactions during clustering, phosphorylation or calcium-dependent modulation. As glycine receptors also display low-affinity for GABA, glycinergic IPSC decay kinetics at co-releasing synapses will be exquisitely sensitive to both glycine and GABA concentration transients in the cleft [105].

7.4.3 *A Role for Zinc at Mixed Synapses*

The divalent cation Zn is thought to play an important role in the brain as a pleiotropic modulator of neurotransmission and is an allosteric modulator of both $GABA_A$ and glycinergic receptors. A releasable pool of Zn is accumulated in vesicles by the transporter ZnT3, and can be revealed in the brain by histochemical procedures like Timm's staining [38]. The highest density of Zn is found at populations of excitatory synapses in the forebrain. However, significant accumulation of vesicular zinc has also been found at inhibitory terminals in regions that display low overall levels of Zn staining, like the retina [3, 85], spinal cord dorsal and ventral horns [18, 37, 176, 177] and cerebellum [38]. Golgi cells of the cerebellum, 70% of which co-release GABA and glycine, express ZnT3 [178, 179] and accumulate Zn in their axonal varicosities [38]. ZnT3 staining of a sparse network of large varicosities is also found in the molecular layer of the cerebellum [179]. These zinc-enriched terminals have been tentatively ascribed to molecular layer interneuron axons [179] but are more closely reminiscent of the axonal plexus of GABA/glycine co-releasing Lugaro cells [51]. Zn was also found to be selectively accumulated in mixed GABA and glycine-containing varicosities of the Lamprey spinal cord [18]. In the mammalian spinal cord Zn stains a subpopulation of GABA-containing varicosities [37, 176, 177] that most likely include mixed synapses. These data suggest that Zn may act as modulator at mixed inhibitory synapses.

$GABA_A$ receptors are potently inhibited by Zn in a subunit-specific manner. Alpha-beta heteromers display an EC50 in the micromolar range, whereas incorporation of a gamma subunit increases the EC50 by about a hundred fold [47]. The nature of the alpha subunits can affect Zn affinity to a lesser extent [24, 47, 56, 57, 90]. Thus the predominant population of $GABA_A$ receptors in the brain, which are clustered at post-synaptic sites through their gamma subunit, is an unlikely target for Zn modulation. Indeed most studies on the inhibition of GABAergic IPSCs by exogenous zinc report EC50 values higher than 10 μM. Extrasynaptic receptors can assemble from alpha beta and delta subunits and these receptors have intermediate EC50 for Zn, depending on the alpha subunit composition [151, 152]. Delta subunits may participate to some mixed synapses, although their level of expression in the brainstem and spinal cord appears very low. In contrast with gamma-containing receptors, Zn inhibition is fully competitive with GABA binding at delta-containing receptors [93]. Thus, at low GABA concentrations, extremely potent inhibition can be achieved with a few micromoles of Zn. Interestingly delta-containing receptors of the dentate gyrus are specifically located at perisynaptic locations, where they are activated by spillover of GABA from the synapse [180], and the resulting slow component of the IPSCs is completely blocked by 10 μM Zn. In summary canonical $GABA_A$ receptors are unlikely to be affected significantly by Zn released at mixed inhibitory synapses, but perisynaptic receptors may be inhibited.

Low micromolar concentrations of Zn were found to potentiate the glycinergic receptors whereas concentrations in excess of 100 μM inhibited glycine-activated currents [99]. The corresponding binding sites for Zn have been identified on the α subunits N-terminal domain of recombinant glycinergic receptors [72, 97, 99, 114, 115, 126]. Because the inhibitory effect has a slow onset [114] and a low EC50 it is not likely to play a physiological role and will not be further discussed. The EC50 for the potentiating effect has been determined in recombinant receptors, using careful chelation of contaminant Zn. Homomeric $\alpha 1$ and $\alpha 2$ receptors display EC50s of 37 nM and 540 nM, respectively [115]. The affinity is decreased about 20 fold in heteromeric $\alpha\beta$ receptors yielding a submicromolar EC50 for $\alpha 1\beta$, the dominant receptor in the adult brain [115]. Zn potentiation of glycine-activated currents results from a shift in the affinity of the receptor for glycine [99] and the potentiating effect varies with the concentration of glycine with a maximum of up to 300%. Occupancy of the high-affinity site by Zn also transforms taurine from a partial agonist of homomeric $\alpha 1$ receptors into a full agonist [97]. Upon binding at the potentiating site, Zn dramatically increases the duration of channel opening bursts [97]. This potentiating effect of Zn parallels the allosteric crosstalk between glycine binding sites (see above) that leads to an increased affinity and to longer channel openings [14, 27]. Because of this mechanism of action, the effect of zinc on synaptic currents cannot easily be deduced from the effects on currents evoked by slow agonist applications. Using fast glycine applications to mimic synaptic activation on zebrafish native receptors, concentrations of zinc as low as 100 nM were found to increase the decay time course of glycine currents several folds [160] while marginally affecting their amplitude. When applied on native and cultured spinal neurons, submicromolar concentrations of zinc were found to potentiate mildly the amplitude of the synaptic currents but greatly slowed their decay [54, 98, 160], increasing the charge transfer several fold [160]. Finally a KI mouse, in which a mutation of glycinergic receptors that suppresses the potentiating Zn binding site was introduced, displays a hyperekplexia phenotype [73]. It is important to note that the motor phenotype of this glycinergic receptor KI animal is the only case in which Zn was demonstrated to subserve a physiological function. KO mice for the ZnT3 transporter display only mild phenotypic changes, possibly due to compensatory mechanisms [34]. In conclusion, buildup of Zn^{2+} concentrations during periods of increased firing of mixed interneurons may adapt the timecourse of IPSPs to ongoing presynaptic activity. By increasing the charge carried by glycinergic IPSCs, zinc modulation would act as a powerful feedback loop [73].

7.5 Conclusion

Mixed inhibition, resulting from the co-release of GABA and glycine, is a prevalent mode of synaptic transmission in the caudal central nervous system. Mixed synapses display an overwhelming variety of structural organization and functional properties. In the few cases where identified synaptic articulations have been studied, their properties seem to be determined by the independent specialization of presynaptic and postsynaptic neurons. The diversity of mixed inhibition is established through developmental processes with a marked tendency to replace GABAergic inhibition, occurring at early stages, with mixed or glycinergic synapses. The glycinergic and GABAergic components of mixed IPSCs systematically display different decay kinetics. In particular, glycinergic receptors behave as allosteric devices with tunable gating kinetics that are controlled by agonist occupancy and a number of activity-dependent parameters. We propose that the adaptability of IPSC kinetics at mixed synapses constitutes the main reason for the co-existence of two inhibitory transmitters acting at similar postsynaptic receptors.

Acknowledgments We thank Dr. Eric Schwartz for critically reading the manuscript. This work was supported by grant ANR-05-neur-030-03, by the CNRS and the INSERM.

References

1. Aguayo LG, van Zundert B, Tapia JC, Carrasco MA, Alvarez FJ (2004) Changes on the properties of glycinergic receptors during neuronal development. Brain Res Rev 47:33–45

2. Akagi H, Miledi R (1988) Heterogeneity of glycinergic receptors and their messenger RNAs in rat brain and spinal cord. Science 242:270–273
3. Akagi T, Kaneda M, Ishii K, Hashikawa T (2001) Differential subcellular localization of zinc in the rat retina. J Histochem Cytochem 49:87–96
4. Alibardi L (2003) Ultrastructural distribution of glycinergic and GABAergic neurons and axon terminals in the rat dorsal cochlear nucleus, with emphasis on granule cell areas. J Anat 203:31–56
5. Armsen W, Himmel B, Betz H, Eulenburg V (2007) The C-terminal PDZ-ligand motif of the neuronal glycine transporter GlyT2 is required for efficient synaptic localization. Mol Cell Neurosci 36:369–380
6. Aubrey KR, Rossi FM, Ruivo R, Alboni S, Bellenchi GC, Le Goff A, Gasnier B, Supplisson S (2007) The transporters GlyT2 and VIAAT cooperate to determine the vesicular glycinergic phenotype. J Neurosci 27:6273–6281
7. Auger C, Marty A (1997) Heterogeneity of functional synaptic parameters among single release sites. Neuron 19:139–150
8. Awatramani GB, Turecek R, Trussell LO (2005) Staggered development of GABAergic and glycinergic transmission in the MNTB. J Neurophysiol 93:819–828
9. Balakrishnan V, Trussell LO (2008) Synaptic inputs to granule cells of the dorsal cochlear nucleus. J Neurophysiol 99:208–219
10. Barbour B, Hausser M (1997) Intersynaptic diffusion of neurotransmitter. TINS 20:377–384
11. Barker JL, Ransom BR (1978) Amino acid pharmacology of mammalian central neurones grown in tissue culture. J Physiol 280:331–354
12. Baurle J, Grusser-Cornehls U (1997) Differential number of glycine- and GABA-immunopositive neurons and terminals in the deep cerebellar nuclei of normal and Purkinje cell degeneration mutant mice. J Comp Neur 382:443–458
13. Beato M, Groot-Kormelink PJ, Colquhoun D, Sivilotti LG (2002) Openings of the rat recombinant alpha 1 homomeric glycinergic receptor as a function of the number of agonist molecules bound. J Gen Physiol 119:443–466
14. Beato M, Groot-Kormelink PJ, Colquhoun D, Sivilotti LG (2004) The activation mechanism of alpha1 homomeric glycinergic receptors. J Neurosci 24:895–906
15. Becker CM, Hoch W, Betz H (1988) Glycinergic receptor heterogeneity in rat spinal cord during postnatal development. EMBO 7:3717–3726
16. Becker CM, Betz H, Schroder H (1993) Expression of inhibitory glycinergic receptors in postnatal rat cerebral cortex. Brain Res 606:220–226
17. Betz H, Kuhse J, Schmieden V, Malosio ML, Langosch D, Prior P, Schmitt B, Kirsch J (1991) How to build a glycinergic postsynaptic membrane. J Cell Sci 15:23–25
18. Birinyi A, Parker D, Antal M, Shupliakov O (2001) Zinc co-localizes with GABA and glycine in synapses in the lamprey spinal cord. J Comp Neurol 433:208–221
19. Bohlhalter S, Mohler H, Fritschy JM (1994) Inhibitory neurotransmission in rat spinal cord: colocalization of glycine- and GABAA-receptors at GABAergic synaptic contacts demonstrated by triple immunofluorescence staining. Brain Res 642:59–69
20. Bormann J (2000) The 'ABC' of GABAergic receptors. TIPS 21:16–19
21. Bowery NG, Smart TG (2006) GABA and glycine as neurotransmitters: a brief history. B J Pharmacol 147 Suppl 1:S109–119
22. Brand A, Behrend O, Marquardt T, McAlpine D, Grothe B (2002) Precise inhibition is essential for microsecond interaural time difference coding. Nature 417:543–547
23. Brickley SG, Cull-Candy SG, Farrant M (1996) Development of a tonic form of synaptic inhibition in rat cerebellar granule cells resulting from persistent activation of GABAA receptors. J Physiol 497:753–759
24. Burgard EC, Tietz EI, Neelands TR, Macdonald RL (1996) Properties of recombinant gamma-aminobutyric acid A receptor isoforms containing the alpha 5 subunit subtype. Mol pharmacol 50:119–127

25. Burger PM, Hell J, Mehl E, Krasel C, Lottspeich F, Jahn R (1991) GABA and glycine in synaptic vesicles: storage and transport characteristics. Neuron 7:287–293
26. Burnstock G (1976) Do some nerve cells release more than one transmitter? Neurosci 1:239–248
27. Burzomato V, Beato M, Groot-Kormelink PJ, Colquhoun D, Sivilotti LG (2004) Single-channel behavior of heteromeric alpha1beta glycinergic receptors: an attempt to detect a conformational change before the channel opens. J Neurosci 24:10924–10940
28. Chattipakorn SC, McMahon LL (2002) Pharmacological characterization of glycine-gated chloride currents recorded in rat hippocampal slices. J Neurophysiol 87:1515–1525
29. Chattipakorn SC, McMahon LL (2003) Strychnine-sensitive glycinergic receptors depress hyperexcitability in rat dentate gyrus. J Neurophysiol 89:1339–1342
30. Chaudhry FA, Reimer RJ, Bellocchio EE, Danbolt NC, Osen KK, Edwards RH, Storm-Mathisen J (1998) The vesicular GABA transporter, VGAT, localizes to synaptic vesicles in sets of glycinergic as well as GABAergic neurons. J Neurosci 18:9733–9750
31. Chen S, Hillman DE (1993) Colocalization of neurotransmitters in the deep cerebellar nuclei. J Neurocytol 22:81–91
32. Chery N, de Koninck Y (1999) Junctional versus extrajunctional glycine and GABA(A) receptor-mediated IPSCs in identified lamina I neurons of the adult rat spinal cord. J Neurosci 19:7342–7355
33. Christensen H, Fykse EM, Fonnum F (1991) Inhibition of gamma-aminobutyrate and glycine uptake into synaptic vesicles. Eur J Pharmacol 207:73–79
34. Cole TB, Robbins CA, Wenzel HJ, Schwartzkroin PA, Palmiter RD (2000) Seizures and neuronal damage in mice lacking vesicular zinc. Epilepsy Res 39:153–169
35. Danglot L, Rostaing P, Triller A, Bessis A (2004) Morphologically identified glycinergic synapses in the hippocampus. Mol Cell Neurosci 27:394–403
36. Danober L, Pape HC (1998) Strychnine-sensitive glycine responses in neurons of the lateral amygdala: an electrophysiological and immunocytochemical characterization. Neurosci 85:427–441
37. Danscher G, Jo SM, Varea E, Wang Z, Cole TB, Schroder HD (2001) Inhibitory zinc-enriched terminals in mouse spinal cord. Neurosci 105:941–947
38. Danscher G, Stoltenberg M (2005) Zinc-specific autometallographic in vivo selenium methods: tracing of zinc-enriched (ZEN) terminals, ZEN pathways, and pools of zinc ions in a multitude of other ZEN cells. J Histochem Cytochem 53:141–153
39. Darstein M, Loschmann PA, Knorle R, Feuerstein TJ (1997) Strychnine-sensitive glycinergic receptors inducing [3H]-acetylcholine release in rat caudatoputamen: a new site of action of ethanol? Naunyn-Schmiedeberg's Arch Pharmacol 356:738–745
40. Darstein M, Landwehrmeyer GB, Kling C, Becker CM, Feuerstein TJ (2000) Strychnine-sensitive glycinergic receptors in rat caudatoputamen are expressed by cholinergic interneurons. Neurosci 96:33–39
41. De Saint Jan D, David-Watine B, Korn H, Bregestovski P (2001) Activation of human alpha1 and alpha2 homomeric glycinergic receptors by taurine and GABA. J Physiol 535:741–755
42. Dean I, Robertson SJ, Edwards FA (2003) Serotonin drives a novel GABAergic synaptic current recorded in rat cerebellar purkinje cells: a Lugaro cell to Purkinje cell synapse. J Neurosci 23:4457–4469
43. Deleuze C, Alonso G, Lefevre IA, Duvoid-Guillou A, Hussy N (2005) Extrasynaptic localization of glycinergic receptors in the rat supraoptic nucleus: further evidence for their involvement in glia-to-neuron communication. Neurosci 133:175–183
44. Diamond J, Roper S (1973) Analysis of Mauthner cell responses to iontophoretically delivered pulses of GABA, glycine and L-glutamate. J Physiol 232:113–128
45. Dieudonné S (1995) Glycinergic synaptic currents in Golgi cells of the rat cerebellum. PNAS 92:1441–1445
46. Dieudonné S, Dumoulin A (2000) Serotonin-driven long-range inhibitory connections in the cerebellar cortex. J Neurosci 20:1837–1848

47. Draguhn A, Verdorn TA, Ewert M, Seeburg PH, Sakmann B (1990) Functional and molecular distinction between recombinant rat GABAA receptor subtypes by Zn2+. Neuron 5:781–788
48. Dudeck O, Lubben S, Eipper S, Knorle R, Kirsch M, Honegger J, Zentner J, Feuerstein TJ (2003) Evidence for strychnine-sensitive glycinergic receptors in human amygdala. Naunyn-Schmiedeberg's Arch Pharmacol 368:181–187
49. Dugue GP, Dumoulin A, Triller A, Dieudonné S (2005) Target-dependent use of co-released inhibitory transmitters at central synapses. J Neurosci 25:6490–6498
50. Dumoulin A, Rostaing P, Bedet C, Levi S, Isambert MF, Henry JP, Triller A, Gasnier B (1999) Presence of the vesicular inhibitory amino acid transporter in GABAergic and glycinergic synaptic terminal boutons. J Cell Sci 112:811–823
51. Dumoulin A, Triller A, Dieudonné S (2001) IPSC kinetics at identified GABAergic and mixed GABAergic and glycinergic synapses onto cerebellar Golgi cells. J Neurosci 21:6045–6057
52. Ebihara S, Takishima T, Shirasaki T, Akaike N (1992) Regional variation of excitatory and inhibitory amino acid-induced responses in rat dissociated CNS neurons. Neurosci Res 14:61–71
53. Ericson M, Molander A, Stomberg R, Soderpalm B (2006) Taurine elevates dopamine levels in the rat nucleus accumbens; antagonism by strychnine. Eur J Neurosci 23:3225–3229
54. Eto K, Arimura Y, Nabekura J, Noda M, Ishibashi H (2007) The effect of zinc on glycinergic inhibitory postsynaptic currents in rat spinal dorsal horn neurons. Brain Res 1161:11–20
55. Faber DS, Korn H (1980) Single-shot channel activation accounts for duration of inhibitory postsynaptic potentials in a central neuron. Science 208:612–615
56. Fisher JL (2002) A histidine residue in the extracellular N-terminal domain of the GABA(A) receptor alpha5 subunit regulates sensitivity to inhibition by zinc. Neuropharmacol 42:922–928
57. Fisher JL, Macdonald RL (1998) The role of an alpha subtype M2-M3 His in regulating inhibition of GABAA receptor current by zinc and other divalent cations. J Neurosci 18:2944–2953
58. Flint AC, Liu X, Kriegstein AR (1998) Nonsynaptic glycinergic receptor activation during early neocortical development. Neuron 20:43–53
59. Frazao R, Nogueira MI, Wassle H (2007) Colocalization of synaptic GABA(C)-receptors with GABA (A)-receptors and glycine-receptors in the rodent central nervous system. Cell Tiss Res 330:1–15
60. Fucile S, de Saint Jan D, David-Watine B, Korn H, Bregestovski P (1999) Comparison of glycine and GABA actions on the zebrafish homomeric glycinergic receptor. J Physiol 517:369–383
61. Fucile S, De Saint Jan D, de Carvalho LP, Bregestovski P (2000) Fast potentiation of glycinergic receptor channels of intracellular calcium in neurons and transfected cells. Neuron 28:571–583
62. Galarreta M, Hestrin S (1997) Properties of GABAA receptors underlying inhibitory synaptic currents in neocortical pyramidal neurons. J Neurosci 17:7220–7227
63. Gao BX, Stricker C, Ziskind-Conhaim L (2001) Transition from GABAergic to glycinergic synaptic transmission in newly formed spinal networks. J Neurophysiol 86:492–502
64. Geerlings A, Nunez E, Lopez-Corcuera B, Aragon C (2001) Calcium- and syntaxin 1-mediated trafficking of the neuronal glycine transporter GLYT2. J Biol Chem 276:17584–17590.
65. Geiman EJ, Zheng W, Fritschy JM, Alvarez FJ (2002) Glycine and GABA(A) receptor subunits on Renshaw cells: relationship with presynaptic neurotransmitters and postsynaptic gephyrin clusters. J Comp Neurol 444:275–289
66. Gentet LJ, Clements JD (2002) Binding site stoichiometry and the effects of phosphorylation on human alpha1 homomeric glycinergic receptors. J Physiol 544:97–106

67. Geurts FJ, De Schutter E, Dieudonné S (2003) Unraveling the cerebellar cortex: cytology and cellular physiology of large-sized interneurons in the granular layer. Cerebellum 2:290–299
68. Ghavanini AA, Mathers DA, Puil E (2005) Glycinergic inhibition in thalamus revealed by synaptic receptor blockade. Neuropharmacol 49:338–349
69. Ghavanini AA, Mathers DA, Kim HS, Puil E (2006) Distinctive glycinergic currents with fast and slow kinetics in thalamus. J Neurophysiol 95:3438–3448
70. Gonzalez-Forero D, Alvarez FJ (2005) Differential postnatal maturation of GABAA, glycinergic receptor, and mixed synaptic currents in Renshaw cells and ventral spinal interneurons. J Neurosci 25:2010–2023
71. Hamann M, Rossi DJ, Attwell D (2002) Tonic and spillover inhibition of granule cells control information flow through cerebellar cortex. Neuron 33:625–633
72. Harvey RJ, Thomas P, James CH, Wilderspin A, Smart TG (1999) Identification of an inhibitory Zn2+ binding site on the human glycinergic receptor alpha1 subunit. J Physiol 520 Pt 1:53–64
73. Hirzel K, Muller U, Latal AT, Hulsmann S, Grudzinska J, Seeliger MW, Betz H, Laube B (2006) Hyperekplexia phenotype of glycinergic receptor alpha1 subunit mutant mice identifies Zn(2+) as an essential endogenous modulator of glycinergic neurotransmission. Neuron 52:679–690
74. Horiuchi M, Loebrich S, Brandstaetter JH, Kneussel M, Betz H (2005) Cellular localization and subcellular distribution of Unc-33-like protein 6, a brain-specific protein of the collapsin response mediator protein family that interacts with the neuronal glycine transporter 2. J Neurochem 94:307–315
75. Hussy N, Deleuze C, Pantaloni A, Desarmenien MG, Moos F (1997) Agonist action of taurine on glycinergic receptors in rat supraoptic magnocellular neurones: possible role in osmoregulation. J Physiol 502:609–621
76. Hussy N, Bres V, Rochette M, Duvoid A, Alonso G, Dayanithi G, Moos FC (2001) Osmoregulation of vasopressin secretion via activation of neurohypophysial nerve terminals glycinergic receptors by glial taurine. J Neurosci 21:7110–7116
77. Ito S, Cherubini E (1991) Strychnine-sensitive glycine responses of neonatal rat hippocampal neurones. J Physiol 440:67–83
78. Jiang Z, Krnjevic K, Wang F, Ye JH (2004) Taurine activates strychnine-sensitive glycinergic receptors in neurons freshly isolated from nucleus accumbens of young rats. J Neurophysiol 91:248–257
79. Jin H, Wu H, Osterhaus G, Wei J, Davis K, Sha D, Floor E, Hsu CC, Kopke RD, Wu JY (2003) Demonstration of functional coupling between gamma -aminobutyric acid (GABA) synthesis and vesicular GABA transport into synaptic vesicles. PNAS 100:4293–4298
80. Jonas P, Bischofberger J, Sandkuhler J (1998) Corelease of two fast neurotransmitters at a central synapse. Science 281:419–424
81. Jones MV, Westbrook GL (1995) Desensitized states prolong GABAA channel responses to brief agonist pulses. Neuron 15:181–191
82. Jones MV, Sahara Y, Dzubay JA, Westbrook GL (1998) Defining affinity with the GABAA receptor. J Neurosci 18:8590–8604
83. Jones MV, Jonas P, Sahara Y, Westbrook GL (2001) Microscopic kinetics and energetics distinguish GABA(A) receptor agonists from antagonists. Biophys J 81:2660–2670
84. Kaneda M, Farrant M, Cull-Candy SG (1995) Whole-cell and single-channel currents activated by GABA and glycine in granule cells of the rat cerebellum. J Physiol 485:419–435
85. Kaneda M, Ishii K, Akagi T, Tatsukawa T, Hashikawa T (2005) Endogenous zinc can be a modulator of glycinergic signaling pathway in the rat retina. J Molec Histol 36:179–185
86. Katsurabayashi S, Kubota H, Higashi H, Akaike N, Ito Y (2004) Distinct profiles of refilling of inhibitory neurotransmitters into presynaptic terminals projecting to spinal neurones in immature rats. J Physiol 560:469–478

87. Kawa K (2003) Glycinergic receptors and glycinergic synaptic transmission in the deep cerebellar nuclei of the rat: a patch-clamp study. J Neurophysiol 90:3490–3500
88. Keller AF, Coull JA, Chery N, Poisbeau P, De Koninck Y (2001) Region-specific developmental specialization of GABA-glycine cosynapses in laminas I-II of the rat spinal dorsal horn. J Neurosci 21:7871–7880
89. Keller AF, Breton JD, Schlichter R, Poisbeau P (2004) Production of 5alpha-reduced neurosteroids is developmentally regulated and shapes GABA(A) miniature IPSCs in lamina II of the spinal cord. J Neurosci 24:907–915
90. Knoflach F, Benke D, Wang Y, Scheurer L, Luddens H, Hamilton BJ, Carter DB, Mohler H, Benson JA (1996) Pharmacological modulation of the diazepam-insensitive recombinant gamma-aminobutyric acidA receptors alpha 4 beta 2 gamma 2 and alpha 6 beta 2 gamma 2. Mol Pharm 50:1253–1261
91. Kolston J, Osen KK, Hackney CM, Ottersen OP, Storm-Mathisen J (1992) An atlas of glycine- and GABA-like immunoreactivity and colocalization in the cochlear nuclear complex of the guinea pig. Anat Embryol 186:443–465
92. Kotak VC, Korada S, Schwartz IR, Sanes DH (1998) A developmental shift from GABAergic to glycinergic transmission in the central auditory system. J Neurosci 18:4646–4655
93. Krishek BJ, Moss SJ, Smart TG (1998) Interaction of H+ and Zn2+ on recombinant and native rat neuronal GABAA receptors. J Physiol 507:639–652
94. Krishtal OA, Osipchuk Yu V, Vrublevsky SV (1988) Properties of glycine-activated conductances in rat brain neurones. Neurosci Lett 84:271–276
95. Laine J, Axelrad H (1996) Morphology of the Golgi-impregnated Lugaro cell in the rat cerebellar cortex: a reappraisal with a description of its axon. J Comp Neurol 375:618–640
96. Laine J, Axelrad H (1998) Lugaro cells target basket and stellate cells in the cerebellar cortex. Neurorep 9:2399–2403
97. Laube B (2002) Potentiation of inhibitory glycinergic neurotransmission by Zn2+: a synergistic interplay between presynaptic P2X2 and postsynaptic glycinergic receptors. Eur J Neurosci 16:1025–1036
98. Laube B, Kuhse J, Rundstrom N, Kirsch J, Schmieden V, Betz H (1995) Modulation by zinc ions of native rat and recombinant human inhibitory glycinergic receptors. J Physiol 483:613–619
99. Laube B, Kuhse J, Betz H (2000) Kinetic and mutational analysis of Zn2+ modulation of recombinant human inhibitory glycinergic receptors. J Physiol 522 Pt 2:215–230
100. Legendre P (1998) A reluctant gating mode of glycinergic receptor channels determines the time course of inhibitory miniature synaptic events in zebrafish hindbrain neurons. J Neurosci 18:2856–2870
101. Legendre P (2001) The glycinergic inhibitory synapse. Cell Mol Life Sci 58:760–793
102. Legendre P, Muller E, Badiu CI, Meier J, Vannier C, Triller A (2002) Desensitization of homomeric alpha1 glycinergic receptor increases with receptor density. Mol Pharmacol 62:817–827
103. Lim R, Alvarez FJ, Walmsley B (2000) GABA mediates presynaptic inhibition at glycinergic synapses in a rat auditory brainstem nucleus. J Physiol 525 Pt 2:447–459
104. Liu QR, Lopez-Corcuera B, Mandiyan S, Nelson H, Nelson N (1993) Cloning and expression of a spinal cord- and brain-specific glycine transporter with novel structural features. J Biol Chem 268:22802–22808
105. Lu T, Rubio ME, Trussell LO (2008) Glycinergic transmission shaped by the co-release of GABA in a mammalian auditory synapse. Neuron In press
106. Malosio ML, Marqueze-Pouey B, Kuhse J, Betz H (1991) Widespread expression of glycinergic receptor subunit mRNAs in the adult and developing rat brain. EMBO 10:2401–2409
107. Mangin JM, Guyon A, Eugene D, Paupardin-Tritsch D, Legendre P (2002) Functional glycinergic receptor maturation in the absence of glycinergic input in dopaminergic neurones of the rat substantia nigra. J Physio 542:685–697

108. Martin G, Siggins GR (2002) Electrophysiological evidence for expression of glycinergic receptors in freshly isolated neurons from nucleus accumbens. J Pharmacol Exp Ther 302:1135–1145
109. McCool BA, Botting SK (2000) Characterization of strychnine-sensitive glycinergic receptors in acutely isolated adult rat basolateral amygdala neurons. Brain Res 859:341–351
110. McIntire SL, Reimer RJ, Schuske K, Edwards RH, Jorgensen EM (1997) Identification and characterization of the vesicular GABA transporter. Nature 389:870–876
111. Meier J (2003) The enigma of transmitter-selective receptor accumulation at developing inhibitory synapses. Cell Tiss Res 311:271–276
112. Meier J, Grantyn R (2004) A gephyrin-related mechanism restraining glycinergic receptor anchoring at GABAergic synapses. J Neurosci 24:1398–1405
113. Meier J, De Chaldee M, Triller A, Vannier C (2000) Functional heterogeneity of gephyrins. Mol Cell Neurosci 16:566–577
114. Miller PS, Beato M, Harvey RJ, Smart TG (2005) Molecular determinants of glycinergic receptor alphabeta subunit sensitivities to Zn2+-mediated inhibition. J Physiol 566:657–670
115. Miller PS, Da Silva HM, Smart TG (2005) Molecular basis for zinc potentiation at strychnine-sensitive glycinergic receptors. J Biol Chem 280:37877–37884
116. Mitchell SJ, Silver RA (2003) Shunting inhibition modulates neuronal gain during synaptic excitation. Neuron 38:433–445
117. Mitchell EA, Gentet LJ, Dempster J, Belelli D (2007) GABAA and glycinergic receptor-mediated transmission in rat lamina II neurones: relevance to the analgesic actions of neuroactive steroids. J Physiol 583:1021–1040
118. Mori M, Gahwiler BH, Gerber U (2002) Beta-alanine and taurine as endogenous agonists at glycinergic receptors in rat hippocampus in vitro. J Physiol 539:191–200
119. Moss SJ, Smart TG (2001) Constructing inhibitory synapses. Nature Rev 2:240–250
120. Mozrzymas JW (2004) Dynamism of GABA(A) receptor activation shapes the "personality" of inhibitory synapses. Neuropharmacol 47:945–960
121. Mugnaini E, Floris A (1994) The unipolar brush cell: a neglected neuron of the mammalian cerebellar cortex. J comparative neurology 339:174–180
122. Mukhtarov M, Ragozzino D, Bregestovski P (2005) Dual Ca2+ modulation of glycinergic synaptic currents in rodent hypoglossal motoneurones. J Physiol 569:817–831
123. Muller E, Triller A, Legendre P (2004) Glycinergic receptors and GABAergic receptor alpha 1 and gamma 2 subunits during the development of mouse hypoglossal nucleus. Eur J Neurosci 20:3286–3300
124. Muller E, Le Corronc H, Triller A, Legendre P (2006) Developmental dissociation of presynaptic inhibitory neurotransmitter and postsynaptic receptor clustering in the hypoglossal nucleus. Mol Cell Neurosci 32:254–273
125. Nabekura J, Katsurabayashi S, Kakazu Y, Shibata S, Matsubara A, Jinno S, Mizoguchi Y, Sasaki A, Ishibashi H (2004) Developmental switch from GABA to glycine release in single central synaptic terminals. Nature Neurosci 7:17–23
126. Nevin ST, Cromer BA, Haddrill JL, Morton CJ, Parker MW, Lynch JW (2003) Insights into the structural basis for zinc inhibition of the glycinergic receptor. J Biol Chem 278:28985–28992
127. Nunzi MG, Shigemoto R, Mugnaini E (2002) Differential expression of calretinin and metabotropic glutamate receptor mGluR1alpha defines subsets of unipolar brush cells in mouse cerebellum. J Comp Neurol 451:189–199
128. Nusser Z, Naylor D, Mody I (2001) Synapse-specific contribution of the variation of transmitter concentration to the decay of inhibitory postsynaptic currents. Biophys J 80:1251–1261
129. O'Brien JA, Berger AJ (1999) Co-transmission of GABA and glycine to brain stem motoneurons. J Neurophysiol 82:1638–1641

130. O'Brien JA, Berger AJ (2001) The nonuniform distribution of the GABA(A) receptor alpha 1 subunit influences inhibitory synaptic transmission to motoneurons within a motor nucleus. J Neurosci 21:8482–8494
131. Oertel D (1991) The role of intrinsic neuronal properties in the encoding of auditory information in the cochlear nuclei. Curr Opin Neurobiol 1:221–228
132. Oertel D, Young ED (2004) What's a cerebellar circuit doing in the auditory system? TINS 27:104–110
133. Ohno K, Koroll M, El Far O, Scholze P, Gomeza J, Betz H (2004) The neuronal glycine transporter 2 interacts with the PDZ domain protein syntenin-1. Mol Cell Neurosci 26:518–529
134. Ottersen OP, Davanger S, Storm-Mathisen J (1987) Glycine-like immunoreactivity in the cerebellum of rat and Senegalese baboon, Papio papio: a comparison with the distribution of GABA-like immunoreactivity and with [3H]glycine and [3H]GABA uptake. Exp Brain Res 66:211–221
135. Ottersen OP, Storm-Mathisen J, Somogyi P (1988) Colocalization of glycine-like and GABA-like immunoreactivities in Golgi cell terminals in the rat cerebellum: a postembedding light and electron microscopic study. Brain Res 450:342–353
136. Pedroarena CM, Kamphausen S (2007) Glycinergic synaptic currents in the deep cerebellar nuclei. Neuropharmacol
137. Poisbeau P, Patte-Mensah C, Keller AF, Barrot M, Breton JD, Luis-Delgado OE, Freund-Mercier MJ, Mensah-Nyagan AG, Schlichter R (2005) Inflammatory pain upregulates spinal inhibition via endogenous neurosteroid production. J Neurosci 25:11768–11776
138. Poyatos I, Ponce J, Aragon C, Gimenez C, Zafra F (1997) The glycine transporter GLYT2 is a reliable marker for glycine-immunoreactive neurons. Brain Res Mol Brain Res 49:63–70
139. Prior P, Schmitt B, Grenningloh G, Pribilla I, Multhaup G, Beyreuther K, Maulet Y, Werner P, Langosch D, Kirsch J, et al. (1992) Primary structure and alternative splice variants of gephyrin, a putative glycinergic receptor-tubulin linker protein. Neuron 8:1161–1170
140. Racca C, Gardiol A, Triller A (1998) Cell-specific dendritic localization of glycinergic receptor alpha subunit messenger RNAs. Neurosci 84:997–1012
141. Rampon C, Luppi PH, Fort P, Peyron C, Jouvet M (1996) Distribution of glycine-immunoreactive cell bodies and fibers in the rat brain. Neurosci 75:737–755
142. Ransom BR, Bullock PN, Nelson PG (1977) Mouse spinal cord in cell culture. III. Neuronal chemosensitivity and its relationship to synaptic activity. J Neurophysiol 40:1163–1177
143. Rigo JM, Badiu CI, Legendre P (2003) Heterogeneity of postsynaptic receptor occupancy fluctuations among glycinergic inhibitory synapses in the zebrafish hindbrain. J Physiol 553:819–832
144. Rossi DJ, Hamann M (1998) Spillover-mediated transmission at inhibitory synapses promoted by high affinity alpha6 subunit GABA(A) receptors and glomerular geometry. Neuron 20:783–795
145. Rozzo A, Armellin M, Franzot J, Chiaruttini C, Nistri A, Tongiorgi E (2002) Expression and dendritic mRNA localization of GABAC receptor rho1 and rho2 subunits in developing rat brain and spinal cord. Eur J Neurosci 15:1747–1758
146. Russier M, Kopysova IL, Ankri N, Ferrand N, Debanne D (2002) GABA and glycine co-release optimizes functional inhibition in rat brainstem motoneurons in vitro. J Physiol 541:123–137
147. Sagne C, El Mestikawy S, Isambert MF, Hamon M, Henry JP, Giros B, Gasnier B (1997) Cloning of a functional vesicular GABA and glycine transporter by screening of genome databases. FEBS Lett 417:177–183
148. Sassoe-Pognetto M, Fritschy JM (2000) Mini-review: gephyrin, a major postsynaptic protein of GABAergic synapses. Eur J Neurosci 12:2205–2210

149. Sato K, Zhang JH, Saika T, Sato M, Tada K, Tohyama M (1991) Localization of glycinergic receptor alpha 1 subunit mRNA-containing neurons in the rat brain: an analysis using in situ hybridization histochemistry. Neurosci 43:381–395
150. Sato K, Kiyama H, Tohyama M (1992) Regional distribution of cells expressing glycinergic receptor alpha 2 subunit mRNA in the rat brain. Brain Res 590:95–108
151. Saxena NC, Macdonald RL (1994) Assembly of GABAA receptor subunits: role of the delta subunit. J Neurosci 14:7077–7086
152. Saxena NC, Macdonald RL (1996) Properties of putative cerebellar gamma-aminobutyric acid A receptor isoforms. Mol Pharmacol 49:567–579
153. Schmieden V, Kuhse J, Betz H (1993) Mutation of glycinergic receptor subunit creates beta-alanine receptor responsive to GABA. Science 262:256–258
154. Sergeeva OA (1998) Comparison of glycine- and GABA-evoked currents in two types of neurons isolated from the rat striatum. Neurosci Lett 243:9–12
155. Sergeeva OA, Haas HL (2001) Expression and function of glycinergic receptors in striatal cholinergic interneurons from rat and mouse. Neurosci 104:1043–1055
156. Simat M, Parpan F, Fritschy JM (2007) Heterogeneity of glycinergic and gabaergic interneurons in the granule cell layer of mouse cerebellum. J Comp Neurol 500:71–83
157. Singer JH, Talley EM, Bayliss DA, Berger AJ (1998) Development of glycinergic synaptic transmission to rat brain stem motoneurons. J Neurophysiol 80:2608–2620
158. Smith AJ, Owens S, Forsythe ID (2000) Characterization of inhibitory and excitatory postsynaptic currents of the rat medial superior olive. J Physiol 529:681–698
159. Spike RC, Watt C, Zafra F, Todd AJ (1997) An ultrastructural study of the glycine transporter GLYT2 and its association with glycine in the superficial laminae of the rat spinal dorsal horn. Neurosci 77:543–551
160. Suwa H, Saint-Amant L, Triller A, Drapeau P, Legendre P (2001) High-affinity zinc potentiation of inhibitory postsynaptic glycinergic currents in the zebrafish hindbrain. J Neurophysiol 85:912–925
161. Takahashi T (2005) Postsynaptic receptor mechanisms underlying developmental speeding of synaptic transmission. Neurosci Res 53:229–240
162. Takahashi T, Momiyama A, Hirai K, Hishinuma F, Akagi H (1992) Functional correlation of fetal and adult forms of glycinergic receptors with developmental changes in inhibitory synaptic receptor channels. Neuron 9:1155–1161
163. Takahashi Y, Shirasaki T, Yamanaka H, Ishibashi H, Akaike N (1994) Physiological roles of glycine and gamma-aminobutyric acid in dissociated neurons of rat visual cortex. Brain Res 640:229–235
164. Taleb O, Betz H (1994) Expression of the human glycinergic receptor alpha 1 subunit in Xenopus oocytes: apparent affinities of agonists increase at high receptor density. EMBO 13:1318–1324
165. Thio LL, Shanmugam A, Isenberg K, Yamada K (2003) Benzodiazepines block alpha2-containing inhibitory glycinergic receptors in embryonic mouse hippocampal neurons. J Neurophysiol 90:89–99
166. Todd AJ (1990) An electron microscope study of glycine-like immunoreactivity in laminae I-III of the spinal dorsal horn of the rat. Neurosci 39:387–394
167. Todd AJ, Sullivan AC (1990) Light microscope study of the coexistence of GABA-like and glycine-like immunoreactivities in the spinal cord of the rat. J Comp Neurol 296:496–505
168. Todd AJ, Watt C, Spike RC, Sieghart W (1996) Colocalization of GABA, glycine, and their receptors at synapses in the rat spinal cord. J Neurosci 16:974–982
169. Tretter V, Jacob TC, Mukherjee J, Fritschy JM, Pangalos MN, Moss SJ (2008) The clustering of GABA(A) receptor subtypes at inhibitory synapses is facilitated via the direct binding of receptor alpha2 subunits to gephyrin. J Neurosci 28:1356–1365
170. Triller A, Choquet D (2003) Synaptic structure and diffusion dynamics of synaptic receptors. Biology of the cell / under the auspices of the European Cell Biol Organ 95:465–476

171. Triller A, Cluzeaud F, Korn H (1987) gamma-Aminobutyric acid-containing terminals can be apposed to glycinergic receptors at central synapses. J Cell Biol 104:947–956
172. Trombley PQ, Hill BJ, Horning MS (1999) Interactions between GABA and glycine at inhibitory amino acid receptors on rat olfactory bulb neurons. J Neurophysiol 82:3417–3422
173. Twyman RE, Macdonald RL (1991) Kinetic properties of the glycinergic receptor main- and sub-conductance states of mouse spinal cord neurones in culture. J Physiol 435:303–331
174. van den Pol AN, Gorcs T (1988) Glycine and glycinergic receptor immunoreactivity in brain and spinal cord. J Neurosci 8:472–492
175. Waldvogel HJ, Baer K, Allen KL, Rees MI, Faull RL (2007) Glycinergic receptors in the striatum, globus pallidus, and substantia nigra of the human brain: an immunohistochemical study. J Comp Neurol 502:1012–1029
176. Wang Z, Danscher G, Kim YK, Dahlstrom A, Mook Jo S (2002) Inhibitory zinc-enriched terminals in the mouse cerebellum: double-immunohistochemistry for zinc transporter 3 and glutamate decarboxylase. Neurosci Lett 321:37–40
177. Wang Z, Danscher G, Mook Jo S, Shi Y, Daa Schroder H (2001) Retrograde tracing of zinc-enriched (ZEN) neuronal somata in rat spinal cord. Brain Res 900:80–87
178. Wang Z, Li JY, Dahlstrom A, Danscher G (2001) Zinc-enriched GABAergic terminals in mouse spinal cord. Brain Res 921:165–172
179. Wang ZY, Stoltenberg M, Huang L, Danscher G, Dahlstrom A, Shi Y, Li JY (2005) Abundant expression of zinc transporters in Bergman glia of mouse cerebellum. Brain Res Bull 64:441–448
180. Wei W, Zhang N, Peng Z, Houser CR, Mody I (2003) Perisynaptic localization of delta subunit-containing GABA(A) receptors and their activation by GABA spillover in the mouse dentate gyrus. J Neurosci 23:10650–10661
181. Wenthold RJ, Huie D, Altschuler RA, Reeks KA (1987) Glycine immunoreactivity localized in the cochlear nucleus and superior olivary complex. Neurosci 22:897–912
182. Werman R, Davidoff RA, Aprison MH (1968) Inhibitory of glycine on spinal neurons in the cat. J Neurophysiol 31:81–95
183. Wilkin GP, Csillag A, Balazs R, Kingsbury AE, Wilson JE, Johnson AL (1981) Localization of high affinity [3H]glycine transport sites in the cerebellar cortex. Brain Res 216:11–33
184. Zafra F, Aragon C, Olivares L, Danbolt NC, Gimenez C, Storm-Mathisen J (1995) Glycine transporters are differentially expressed among CNS cells. J Neurosci 15:3952–3969
185. Zeilhofer HU, Studler B, Arabadzisz D, Schweizer C, Ahmadi S, Layh B, Bosl MR, Fritschy JM (2005) Glycinergic neurons expressing enhanced green fluorescent protein in bacterial artificial chromosome transgenic mice. J Comp Neurol 482:123–141
186. Zhang D, Pan ZH, Awobuluyi M, Lipton SA (2001) Structure and function of GABA(C) receptors a comparison of native versus recombinant receptors. TIPS 22:121–132
187. Zhang LH, Gong N, Fei D, Xu L, Xu TL (2008) Glycine Uptake Regulates Hippocampal Network Activity via Glycinergic receptor-Mediated Tonic Inhibition. Neuropsychopharmacol 33:701–711

Chapter 8
Glutamate Co-Release by Monoamine Neurons

Louis-Eric Trudeau, Grégory Dal Bo, and José Alfredo Mendez

Abstract A wide range of indirect data obtained during the last two decades has suggested that monoamine neurons may co-release neurotransmitters. Ultrastructural investigations have consistently reported that these neurons, including dopamine, serotonin and norepinephrine neurons establish both junctional (i.e. synaptic) and non-junctional (i.e. non-synaptic) axon terminals. The hypothesis that some of these terminals can mediate synaptic glutamate release has received strong support from cell culture studies as well as indirect support from electrophysiological recordings obtained in slice preparations and in intact animals. The molecular identification of vesicular glutamate transporters (VGluTs) in recent years has provided a new impetus and a new strategy to confirm the glutamatergic phenotype of neurons. A number of recent results now confirm that while many serotonin neurons express VGluT3, a subset of norepinephrine and dopamine neurons express VGluT2, thus supporting the hypothesis of glutamate co-transmission. The possibility that the neurotransmitter repertoire of central monoamine neurons may be plastic during development and in the context of activity-dependent neuronal plasticity or disease is now a major direction of current research.

8.1 General Introduction

Dopamine (DA)- and noradrenaline (NA)-containing neurons of the ventral mesencephalon play an important regulatory role in motivated behavior. Subtle disruption of their functioning is thought to accompany diseases such as schizophrenia and attention deficit and hyperactivity disorder (ADHD). They are also a primary target of drugs of abuse that generally enhance DA secretion in the brain. Degeneration of DA and NA neurons is also a primary hallmark of Parkinson's disease. Another important population of CNS monoamine neurons are serotonin (5-HT)-containing neurons of the raphe nuclei. These

L.-E. Trudeau (✉)
Department of pharmacology, CNS Research Group, Faculty of Medicine, Université de Montréal, Montrial, Quévec, Canada

R. Gutierrez (ed.), *Co-Existence and Co-Release of Classical Neurotransmitters*,
DOI 10.1007/978-0-387-09622-3_8, © Springer Science+Business Media, LLC 2009

neurons are thought to play multiple roles in behavioral activation, sleep and emotional processes. Although DA, NA and 5-HT are the primary neurotransmitters of these monoamine neurons, these cells have also long been known to also contain other neurotransmitters such as neuropeptides like cholecystokinin (CCK) or neurotensin (NT). However, very little is known concerning the specific physiological roles of such peptide co-release. In addition to neuropeptides, recent work has also raised the intriguing possibility that some monoamine neurons use glutamate as a co-transmitter. In this chapter, we will provide an overview of this possibility and will examine in detail the published literature supporting this hypothesis.

8.2 Morphological Heterogeneity of Monoaminergic Axon Terminals

In the context of the idea that monoamine neurons release multiple neurotransmitters, it seems important to first highlight the fact that the ultrastructural investigations of the axon terminals of these neurons in the striatum and cerebral cortex showed that these neurons establish two morphologically distinctive types of axon terminals. It was found that DA-, NA- and 5-HT-containing axon terminals in the striatum and cerebral cortex establish a large contingent of terminals that display no clearly identifiable synaptic specializations, meaning that no synaptic cleft and postsynaptic density can be detected, even in serial sections (Descarries et al., 1975; Descarries et al., 1977; Beaudet and Descarries, 1978; Descarries et al., 1980; Pickel et al., 1981; Voorn et al., 1986; Descarries et al., 1996; Antonopoulos et al., 2002). Such terminals have been termed “free nerve endings”, “asynaptic” axon terminals or “non-synaptic” terminals and are proposed to be involved in “diffuse” or “volume” transmission (Descarries and Mechawar, 2000; Descarries et al., 2007). The significance of this dual character of monoamine neuron axon terminals is presently unclear. However, one possibility is that they may serve different functions. For example, while the asynaptic terminals may be sites where dopamine is the main or only transmitter released, the synaptic junctions could be sites where other neurotransmitters such as glutamate are released, either alone or together with dopamine (Sulzer et al., 1998; Trudeau, 2004; Descarries et al., 2007).

8.3 Initial Electrophysiological and Anatomical Work Suggesting the Presence of Glutamate in Monoamine Neurons

The first indirect evidence for glutamate as a co-transmitter in monoamine neurons was provided by pioneering studies of Ottersen and Storm-Mathisen (Storm-Mathisen et al., 1983; Ottersen and Storm-Mathisen, 1984) who developed an immunocytochemical method to detect glutamate in fixed brain tissue.

They demonstrated the presence of glutamate-like immunoreactivity in presumed DA neurons of the substantia nigra as well as in presumed 5-HT neurons in the dorsal raphe. Although suggestive, these findings suffered from the limitation that the anti-glutamate immunostaining also revealed metabolic pools of glutamate, which complicated the distinction between *bona fide* glutamatergic neurons and other non-glutamate neurons displaying detectable levels of glutamate of metabolic origin. Others subsequently confirmed the presence of glutamate-like immunoreactivity in 5-HT neurons of the medulla oblongata of the rat (Nicholas et al., 1992) and in NA neurons of the locus coeruleus of the mouse and rat (Fung et al., 1994; Liu et al., 1995).

More convincing evidence was subsequently provided by evaluating the distribution in the brain of glutaminase, an enzyme thought to be critical for the synthesis of neurotransmitter pools of glutamate. Double-labelling experiments for catecholaminergic markers and glutaminase showed that in the rat, numerous DA neurons of the substantia nigra, NA neurons of the locus coeruleus and 5-HT neurons of raphe nuclei are strongly immunopositive for glutaminase (Kaneko et al., 1990). The presence of glutaminase in mesencephalic DA neurons was confirmed by others in the rat and extended in the same study to the monkey brain (Sulzer et al., 1998). Although these findings are compatible with the idea that some of these neurons contain abundant levels of neurotransmitter glutamate, they come short of demonstrating that glutamate is packaged in synaptic vesicles in these neurons and is in fact used as a co-transmitter. As detailed below, only the presence of a vesicular glutamate transporter can ascertain that pools of glutamate are indeed packaged and concentrated in synaptic vesicles.

Indirect electrophysiological evidence for the release of glutamate by monoamine neurons has also been reported by a number of groups. For example, extracellular stimulation of the substantia nigra in the rat or the cat has been shown to induce short-latency excitatory synaptic response in neurons of the striatum (Hull et al., 1970; Hull et al., 1973; Kitai et al., 1975; Wilson et al., 1982). Similar responses have been reported in other brain regions after electrical stimulation of presumed 5-HT neurons of the dorsal raphe (Park et al., 1982; Holtman et al., 1986; Fung and Barnes, 1989) and presumed NA neurons of the locus coeruleus (Fung et al., 1994). These studies support the idea that monoamine neurons can produce rapid excitatory synaptic signals in target neurons. However, the use of extracellular stimulation of non-identified neurons produced results that could potentially be explained in part by stimulation of non-monoaminergic glutamate neurons located in these nuclei. The contribution of fibres of passage originating from other nearby structures could also not be excluded in most of these studies.

More recently, convincing electrophysiological data were obtained using a brain slice model and in vivo preparations. In the first of these studies (Chuhma et al., 2004), a thick angled brain slice was prepared from mice in such a way as to maintain the integrity of mesencephalic DA neurons, of a subset of ventral striatal target neurons as well as part of the medial forebrain bundle containing some of the projecting axons. In this preparation, extracellular electrical stimulation of presumed DA neurons in the VTA evoked short-latency excitatory synaptic responses in striatal medium spiny neurons. Arguing in favor of the fact that the responses were caused by the firing of DA neurons, local application of quinpirole, a D2 receptor agonist known to hyperpolarize DA neurons, strongly reduced the amplitude of the synaptic responses. Although quite convincing, a caveat of this experiment is that the presence of purely glutamatergic neurons in the VTA could not be excluded. In the light of recent findings suggesting the presence of such neurons (Yamaguchi et al., 2007) and the possible presence of non-DA neurons responding to D2 agonists in the VTA (Margolis et al., 2006), the purely dopaminergic origin of the fast glutamate responses evoked in ventral striatal neurons in this study remains uncertain. Finally, recent data obtained in the intact anesthetised rat reported the presence of fast, glutamate-mediated synaptic responses evoked in prefrontal cortex neurons by extracellular stimulation of the VTA (Lapish et al., 2006). Interestingly, 6-hydroxydopamine lesions of the VTA essentially eliminated these synaptic responses, arguing in favor of the idea that the responses were indeed generated by DA neurons and not by fibres of passage or other intrinsic glutamatergic neurons.

8.4 Initial Microculture Studies Showing Glutamate Co-release by 5-HT and DA Neurons

In the absence of paired recordings between a monoamine neuron and a monosynaptically connected target neuron in brain slice or in vivo, the most direct evidence to date for glutamate co-release by monoamine neurons has been obtained from primary culture preparations. In groundbreaking work, Johnson showed that over 60% of single 5-HT neurons of the rat raphe nuclei in isolated culture established glutamatergic synaptic contacts on their own dendrites or on local target neurons (Johnson, 1994a; Johnson, 1994b; Johnson and Yee, 1995). It is of interest to mention that in this same report, 5-HT-receptor-mediated slow synaptic responses were also observed, an observation that suggested the occurrence of co-release of the two transmitters. A similar situation has not yet been reported for other classes of monoamine neurons.

Highly complementary data were subsequently reported for DA neurons by the groups of Sulzer, Rayport and Trudeau. The first report demonstrating the establishment of glutamatergic synaptic contacts by isolated rat DA neurons in culture was published by Sulzer and collaborators (Sulzer et al., 1998). Using single neuron cultures, they showed that approximately 60% of single DA neurons generated glutamate-mediated fast synaptic responses when stimulated by an action potential. This was not accompanied by a slow DA-mediated synaptic response, unlike what was reported in 5-HT neurons (Johnson, 1994a). However, in response to paired stimulation, the second glutamate-mediated synaptic response was smaller, an effect that was absent in the presence of a D2 receptor antagonist or after depletion of vesicular DA stores with reserpine. These findings were interpreted as being caused by the presynaptic action of DA, thought to be co-released with glutamate. The establishment of glutamate synapses by single DA neurons was confirmed by Bourque et al., who showed in addition that the glutamatergic phenotype of DA neurons is enhanced by exposure to the growth factor GDNF (Bourque and Trudeau, 2000), through a mechanism that has not yet been fully characterized. It is also interesting to note that glutamatergic synapses established by isolated DA neurons are negatively regulated by D2 receptors, analogous to the negative regulation of DA release by D2 autoreceptors (Sulzer et al., 1998; Congar et al., 2002). The ability to measure miniature glutamate-mediated synaptic events in isolated DA neurons has permitted detailed analysis of the presynaptic action of D2 receptors in DA neurons and shown that this receptor negatively regulates the secretory process through a mechanism implicating Kv potassium channels (Congar et al., 2002).

Because these studies were carried out in single isolated neurons, a preparation which leads to the formation of autapses (i.e., synapses formed by a neuron onto its own dendrites or cell body), a valid question is whether DA neurons also establish glutamatergic synapses onto their normal target neurons. This issue was addressed in an experiment in which DA neurons were co-cultured together with ventral striatal GABA neurons (Joyce and Rayport, 2000). In this study, paired recordings showed that stimulation of presumed DA neurons evoked glutamate-mediated postsynaptic responses in striatal neurons. Since the dopaminergic phenotype of presynaptic neurons was not systematically confirmed in all recordings in this report, the relative proportion of DA neurons able to establish glutamate synapses on striatal neurons is not known. However, the data nonetheless argue strongly that synaptic glutamate release by DA neurons is not a phenomenon restricted to single neuron cultures and autapse formation. In this context, it is also interesting to note that using an explant culture system, Plenz and Kitai showed that extracellular stimulation of a ventral mesencephalic slice containing DA neurons can evoke fast excitatory synaptic responses in medium spiny neurons present in a nearby striatal slice (Plenz and Kitai, 1996).

A detailed mechanism explaining the ability of a subset of monoamine neurons to use glutamate as a cotransmitter is not presently available. However, as explained in the section below, expression of a vesicular glutamate transporter by some of these neurons may provide the basis for this form of co-transmission.

8.5 Discovery of Vesicular Glutamate Transporters

While glutamate has long been known to be one of the main and most abundant neurotransmitter in the central nervous system, the morphological identification of glutamatergic neurons has been difficult until recently. Glutamate is synthesized in the presynaptic cytoplasm by phosphate-activated glutaminase (PAG) before being stored in synaptic vesicles in axon terminals prior to release by exocytosis at synapses. However, the immunodetection of glutamate and PAG in neurons is not sufficient to confirm with high certainty that a given neuron is indeed glutamatergic. For example, PAG and glutamate are found in GABAergic cells and astrocytes (Kaneko and Mizuno, 1994; Kaneko and Fujiyama, 2002). This problem in phenotypic identification of neurons was resolved recently with the characterization of three VGluTs. The first was initially identified as a brain-specific Na+-dependent, inorganic phosphate transporter (BNPI) (Ni et al., 1994). It was reported to be abundantly expressed at excitatory synapses in the brain and localized to synaptic vesicles, rather than at the plasma membrane where Na^+-dependent transporters are normally found (Bellocchio et al., 1998; Takamori et al., 2000; Bellocchio et al., 2000). Moreover, cells transfected with BNPI showed glutamate uptake in a vesicular compartment, and BNPI expression in cultured hippocampal GABA neurons allowed these cells to release glutamate and induce excitatory post-synaptic currents in target cells that can be blocked by antagonists of AMPA and NMDA-receptors (Takamori et al., 2000). BNPI thus functions as a vesicular glutamate transporter and was renamed VGluT1. It is now generally agreed that the presence of a VGluT in neurons is strong evidence for the fact that these neurons use glutamate as a neurotransmitter.

Closely after the discovery of VGluT1, a second VGluT (VGluT2) was identified and showed 82% identity with VGluT1. Like VGluT1, VGlut2 was found to be present in the membrane of synaptic-like vesicles in axons terminals (Hisano et al., 2000; Fujiyama et al., 2001). The first investigations of the distribution of VGluT1 and VGluT2 mRNA and protein in the mammalian brain revealed a striking complementary distribution of the two transporters (Herzog et al., 2001; Fujiyama et al., 2001; Fremeau et al., 2001). VGluT1 is predominantly expressed in cortical structures including the hippocampus, cerebral and cerebellar cortices, and layers I-III of the neocortex. In comparison, VGluT2 is mainly found in sub-cortical structures, including diencephalic nuclei, the brainstem and telencephalic regions. The subcellular localization of VGluT1 and VGluT2 protein revealed that the expression is confined to axon terminals, with little if any signal found in the cell body and dendrites. Considering this fact, it is clear that except under rare conditions such as in single-neuron cultures, the identification of a given neuron as glutamatergic cannot be performed by the simple immunocytochemical localization of VGluT1 or VGluT2. However, the use of in situ hybridization or single-cell RT-PCR to detect VGluT1 or VGluT2 mRNA is an excellent strategy for the phenotypic characterization of potential glutamatergic neurons. In situ hybridization studies confirmed the complementary distribution of VGluT1 and 2 mRNA in the brain. While VGluT1 mRNA is mainly expressed in the neocortex, hippocampus, cerebral and cerebellar cortices, VGluT2 mRNA is highly expressed in diencephalic, thalamic and hypothalamic nuclei, as well as in the brainstem and in deep cerebellar nuclei (Hisano et al., 2000; Herzog et al., 2001; Fremeau et al., 2001; Varoqui et al., 2002; Kaneko and Fujiyama, 2002; Kaneko et al., 2002).

Considering the close and complementary distribution of VGluT1 and VGluT2, comprising most known glutamatergic cell groups, the identification of a third VGluT in 2002 was somewhat surprising. VGluT3 presents 75% of homology with VGluT1 and VGluT2, and shows a much less extensive distribution in sparse populations of neurons that had not previously been considered to be glutamatergic. VGluT3 is found in striatal cholinergic interneurons of the striatum, in serotoninergic neurons of the raphé nuclei, and in GABAergic basket cells of rat hippocampus (Schafer et al., 2002; Fremeau et al., 2002; Gras et al., 2002; Somogyi et al., 2004; Herzog et al., 2004), suggesting that neurons expressing VGluT3 probably release glutamate as a co-transmitter in addition to another classical neurotransmitter. Another interesting difference between VGluT3 and the two other VGluTs is that in some neurons, VGluT3 protein is also found in cell bodies and dendrites, presuming mediating somatodendritic glutamate release (Fremeau et al., 2002; Harkany et al., 2004). Although VGluTs are most abundantly expressed in neurons, several studies reported expression of these proteins in astrocytes and other cell types. For example, VGluT1, VGluT2 and VGluT3 have been detected in freshly dissociated and primary cultured astrocytes, as well as in vivo, by single-cell RT-PCR, immunocytochemistry and electronic microscopy (Morimoto et al., 2003; Zhang et al., 2004; Montana et al., 2004; Bezzi et al., 2004). These results suggest that such transporters could be involved in glutamate release by astrocytes.

8.6 Presence of Vesicular Glutamate Transporters in Monoamine Neurons

The first direct morphological evidence for the presence of a VGluT in catecholaminergic neurons was established by Stornetta et al., (Stornetta et al., 2002a; Stornetta et al., 2002b). Using in situ hybridization to localize VGluT2 mRNA coupled with immunodetection of tyrosine hydroxylase (TH), the authors showed that the majority (75 to 80%) of adrenergic neurons in the C1, C2 and C3 subgroups and of A2 noradrenergic neurons contain VGluT2 mRNA (Stornetta et al., 2002a). Further experiments showed that $TH^{+}/VGluT2^{+}$ neurons project to sympathetic preganglionic neurons of the spinal cord (Stornetta et al., 2002b; Nakamura et al., 2004) and to neurons of the retrotrapezoid nucleus involved in chemosensitivity (Rosin et al., 2006).

In addition to the demonstration of VGluT2 in adrenergic and noradrenergic neurons, solid data also exists showing the presence of VGluT3 in 5-HT neurons of the raphe nuclei. A combination of immunocytochemistry to detect tryptophan hydroxylase and in situ hybridization to detect VGluT3 mRNA, demonstrated that 5-HT neurons express this transporter (Fremeau et al., 2002; Gras et al., 2002; Herzog et al., 2004). Compatible with the possibility of co-release of glutamate and 5-HT, VGluT3 protein was also reported to be highly colocalized with vesicular monoamine transporter (VMAT2) in axon terminals in the hippocampus of P15 rats (but less so in adults rats) (Boulland et al., 2004). As these terminals do not express TH, they were suggested to be serotoninergic (Boulland et al., 2004). Using confocal microscopy, it was also demonstrated that VGluT3-positive terminals on sympathetic preganglionic neurons contain 5-HT, confirming their serotoninergic nature (Nakamura et al., 2004).

The possible expression of VGluTs in DA neurons has also attracted great interest during the last few years. An initial report used the northern blot technique to reveal the presence of VGluT2 mRNA in the substantia nigra (Aihara et al., 2000). However, the cellular origin of this signal was not investigated in this initial study. VGluT3 mRNA has also been suggested to be present in DA neurons of the VTA and substantia nigra (Fremeau et al., 2002). However, although VGluT3 protein is abundantly expressed in these nuclei in axon terminals that arise from other nuclei including the raphe (Herzog et al., 2004), the observation of VGluT3 mRNA in DA neurons has not been confirmed by other groups.

To examine this question in more detail, we examined the expression of the three VGluTs in rat mesencephalic DA neurons in primary culture. The use of single neuron cultures allowed the confirmation of the cellular origin of the signal detected by immunocytochemistry. We found by double immunocytochemistry that approximately 80% of TH-positive DA neurons selectively express the VGluT2 isoform (Dal Bo et al., 2004). The selectivity of this expression was validated by single cell RT-PCR experiments, which confirmed the absence of VGluT1 or VGluT3 mRNA in these same cells (Dal Bo et al., 2004). This first unequivocal demonstration of the expression of a VGluT in DA neurons confirmed that VGluT2 is the only isoform present and provided a convenient explanation of the ability of these neurons to establish glutamatergic synapses.

However, additional studies were required to evaluate the expression profile of VGluTs in DA neurons in vivo. A first suggestion was made that VGluT2-positive cells are present in the VTA and caudal linear nucleus (CLi). Retrograde labelling with fluorogold applied in the prefrontal cortex, followed by in situ hybridization against VGluT2 mRNA, showed evidence for VGluT2 expression in VTA and CLi (Hur and Zaborszky, 2005). However, although DA neurons are known to project to the prefrontal cortex, a limitation of this study is that there was no demonstration that the VGluT2-expressing neurons found were dopaminergic. Using a double in situ hybridization technique to localize TH and VGluT2 mRNA, we reported in 2005 the presence of double-labeled neurons in the VTA of postnatal (P5 to P15) juvenile rat (Dal Bo et al., 2005, Dal Bo et al., in press). The proportion of double-labeled neurons was modest, representing approximately 1 to 2% of DA neurons in the SN and VTA, respectively. Compatible with this initial observation, another group used a combination of in situ hybridization against VGluT2 mRNA and immunocytochemistry against TH to examine VGluT2 expression in the ventral mesencephalon of adult rats (Kawano et al., 2006). Numerous double-labelled neurons were found in the VTA, while very few were found in the substantia nigra and retrorubral nucleus. Within the VTA, the distribution of VGluT2-positive DA neurons was different according the subregion examined. Areas close to the midline such as the CLi and rostral linear nucleus (RLi), presented a higher proportion of double-labelled neurons (21 to 53%), while parabrachial pigmented and paranigral nuclei showed weaker proportions (3 to 5%) (Kawano et al., 2006). VGluT2-expressing DA neurons were also found occasionally in other dopaminergic regions such as the A11, A13, and A15 subgroups, while no such neurons were found in the A12, A14, A16, and A17 subgroups (Kawano et al., 2006). Finally, another group also reported the presence of VGluT2-expressing DA neurons using double in situ hybridization (Yamaguchi et al., 2007). However, in this study, the proportion of DA neurons expressing VGluT2 mRNA was reported to be very small (7 out of 6357 DA neurons) and a large number of neurons containing VGluT2 but not TH mRNA was found, suggesting for the first time the presence of purely glutamatergic neurons in the ventral mesencephalon. The reason explaining the different proportion of DA neurons found to contain VGluT2 mRNA in the three studies is unclear. However, as in situ hybridization may not be ideal to detect mRNA present in low copy number, we re-examined this question using single-cell RT-PCR. For this experiment, we took advantage of mice expressing GFP under the control of the TH promoter (Sawamoto et al., 2001; Jomphe et al., 2005), thus facilitating the identification and isolation of DA neurons. Freshly dissociated DA neurons were thus examined in mice of different postnatal ages. We found that at P0, 25% (17/68) of DA neurons expressed VGLUT-2 mRNA, while this proportion decreased to 14% (8/58) at P45 (Mendez et al., 2005, Mendez et al., 2008). This higher proportion of DA neurons containing VGluT2 mRNA is compatible with the idea that numerous DA neurons may contain low levels of VGluT2 mRNA that may escape detection using in situ hybridization, perhaps leading to an under-estimation of double-labelled neurons and an over-estimation of the number of purely glutamatergic neurons. It should be pointed out that, in this study, VGluT1 and VGluT3 mRNA were never detected in DA neurons confirming that DA neurons only express the VGluT2 subtype.

In addition to the presence of mRNA, the presence of VGluT2 protein in dopaminergic axon terminals was examined. Using confocal microscopy, Kawano et al. (2006) examined the nucleus accumbens (nAcb) and reported the presence of VGluT2 immunoreactivity in dopaminergic axon terminals. However, considering the limited spatial resolution of confocal microscopy and the relative abundance of VGluT2-containing axon terminals in the striatum that originate from other nuclei including the thalamus, the possibility of false positive double-labeling could not be excluded. Only electron microscopy has sufficient spatial resolution to confirm with a high degree of certainty the presence of VGluT2 protein in dopaminergic axon terminals. Using immunoelectron microscopy and double-labelling for VGluT2, detected with immunogold, and TH labelled with immunoperoxydase, we confirmed the existence of VGluT2-containing dopaminergic axon terminals in the striatum of P15 and P90 rats, and suggested a lower occurrence in older rats (Bérubé-Carriere et al., 2006; Bérubé-Carriere et al., 2007).

8.7 Regulation of the Expression of Vesicular Glutamate Transporters in Neurons

As discussed previously, while VGluT2 protein can be found in a majority of DA neurons in culture, only a small subset express the transporter under basal conditions in juvenile and adult animals. This provides an indirect indication that expression of VGluTs in monoamine neurons and in neurons in general, could perhaps be regulated through various signals. In fact, early evidence for the inducibility of VGluT expression came from the cloning of VGluT1 and VGluT2. Whereas VGluT1, identified as BNPI was found to be upregulated in response to subtoxic treatment of cerebellar granule cells with NMDA (Ni et al., 1994), VGluT2, identified as a cDNA fragment, AB20, was reported to be upregulated under activin A/betacellulin treatment to induce neuronal–like/ neuroendocrine cell differentiation of AR42J cells (Mashima et al., 1999; Aihara et al., 2000). Although in *Drosophila* it has been proposed that a single glutamate transporter is sufficient to fill a synaptic vesicle (Daniels et al., 2006), there is compelling evidence showing that an increase in the number of VGluTs per vesicle increases the amount of glutamate packaged and released by a single vesicle (Daniels et al., 2004). Taken together with the findings that quantal transmission does not cause saturation of glutamate receptors (Liu et al., 1999; Yamashita et al., 2003), these findings favor the notion that modulating the expression of VGluTs could have an important physiological impact. In addition, up-regulation of a VGluT in a neuron already releasing another neurotransmitter could in principle modify the neurotransmitter repertoire of a neuron or be implicated in various pathophysiological mechanisms. Several groups have therefore started to evaluate possible changes in VGluT expression in disease and pathological models. For example, compared to matched controls, VGluT1 protein has been found to be elevated in the molecular layer of the dentate gyrus in hippocampal tissue obtained from individuals diagnosed with schizophrenia (Talbot et al., 2004). However, VGluT1 mRNA was found to be decreased in both hippocampus and in the dorsolateral prefrontal cortex of schizophrenia patients in another study (Eastwood and Harrison, 2005). Finally, another group reported that the abundance of VGluT2 mRNA is increased in the thalamus of schizophrenia patients (Smith et al., 2001). Evidence has also been provided for alterations of VGluT expression in Parkinson disease (PD). Besides the clear loss of DA neurons, an imbalance between the so called “direct” and “indirect” pathways of the basal ganglia is thought to play an important role in the symptoms of this disease. Interestingly, the levels of both VGluT1 and VGluT2 proteins are increased in the putamen of PD patients whereas VGluT1 expression is reduced in both the prefrontal and temporal cortices (Kashani et al., 2007a). Whether this is causally related to PD or a response to DA denervation remains uncertain. However, 6-OHDA lesion experiments have shown no such increase of any VGluT protein in the whole striatum after dopaminergic denervation (Robelet et al., 2004; Chung et al., 2006–2007). Another disease in which a modulation of VGluT expression has been reported is Alzheimer's disease (AD). Whereas VGluT1 is decreased in both the parietal and occipital cortices in AD patients (Kirvell et al., 2006), both VGluT1 and VGluT2 are decreased in the prefrontal dorsolateral cortex (Kashani et al., 2007b). Strikingly, no changes were reported to occur in the temporal cortex, an area that is particularly affected in AD.

VGluT expression has also been shown to be modulated in response to therapeutic agents such as antidepressants or antiepileptic drugs. For example, chronic treatment of rodents with the selective 5-HT reuptake inhibitors fluoxetine and paroxetine as well as desipramine, a norepinephrine reuptake inhibitor, increased expression of VGluT1 mRNA in the anterior cortex as well as in the CA1, CA3 and dentate gyrus of the hippocampus (Tordera et al., 2005; Moutsimilli et al., 2005) with a concomitant increase in VGluT1 protein in the thalamus. In the same report, the antipsychotic clozapine and the mood stabilizer lithium had a moderate but significant impact on both VGluT1 mRNA and protein expression in the same areas. It is noteworthy that these treatments induced no robust changes in VGluT2 or VGluT3 mRNA levels and that the increase in VGluT1 expression was not present in response to acute treatments. This later observation suggests that these changes are likely to be delayed adaptive responses to the drug treatments. In the Mongolian gerbil model of epilepsy, the seizure-sensitive (SS) gerbils showed a marked increased of both VGluT1 and VGluT2 proteins in the dentate gyrus of the hippocampus (Kang et al., 2005). These increases were reduced in response to treatment with the antiepileptic drug valproic acid, an effect that was not reproduced with vigabatrin, an agent that belongs to a different class (Kang et al., 2005). Similarly, riluzole was found to reduce the elevated expression of VGluT1 protein found in the hippocampus of rats in the pilocarpine-induced epilepsy model (Kim et al., 2007). Interestingly both valproic acid and riluzole are Na^+ channel blocker, therefore suggesting that decreases in neuronal firing may be involved in the effects of these agents on VGluT expression.

The expression of VGluTs has also been reported to be modified under other conditions. For example, levels of VGluT1 and VGluT2 have been reported to be induced in the hippocampus of rats after hypoxic-ischemia insults (Kim et al., 2005). Repetitive electroconvulsive shock (ECS) induces VGluT1 mRNA expression in the cortex and dentate gyrus (Tordera et al., 2005), an effect that is dependent on the activation of monoamines. Osmotic challenges induced by water deprivation or salt loading also induce VGluT2 mRNA in neurons of the hypothalamic-neurohypophysial system (Kawasaki et al., 2005; Kawano et al., 2006; Kawasaki et al., 2006). Finally, the expression of VGluT2 mRNA is increased in peptide-hormone secreting cells of the adenohypophysis of ovariectomized rats treated with estradiol or after induction of a state of hypothyroidism (Hrabovszky et al., 2006).

As mentioned above, VGluTs are also expressed in monoamine neurons and examples suggesting the regulation of glutamate cotransmission in these cell populations are beginning to be examined. In particular, in C1 adrenergic neurons of the rostral ventral lateral medulla, an area devoted to the control of arterial pressure, VGluT2 is upregulated in hydralazine-induced hypotensive rats (Stornetta et al., 2002b). Although the increase is not exclusive of these adrenergic neurons (because other non-TH neurons also respond to the same stimulus by an increase in VGluT2), two aspects of this regulatory response are worth pointing out: (a) the up-regulation is relatively fast, since the increase in VGluT2 occurs within approximately 2 h after the induction of hypotension and (b) the increase of VGluT2 is accompanied by nuclear translocation of c-Fos. Although the VGluT2 promoter has not yet been characterized, a computational analysis to identify potential transcription factor binding sites using TFSEARCH, TRANSFAC and Matinspector reveals the presence of putative AP-1 and CREB binding sites within the 5' region of VGluT2 reported in the genebank database. It is therefore possible that after the induction of hypotension, the baroreceptors lead to VGluT2 up-regulation through signal pathways involving CREB activation and subsequent AP-1 up-regulation.

Finally, we recently reported two examples of the regulation of VGluT2 expression in mesencephalic DA neurons. When evaluating the expression of VGluT2 in isolated DA neurons in culture, we noted that the proportion of DA neurons expressing VGluT2 increased with time (Dal Bo et al., 2004). However, under culture conditions that allowed contact with GABAergic neurons, this induction was not observed, thus suggesting that interaction with GABAergic neurons normally mediates a negative regulatory signal influencing VGluT2 expression in DA neurons (Mendez et al., 2005; Mendez et al., 2008). We have also obtained evidence suggesting the possibility of an up-regulation of VGluT2 expression in surviving DA neurons after a partial neonatal 6-OHDA lesion (Dal Bo et al., 2005, Dal Bo et al., in press). Because the increase in the number of DA neurons expressing VGluT2 could not be simply explained by the preferential survival of DA neurons that already expressed VGluT2 prior to the lesion, it is possible that the induction represented a form of neuroadaptive derepression of the glutamatergic phenotype of DA neurons.

8.8 Conclusions and Future Directions

Although the expression of VGluTs in monoamine neurons is now well established, the physiological functions and implications of this ability to use glutamate as a co-transmitter have remained elusive. Identification of the physiological and pathological roles of such cotransmission will require approaches such as conditional gene targeting to prevent expression of VGluTs in select monoamine neuron populations. It will also be important to understand the molecular mechanisms of induction of VGluT in monoamine and as well as in other neuronal populations. The recent availability of transgenic mouse lines expressing a bacterial artificial chromosome (BAC) vector carrying the promoter region of VGluTs fused to the reporter gene EGFP should prove to be quite useful to reach this objective (Gong et al., 2003). It will also be important to characterize in detail the promoter regions of the VGluT1-2 and 3 genes. Another major question that is still unanswered is whether or not the glutamate that is coreleased by monoamine neurons in vivo is destined to mediate fast information transfer at synapses as opposed to other slower, modulatory roles of glutamate. Although this is likely to be extremely difficult, paired recordings between a VGluT-expressing monoamine neuron and a target neuron in a brain slice preparation would seem like the only way to settle this issue.

Acknowledgments L-E Trudeau is a senior scholar of the Fonds de la Recherche en Santé du Québec and receives support from the Canadian Institutes of Health Research. Support from the National Alliance for Research on Schizophrenia and Depression is also acknowledged for some of the work discussed in the chapter. José Alfredo Mendez is supported by the Jasper postdoctoral fellowship of the Groupe de Recherche sur le système nerveux central.

References

Aihara Y, Mashima H, Onda H, Hisano S, Kasuya H, Hori T, Yamada S, Tomura H, Yamada Y, Inoue I, Kojima I, Takeda J (2000) Molecular cloning of a novel brain-type Na(+)-dependent inorganic phosphate cotransporter. J Neurochem 74:2622–2625

Antonopoulos J, Dori I, Dinopoulos A, Chiotelli M, Parnavelas JG (2002) Postnatal development of the dopaminergic system of the striatum in the rat. Neuroscience 110:245–256

Beaudet A, Descarries L (1978) The monoamine innervation of rat cerebral cortex: synaptic and nonsynaptic axon terminals. Neuroscience 3:851–860

Bellocchio EE, Hu H, Pohorille A, Chan J, Pickel VM, Edwards RH (1998) The localization of the brain-specific inorganic phosphate transporter suggests a specific presynaptic role in glutamatergic transmission. J Neurosci 18:8648–8659

Bellocchio EE, Reimer RJ, Fremeau RT Jr, Edwards RH (2000) Uptake of glutamate into synaptic vesicles by an inorganic phosphate transporter. Science 289:957–960

Bezzi P, Gundersen V, Galbete JL, Seifert G, Steinhauser C, Pilati E, Volterra A (2004) Astrocytes contain a vesicular compartment that is competent for regulated exocytosis of glutamate. Nat Neurosci 7:613–620

Boulland JL, Qureshi T, Seal RP, Rafiki A, Gundersen V, Bergersen LH, Fremeau RT Jr, Edwards RH, Storm-Mathisen J, Chaudhry FA (2004) Expression of the vesicular glutamate transporters during development indicates the widespread corelease of multiple neurotransmitters. J Comp Neurol 480:264–280

Bourque MJ, Trudeau LE (2000) GDNF enhances the synaptic efficacy of dopaminergic neurons in culture. Eur J Neurosci 12:3172–180

Bérubé-Carriere N, Riad M, Dal Bo G, Trudeau L-E, Descarries L Colocalisation of dopamine and glutamate in axon terminals of VTA neurons innervating the nucleus accumbens. Abstract Viewer/Itinerary Planner. Washington, DC: Society for Neuroscience. Online. 722.11. 2006

Bérubé-Carriere N, Riad M, Dal Bo G, Trudeau L-E, Descarries L Vesicular glutamate transporter 2 in dopamine axon terminals of the nucleus accumbens. Abstract Viewer/Itinerary Planner. Washington, DC: Society for Neuroscience. Online. 38.9. 2007

Chuhma N, Zhang H, Masson J, Zhuang X, Sulzer D, Hen R, Rayport S (2004) Dopamine neurons mediate a fast excitatory signal via their glutamatergic synapses. J Neurosci 24:972–981

Chung EK, Chen LW, Chan YS, Yung KK (2006–2007) Up-regulation in expression of vesicular glutamate transporter 3 in substantia nigra but not in striatum of 6-hydroxydopamine-lesioned rats. Neurosignals 15:238–248

Congar P, Bergevin A, Trudeau LE (2002) D2 receptors inhibit the secretory process downstream from calcium influx in dopaminergic neurons: implication of k(+) channels. J Neurophysiol 87:1046–1056

Dal Bo G, Bérubé-Carrière N, Mendez JA, Leo D, Riad M, Descarries L, Lévesque D, Trudeau L-E Enhanced glutamatergic phenotype of mesencephalic dopamine neurons after neonatal 6-hydroxydopamine lesion. Neuroscience (in press).

Dal Bo G, Lévesque D, Trudeau L-E A 6-OHDA lesion up-regulates VGLUT2 expression in dopaminergic neurons of the VTA. Abstract Viewer/Itinerary Planner. Washington, DC: Society for Neuroscience. Online. 155. 6. 2005

Dal Bo G, St-Gelais F, Danik M, Williams S, Cotton M, Trudeau LE (2004) Dopamine neurons in culture express VGLUT2 explaining their capacity to release glutamate at synapses in addition to dopamine. J Neurochem 88:1398–1405

Daniels RW, Collins CA, Chen K, Gelfand MV, Featherstone DE, DiAntonio A (2006) A single vesicular glutamate transporter is sufficient to fill a synaptic vesicle. Neuron 49:11–16

Daniels RW, Collins CA, Gelfand MV, Dant J, Brooks ES, Krantz DE, DiAntonio A (2004) Increased expression of the Drosophila vesicular glutamate transporter leads to excess glutamate release and a compensatory decrease in quantal content. J Neurosci 24:10466–10474

Descarries L, Beaudet A, Watkins KC (1975) Serotonin nerve terminals in adult rat neocortex. Brain Res 100:563–588

Descarries L, Bérubé-Carrière N, Riad M, Bo GD, Mendez JA, Trudeau LE (2007) Glutamate in dopamine neurons: Synaptic versus diffuse transmission. Brain Res Rev.

Descarries L, Bosler O, Berthelet F, Des Rosiers MH (1980) Dopaminergic nerve endings visualised by high-resolution autoradiography in adult rat neostriatum. Nature 284:620–622

Descarries L, Mechawar N (2000) Ultrastructural evidence for diffuse transmission by monoamine and acetylcholine neurons of the central nervous system. Prog Brain Res 125:27–47

Descarries L, Watkins KC, Garcia S, Bosler O, Doucet G (1996) Dual character, asynaptic and synaptic, of the dopamine innervation in adult rat neostriatum: a quantitative autoradiographic and immunocytochemical analysis. J Comp Neurol 375:167–186

Descarries L, Watkins KC, Lapierre Y (1977) Noradrenergic axon terminals in the cerebral cortex of rat. III. Topometric ultrastructural analysis. Brain Res 133:197–222

Eastwood SL, Harrison PJ (2005) Decreased expression of vesicular glutamate transporter 1 and complexin II mRNAs in schizophrenia: further evidence for a synaptic pathology affecting glutamate neurons. Schizophr Res 73:159–172

Fremeau RT Jr, Burman J, Qureshi T, Tran CH, Proctor J, Johnson J, Zhang H, Sulzer D, Copenhagen DR, Storm-Mathisen J, Reimer RJ, Chaudhry FA, Edwards RH (2002) The identification of vesicular glutamate transporter 3 suggests novel modes of signaling by glutamate. Proc Natl Acad Sci U S A 99:14488–14493

Fremeau RT Jr, Troyer MD, Pahner I, Nygaard GO, Tran CH, Reimer RJ, Bellocchio EE, Fortin D, Storm-Mathisen J, Edwards RH (2001) The expression of vesicular glutamate transporters defines two classes of excitatory synapse. Neuron 31:247–260

Fujiyama F, Furuta T, Kaneko T (2001) Immunocytochemical localization of candidates for vesicular glutamate transporters in the rat cerebral cortex. J Comp Neurol 435:379–387
Fung SI, Chan JY, Manzoni D, White SR, Lai YY, Strahlendorf HK, Zhuo H, Liu RH, Reddy VK, Barnes CD (1994) Co-transmitter-mediated locus coeruleus action on motoneurons. Brain Res Bull 35:423–432
Fung SJ, Barnes CD (1989) Raphe-produced excitation of spinal cord motoneurons in the cat. Neurosci Lett 103:185–190
Fung SJ, Reddy VK, Liu RH, Wang Z, Barnes CD (1994) Existence of glutamate in noradrenergic locus coeruleus neurons of rodents. Brain Res Bull 35:505–512
Gong S, Zheng C, Doughty ML, Losos K, Didkovsky N, Schambra UB, Nowak NJ, Joyner A, Leblanc G, Hatten ME, Heintz N (2003) A gene expression atlas of the central nervous system based on bacterial artificial chromosomes. Nature 425:917–925
Gras C, Herzog E, Bellenchi GC, Bernard V, Ravassard P, Pohl M, Gasnier B, Giros B, El Mestikawy S (2002) A third vesicular glutamate transporter expressed by cholinergic and serotoninergic neurons. J Neurosci 22:5442–5451
Harkany T, Holmgren C, Hartig W, Qureshi T, Chaudhry FA, Storm-Mathisen J, Dobszay MB, Berghuis P, Schulte G, Sousa KM, Fremeau RT Jr, Edwards RH, Mackie K, Ernfors P, Zilberter Y (2004) Endocannabinoid-independent retrograde signaling at inhibitory synapses in layer 2/3 of neocortex: involvement of vesicular glutamate transporter 3. J Neurosci 24:4978–988
Herzog E, Bellenchi GC, Gras C, Bernard V, Ravassard P, Bedet C, Gasnier B, Giros B, El Mestikawy S (2001) The existence of a second vesicular glutamate transporter specifies subpopulations of glutamatergic neurons. J Neurosci 21:RC181
Herzog E, Gilchrist J, Gras C, Muzerelle A, Ravassard P, Giros B, Gaspar P, El Mestikawy S (2004) Localization of VGLUT3, the vesicular glutamate transporter type 3, in the rat brain. Neuroscience 123:983–1002
Hisano S, Hoshi K, Ikeda Y, Maruyama D, Kanemoto M, Ichijo H, Kojima I, Takeda J, Nogami H (2000) Regional expression of a gene encoding a neuron-specific Na(+)-dependent inorganic phosphate cotransporter (DNPI) in the rat forebrain. Brain Res Mol Brain Res 83:34–43
Holtman JR Jr, Dick TE, Berger AJ (1986) Involvement of serotonin in the excitation of phrenic motoneurons evoked by stimulation of the raphe obscurus. J Neurosci 6:1185–1193
Hrabovszky E, Kallo I, Turi GF, May K, Wittmann G, Fekete C, Liposits Z (2006) Expression of vesicular glutamate transporter-2 in gonadotrope and thyrotrope cells of the rat pituitary. Regulation by estrogen and thyroid hormone status. Endocrinology 147:3818–3825
Hull CD, Bernardi G, Buchwald NA (1970) Intracellular responses of caudate neurons to brain stem stimulation. Brain Res 22:163–179
Hull CD, Bernardi G, Price DD, Buchwald NA (1973) Intracellular responses of caudate neurons to temporally and spatially combined stimuli. Exp Neurol 38:324–336
Hur EE, Zaborszky L (2005) Vglut2 afferents to the medial prefrontal and primary somatosensory cortices: a combined retrograde tracing in situ hybridization. J Comp Neurol 483:351–373
Johnson MD (1994a) Synaptic glutamate release by postnatal rat serotonergic neurons in microculture. Neuron 12:433–442
Johnson MD (1994b) Electrophysiological and histochemical properties of postnatal rat serotonergic neurons in dissociated cell culture. Neuroscience 63:775–787
Johnson MD, Yee AG (1995) Ultrastructure of electrophysiologically characterized synapses formed by serotonergic raphe neurons in culture. Neuroscience 67:609–623
Jomphe C, Bourque MJ, Fortin GD, St-Gelais F, Okano H, Kobayashi K, Trudeau LE (2005) Use of TH-EGFP transgenic mice as a source of identified dopaminergic neurons for physiological studies in postnatal cell culture. J Neurosci Methods 146:1–12

Joyce MP, Rayport S (2000) Mesoaccumbens dopamine neuron synapses reconstructed in vitro are glutamatergic. Neuroscience 99:445–456
Kaneko T, Akiyama H, Nagatsu I, Mizuno N (1990) Immunohistochemical demonstration of glutaminase in catecholaminergic and serotoninergic neurons of rat brain. Brain Res 507:151–154
Kaneko T, Fujiyama F (2002) Complementary distribution of vesicular glutamate transporters in the central nervous system. Neurosci Res 42:243–250
Kaneko T, Fujiyama F, Hioki H (2002) Immunohistochemical localization of candidates for vesicular glutamate transporters in the rat brain. J Comp Neurol 444:39–62
Kaneko T, Mizuno N (1994) Glutamate-synthesizing enzymes in GABAergic neurons of the neocortex: a double immunofluorescence study in the rat. Neuroscience 61:839–1849
Kang TC, Kim DS, Kwak SE, Kim JE, Kim DW, Kang JH, Won MH, Kwon OS, Choi SY (2005) Valproic acid reduces enhanced vesicular glutamate transporter immunoreactivities in the dentate gyrus of the seizure prone gerbil. Neuropharmacology 49:912–921
Kashani A, Betancur C, Giros B, Hirsch E, El Mestikawy S (2007a) Altered expression of vesicular glutamate transporters VGLUT1 and VGLUT2 in Parkinson disease. Neurobiol Aging 28:568–578
Kashani A, Lepicard E, Poirel O, Videau C, David JP, Fallet-Bianco C, Simon A, Delacourte A, Giros B, Epelbaum J, Betancur C, El Mestikawy S (2007b) Loss of VGLUT1 and VGLUT2 in the prefrontal cortex is correlated with cognitive decline in Alzheimer disease. Neurobiol Aging
Kawano M, Kawasaki A, Sakata-Haga H, Fukui Y, Kawano H, Nogami H, Hisano S (2006) Particular subpopulations of midbrain and hypothalamic dopamine neurons express vesicular glutamate transporter 2 in the rat brain. J Comp Neurol 498:581–592
Kawasaki A, Hoshi K, Kawano M, Nogami H, Yoshikawa H, Hisano S (2005) Up-regulation of VGLUT2 expression in hypothalamic-neurohypophysial neurons of the rat following osmotic challenge. Eur J Neurosci 22:672–680
Kawasaki A, Shutoh F, Nogami H, Hisano S (2006) VGLUT2 expression is up-regulated in neurohypophysial vasopressin neurons of the rat after osmotic stimulation. Neurosci Res.
Kim DS, Kwak SE, Kim JE, Won MH, Choi HC, Song HK, Kwon OS, Kim YI, Choi SY, Kang TC (2005) Bilateral enhancement of excitation via up-regulation of vesicular glutamate transporter subtype 1, not subtype 2, immunoreactivity in the unilateral hypoxic epilepsy model. Brain Res 1055:122–130
Kim JE, Kim DS, Kwak SE, Choi HC, Song HK, Choi SY, Kwon OS, Kim YI, Kang TC (2007) Anti-glutamatergic effect of riluzole: comparison with valproic acid. Neuroscience 147:136–145
Kirvell SL, Esiri M, Francis PT (2006) Down-regulation of vesicular glutamate transporters precedes cell loss and pathology in Alzheimer's disease. J Neurochem 98:939–950
Kitai ST, Wagner A, Precht W, Ono T (1975) Nigro-caudate and caudato-nigral relationship: an electrophysiological study. Brain Res 85:44–48
Lapish CC, Seamans JK, Judson Chandler L (2006) Glutamate-dopamine cotransmission and reward processing in addiction. Alcohol Clin Exp Res 30:1451–1465
Liu G, Choi S, Tsien RW (1999) Variability of neurotransmitter concentration and nonsaturation of postsynaptic AMPA receptors at synapses in hippocampal cultures and slices. Neuron 22:395–409
Liu RH, Fung SJ, Reddy VK, Barnes CD (1995) Localization of glutamatergic neurons in the dorsolateral pontine tegmentum projecting to the spinal cord of the cat with a proposed role of glutamate on lumbar motoneuron activity. Neuroscience 64:193–208
Margolis EB, Lock H, Hjelmstad GO, Fields HL (2006) The ventral tegmental area revisited: is there an electrophysiological marker for dopaminergic neurons? J Physiol 577:907–924

Mashima H, Yamada S, Tajima T, Seno M, Yamada H, Takeda J, Kojima I (1999) Genes expressed during the differentiation of pancreatic AR42J cells into insulin-secreting cells. Diabetes 48:304–309

Mendez JA, Bourque M-J, Bourdeau ML, Danik M, Williams S, Lacaille J-C, Trudeau L-E (2008) Contact-dependent regulation of the neurotransmitter phenotype of dopamine neurons. J Neurosci 28:6309–6318.

Mendez JA, Bourque M-J, Trudeau L-E (2005) Evidence for the expression of VGLUT2 mRNA in acutely dissociated TH-GFP mouse dopamine neurons. Abstract Viewer/Itinerary Planner. Washington, DC: Society for Neuroscience. Online. 155. 3. 2005.

Montana V, Ni Y, Sunjara V, Hua X, Parpura V (2004) Vesicular glutamate transporter-dependent glutamate release from astrocytes. J Neurosci 24:2633–642

Morimoto R, Hayashi M, Yatsushiro S, Otsuka M, Yamamoto A, Moriyama Y (2003) Co-expression of vesicular glutamate transporters (VGLUT1 and VGLUT2) and their association with synaptic-like microvesicles in rat pinealocytes. J Neurochem 84:382–391

Moutsimilli L, Farley S, Dumas S, El Mestikawy S, Giros B, Tzavara ET (2005) Selective cortical VGLUT1 increase as a marker for antidepressant activity. Neuropharmacology 49:890–900

Nakamura K, Wu SX, Fujiyama F, Okamoto K, Hioki H, Kaneko T (2004) Independent inputs by VGLUT2- and VGLUT3-positive glutamatergic terminals onto rat sympathetic preganglionic neurons. Neuroreport 15:431–436

Ni B, Rosteck PR Jr, Nadi NS, Paul SM (1994) Cloning and expression of a cDNA encoding a brain-specific Na(+)- dependent inorganic phosphate cotransporter. Proc Natl Acad Sci U S A 91:5607–5611

Nicholas AP, Pieribone VA, Arvidsson U, Hokfelt T (1992) Serotonin-, substance P- and glutamate/aspartate-like immunoreactivities in medullo-spinal pathways of rat and primate. Neuroscience 48:545–459

Ottersen OP, Storm-Mathisen J (1984) Glutamate- and GABA-containing neurons in the mouse and rat brain, as demonstrated with a new immunocytochemical technique. J Comp Neurol 229:374–392

Park MR, Gonzales-Vegas JA, Kitai ST (1982) Serotonergic excitation from dorsal raphe stimulation recorded intracellularly from rat caudate-putamen. Brain Res 243:49–58

Pickel VM, Beckley SC, Joh TH, Reis DJ (1981) Ultrastructural immunocytochemical localization of tyrosine hydroxylase in the neostriatum. Brain Res 225:373–385

Plenz D, Kitai ST (1996) Organotypic cortex-striatum-mesencephalon cultures: the nigrostriatal pathway. Neurosci Lett 209:177–180

Robelet S, Melon C, Guillet B, Salin P, Kerkerian-Le Goff L (2004) Chronic L-DOPA treatment increases extracellular glutamate levels and GLT1 expression in the basal ganglia in a rat model of Parkinson's disease. Eur J Neurosci 20:1255–1266

Rosin DL, Chang DA, Guyenet PG (2006) Afferent and efferent connections of the rat retrotrapezoid nucleus. J Comp Neurol 499:64–89

Sawamoto K, Nakao N, Kobayashi K, Matsushita N, Takahashi H, Kakishita K, Yamamoto A, Yoshizaki T, Terashima T, Murakami F, Itakura T, Okano H (2001) Visualization, direct isolation, and transplantation of midbrain dopaminergic neurons. Proc Natl Acad Sci U S A 98:6423–6428

Schafer MK, Varoqui H, Defamie N, Weihe E, Erickson JD (2002) Molecular cloning and functional identification of mouse vesicular glutamate transporter 3 and its expression in subsets of novel excitatory neurons. J Biol Chem 277:50734–48

Smith RE, Haroutunian V, Davis KL, Meador-Woodruff JH (2001) Vesicular glutamate transporter transcript expression in the thalamus in schizophrenia. Neuroreport 12:2885–2887

Somogyi J, Baude A, Omori Y, Shimizu H, El Mestikawy S, Fukaya M, Shigemoto R, Watanabe M, Somogyi P (2004) GABAergic basket cells expressing cholecystokinin contain vesicular glutamate transporter type 3 (VGLUT3) in their synaptic terminals in hippocampus and isocortex of the rat. Eur J Neurosci 19:552–569

Storm-Mathisen J, Leknes AK, Bore AT, Vaaland JL, Edminson P, Haug FM, Ottersen OP (1983) First visualization of glutamate and GABA in neurones by immunocytochemistry. Nature 301:517–520

Stornetta RL, Sevigny CP, Guyenet PG (2002a) Vesicular glutamate transporter DNPI/VGLUT2 mRNA is present in C1 and several other groups of brainstem catecholaminergic neurons. J Comp Neurol 444:191–206

Stornetta RL, Sevigny CP, Schreihofer AM, Rosin DL, Guyenet PG (2002b) Vesicular glutamate transporter DNPI/VGLUT2 is expressed by both C1 adrenergic and nonaminergic presympathetic vasomotor neurons of the rat medulla. J Comp Neurol 444:207–220

Sulzer D, Joyce MP, Lin L, Geldwert D, Haber SN, Hattori T, Rayport S (1998) Dopamine neurons make glutamatergic synapses in vitro. J Neurosci 18:4588–4602

Takamori S, Rhee JS, Rosenmund C, Jahn R (2000) Identification of a vesicular glutamate transporter that defines a glutamatergic phenotype in neurons. Nature 407:189–194

Talbot K, Eidem WL, Tinsley CL, Benson MA, Thompson EW, Smith RJ, Hahn CG, Siegel SJ, Trojanowski JQ, Gur RE, Blake DJ, Arnold SE (2004) Dysbindin-1 is reduced in intrinsic, glutamatergic terminals of the hippocampal formation in schizophrenia. J Clin Invest 113:1353–1363

Tordera RM, Pei Q, Sharp T (2005) Evidence for increased expression of the vesicular glutamate transporter, VGLUT1, by a course of antidepressant treatment. J Neurochem 94:875–883

Trudeau LE (2004) Glutamate co-transmission as an emerging concept in monoamine neuron function. J Psychiatry Neurosci 29:296–310

Varoqui H, Schafer MK, Zhu H, Weihe E, Erickson JD (2002) Identification of the differentiation-associated Na+/PI transporter as a novel vesicular glutamate transporter expressed in a distinct set of glutamatergic synapses. J Neurosci 22:142–155

Voorn P, Jorritsma-Byham B, Van Dijk C, Buijs RM (1986) The dopaminergic innervation of the ventral striatum in the rat: a light- and electron-microscopical study with antibodies against dopamine. J Comp Neurol 251:84–99

Wilson CJ, Chang HT, Kitai ST (1982) Origins of postsynaptic potentials evoked in identified rat neostriatal neurons by stimulation in substantia nigra. Exp Brain Res 45:157–167

Yamaguchi T, Sheen W, Morales M (2007) Glutamatergic neurons are present in the rat ventral tegmental area. Eur J Neurosci 25:106–118

Yamashita T, Ishikawa T, Takahashi T (2003) Developmental increase in vesicular glutamate content does not cause saturation of AMPA receptors at the calyx of held synapse. J Neurosci 23:3633–638

Zhang Q, Pangrsic T, Kreft M, Krzan M, Li N, Sul JY, Halassa M, Van Bockstaele E, Zorec R, Haydon PG (2004) Fusion-related release of glutamate from astrocytes. J Biol Chem 279:12724–12733

Chapter 9
Dopamine and Serotonin Crosstalk Within the Dopaminergic and Serotonergic Systems

Fu-Ming Zhou and John A. Dani

Abstract Dopamine (DA) and serotonin (5-hydroxytryptamine, 5-HT) fibers broadly innervate the brain. DA transporters (DATs) and 5-HT transporters (SERTs) determine the temporal and spatial extent of the signals, but the transporters are not completely selective. Under circumstances that elevate extracellular 5-HT (such as treatment with antidepressants that block SERTs), DATs expressed on the dense DA terminals of the striatum uptake 5-HT. Subsequently, 5-HT enters DA synaptic vesicles, and DA and 5-HT are co-released. Circumstances also can favor DA loading into 5-HT neurons and terminals. For example, during L-dopa treatment of Parkinson's disease, 5-HT neurons uptake L-dopa that is subsequently converted to DA, leading to co-release. DA and 5-HT neurons promiscuously uptake each other's neurotransmitter and co-release occurs under certain conditions. This process induces DA and 5-HT co-signaling mechanisms that have important implications for neuropsychiatric disorders and their treatments.

9.1 Introduction

Dopamine (DA) is synthesized from the dietary amino acid tyrosine (Fig. 9.1a). In late 1950 s, it was shown that DA serves important functional roles in the basal ganglia (Carlsson et al. 1958; Bertler and Rosengren 1959). The presence and organization of DA neurons and their projections in the brain were established in the early 1960 s (Carlsson 2001, 2002a,b). Using the then newly invented formaldehyde fluorescence method, the most important DA cell groups and their main projections were mapped (Carlsson et al. 1962; Falck 1962; Dahlstrom and Fuxe 1964). These original findings were confirmed, extended, and refined with more sensitive immunohistochemical methods (Bjorklund and Lindvall 1984, 1986).

J.A. Dani (✉)
Department of Neuroscience, Menninger Department of Psychiatry and Behavioral Science, Baylor College of Medicine, Houston, TX 77030
e-mail: jdani@bcm.tmc.edu

R. Gutierrez (ed.), *Co-Existence and Co-Release of Classical Neurotransmitters*,
DOI 10.1007/978-0-387-09622-3_9,

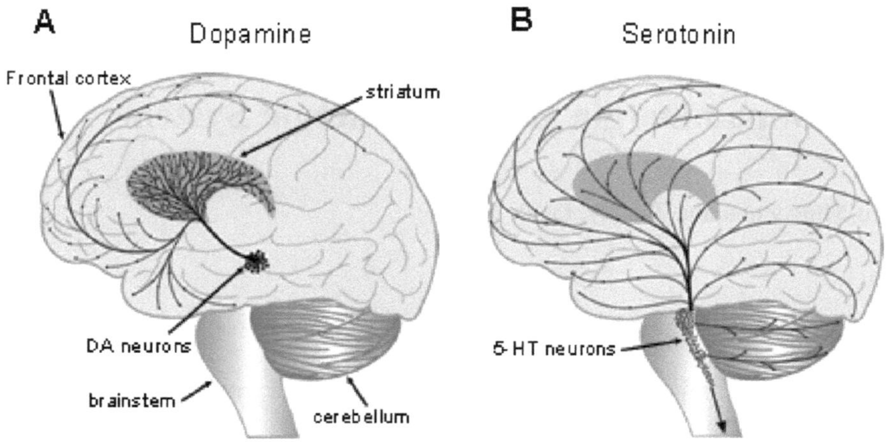

Fig. 9.1 A simplified representation of DA and 5-HT projections in the human brain. **(a)** The cell bodies of the mesoforebrain dopamine system are located mainly in the substantia nigra pars compacta (SNc, A9) and the ventral tegmental area (VTA, A10) of the midbrain. SNc dopamine neurons mainly project to the dorsal striatum, and VTA dopamine neurons mainly project to the mesocorticolimbic areas such as the nucleus accumbens, amygdala, and prefrontal cortex. **(b)** The cell bodies of the serotonin system are located in the raphe nuclei. These serotonin neurons project to virtually all brain areas in a relatively even manner, though some regional variation exists. *From Zhou and Dani*, 2006. (*See* Color Plate 8)

Soon after its discovery in the basal ganglia, DA was proposed to be involved in motor control and Parkinson's disease (Carlsson 1959). This proposal gained rapid support in the 1960 s from evidence that DA neurons were significantly lost in Parkinson's disease (see Hornykiewicz 2001, 2002). Following these discoveries, DA replacement therapy with L-3,4-dihydroxyphenylalanine (L-dopa, an immediate DA precursor) became the standard treatment for Parkinson's disease (Hornykiewicz 2001, 2002; Carlsson 2002a; Olanow 2004). Almost simultaneously, it was hypothesized that the DA system was involved in the pathophysiology and treatment of psychosis and schizophrenia (Carlsson and Lindqvist 1963; van Rossum 1966). By the mid-1970 s, the antipsychotic drugs were found to be potent inhibitors of the DA D_2 class of receptors (Seeman and Lee 1975; Creese et al. 1976; Seeman 1980). These findings established the mesoforebrain DA system as a site of pathophysiological changes and as a key target for treatment (Kapur and Mamo 2003; Goto and Grace 2007; Guillin et al. 2007).

Serotonin (5-HT) is synthesized from the dietary amino acid tryptophan (Fig. 9.1b). Its presence in the brain was discovered in the 1950 s (Twarog and Page 1953; Pletscher et al. 1956; see Whitaker-Azmitia 1999 and Carlsson 2001 for historical accounts). Using the formaldehyde fluorescence method, 5-HT-containing neurons were mapped to raphe nuclei in the brainstem at the same time DA neurons were localized in the midbrain (Carlsson et al. 1962; Dahlstrom and Fuxe 1964; Anden et al. 1965). The broad, diffusive 5-HT projections in the brain were revealed in greater detail in the late 1970 s using more sensitive immunohistochemical methods (Steinbusch et al. 1978; Steinbusch 1981).

Research on brain 5-HT was also intimately related to the birth and development of modern neuropsychopharmacology. Soon after its discovery in the brain in the early 1950 s, the all 5-HT system was quickly proposed to be linked to the behavioral calming or tranquillizing effects of reserpine that was used to treat aspects of agitation in psychotic patients (Brodie et al. 1955; Aghajanian and Marek 1999). Reserpine is now known to inhibit the vesicular monoamine transporter (VMAT), thereby releasing and depleting 5-HT and other monoamine storage (Pletscher et al. 1956; see Yelin and Schuldiner 2002 for review). The brain 5-HT system was shown to be a critical target of hallucinogens such as lysergic acid diethylamide (LSD) (Woolley and Shaw 1954; Shore et al. 1955; Pletscher et al. 1956; Aghajanian and Marek 1999). The success of selective 5-HT reuptake inhibitors as antidepressants further demonstrates the importance of the 5-HT system in brain functions (Wong et al. 1995, 2005; Carlsson 2001, 2002b; Baldessarini 2006; Zhou and Dani 2006).

9.2 Mesostriatal Dopamine System

9.2.1 Basic Anatomy

In mammals, the main DA cell groups are located in the midbrain. Those in the substantia nigra (A9 group) and ventral tegmental area (A10 group) are relatively large and make prominent projections, and their functions are altered in brain disorders (Grace 1991; Damier et al. 1999; Obeso et al. 2000; Hornykiewicz 2001; Carlsson 2002a); Schultz 2002, 2007; Olanow 2004; Hyman et al. 2006; Tanaka 2006; Di Chiara and Bassareo 2007; Goto and Grace 2007; Guillin et al. 2007; Sulzer 2007. Each side of the midbrain DA areas has about 12,000 DA neurons in mice, around 22,000 in rats, about 240,000 in monkeys, and about 500,000 in young humans (German and Manaye 1993; Nelson et al. 1996; Oorschot 1996; Emborg et al. 1998; Chu et al. 2002; see Bjorklund and Lindvall 1984, 1986; Bentivoglio and Morelli 2005; Bjorklund and Dunnett 2007 for review).

Midbrain DA neurons in the substantia nigra pars compacta (SNc) and the ventral tegmental area (VTA) project to many targets (Fig. 9.1a), but most densely innervate the striatum, forming the mesostriatal DA system (Bjorklund and Lindvall 1984, 1986; Haber et al. 2000; Prensa and Parent 2001; Haber 2003; Bentivoglio and Morelli 2005; Bjorklund and Dunnett 2007). The striatum is often divided into the dorsal and ventral portions. The dorsal striatum comprises the caudate nucleus and the putamen. The ventral striatum includes the ventral conjunction of the caudate and putamen, the nucleus accumbens (NAc), and portions of the olfactory tubercle. DA neurons of the VTA and in a minor portion of the SNc also project to limbic and cortical areas (such as the amygdala, the entorhinal cortex, and prefrontal cortex), forming the mesolimbocortical DA system (Bjorklund and Lindvall 1984, 1986; Haber 2003; Bentivoglio and Morelli 2005; Bjorklund and Dunnett 2007).

Quantitative anatomical analyses have estimated that in the striatum the average density of DA synapses or varicosities is $0.1/\mu m^3$ out of the total synapse or varicosity density of $1.2/\mu m^3$ (Doucet et al. 1986; Groves et al. 1994; Descarries and Mechawar 2000; Arbuthnott and Wickens 2007). The average distance between 2 DA varicosities on a DA axon is about 4 μm while the average nearest-neighbor distance in a volume between DA varicosities is 1.2 μm because of the distribution of DA varicosities arising from different neighboring DA axons in 3D space (Arbuthnott and Wickens 2007). Because DA often spills over out of synapses or diffuses from release sites for about 5 μm before being uptaken (Peters and Michael 2000; Cragg and Rice 2004), such a high packing density of DA varicosities probably ensures that each point in the striatum receives DA from multiple release sites (Fig. 9.2).

9.2.2 DA Receptors in the Mesostriatal DA System

Molecular cloning and functional studies have identified five subtypes of DA receptors that all couple to G-proteins (Neve and Neve 1997; Missale et al. 1998; Bentivoglio and Morelli 2005; Herve and Girault 2005; Alexander et al.

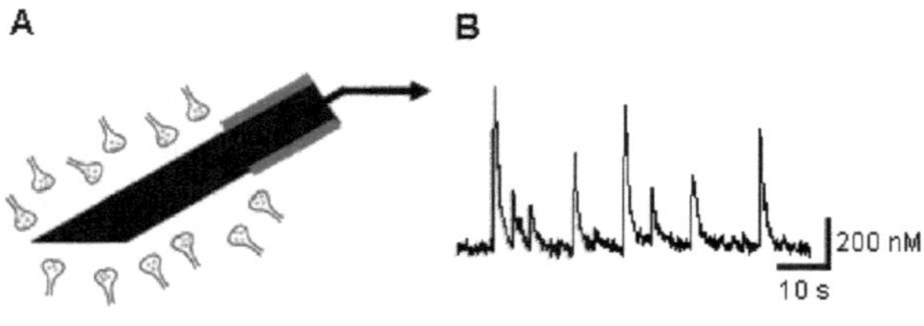

Fig. 9.2 DA released measured with carbon-fiber electrodes using fast-scan cyclic voltammetry. **(a)** Illustration of carbon-fiber electrode used for fast-scan cyclic voltammetry recordings of vesicular DA release in the striatum. Not drawn to scale. **(b)** The trace shows action potential-dependent spontaneous vesicular DA release recorded in the striatum using voltammetry. Action potential-independent single DA vesicle release events (minis) are below the detection sensitivity for voltammetry. Thus, these relatively large DA spikes arise from synchronized release of multiple axon terminals and varicosities originating from the same parent DA axon fibers in the vicinity of the carbon-fiber electrode. *Unpublished data of Zhou and Dani.* (*See* Color Plate 8)

2007). D_1 and D_5 subtypes display the pharmacological and physiological properties of the original D_1 classification (Kebabian and Calne 1979) and are called D_1-type receptors. D_2, D_3, and D_4 subtypes correspond to the original D_2 classification and are called D_2-type receptors. The D_1-type receptors classically couple to $G_{s/olf}$ and activate adenylyl cyclase, and the D_2-type receptors couple to $G_{i/o}$ and inhibit adenylyl cyclase (Kebabian and Calne 1979; Missale et al. 1998; Neve et al. 2004; Alexander et al. 2007). Active ongoing research continues to reveal other G-protein pathways. For instance, D_1-type receptors couple to Gq and activate phospholipase C (PLC), thereby stimulating inositol triphosphate (IP3) and diacylglycerol (DAG)-mediated signaling mechanisms (Wang et al. 1995; Jin et al. 2001; Mannoury la Cour et al. 2007; Rashid et al. 2007a). The DA-induced depolarization in cortical interneurons and the DA-induced increase in spiking in medium spiny neurons in the nucleus accumbens involve co-activation of D_1-type and D_2-type receptors (Hopf et al. 2003; Wu and Hablitz 2005; Inoue et al. 2007). D_1 and D_2 receptors also synergistically activate G_q and stimulate a PLC-mediated Ca^{2+} signaling mechanism (Lee et al. 2004; Rashid et al. 2007a,b).

Within the midbrain DA regions, γ-aminobutyric acid (GABA) and glutamate afferent terminals express D_1 receptors that regulate transmitter release (Cameron and Williams 1993; Levey et al. 1993; Missale et al. 1998; Miyazaki and Lacey 1998; Ibanez-Sandoval et al. 2006). In both rodents and primates, including humans, DA neurons in the SNc and the VTA strongly express D_2 receptors (Levey et al. 1993; Yung et al. 1995; Ariano 1997; Khan et al. 1998). These expressed D_2 receptors serve as inhibitory autoreceptors that increase a K^+ conductance (Lacey et al. 1987, 1988; Lacey 1993; White 1996; Mercuri et al. 1997; Diana and Tepper 2002).

D_1 and D_2 receptors are highly expressed in the neuronal populations in the striatum and nucleus accumbens (Meador-Woodruff et al. 1991; Mansour et al. 1992; Levey et al. 1993; Bergson et al. 1995; Yung et al. 1995; Surmeier et al. 1996, 2007; Ariano 1997). While some evidence suggests a certain level of D_1 and D_2 co-expression (Aizman et al. 2000), there is general segregation among the cell types (Gerfen 2000; Surmeier et al. 2007). For example, in rodents, the striatopallidal medium spiny projection neurons express nearly exclusively D_2 receptors, and the striatonigral medium spiny projection neurons express D_1 receptors (Gerfen 1992; Surmeier et al. 1996; Gong et al. 2003; Day et al. 2006; Gerfen 2006). D_2 receptors also are widely expressed on DA, glutamate, and other afferent axon terminals in the striatum and nucleus accumbens where they inhibit transmitter release upon activation (Hersch et al. 1995; Yung et al. 1995; Khan et al. 1998; Missale et al. 1998; Usiello et al. 2000; Schmitz et al. 2003; Bamford et al. 2004; Wu et al. 2006).

Acting on different types of DA receptors, DA affects neuronal activity by regulating intrinsic membrane ion channels, neurotransmitter receptors, and neurotransmitter release (Nicola et al. 2000; Neve et al. 2004; Surmerier et al. 2007). Generally, activation of the D_1-type receptors tends to increase neuronal excitability while activation of D_2-type receptors often tends to exert the opposite effects (Gonon 1997; West and Grace 2002), but more complex effects via receptor synergism are possible and more common than initially anticipated (Hopf et al. 2003; Inoue et al. 2007; Rashid et al. 2007a,b). Activation of D_1 receptors enhances glutamatergic synaptic transmission by increasing the functionality of glutamate receptors (Snyder et al. 2000; Dunah and Standaert 2001; Mangiavacchi and Wolf 2004; Sun et al. 2005; Gao et al. 2006; Hallett et al. 2006). When the medium spiny neurons in the striatum and nucleus accumbens are in the so called up-state with the membrane potential around –60 mV, D_1 receptor activation increases cell excitability by concerted effects on multiple intrinsic channels (Hernandez-Lopez et al. 2000). When the cell is in the down-state with the membrane potential around –80 mV, D_1 receptor activation may have the opposite effect on cell excitability (Hernandez-Lopez et al. 2000; Surmeier et al. 2007). On the other hand, D_2 receptor activation often reduces glutamate receptor function (Liu et al. 2006) and membrane excitability by inhibiting Na^+ and Ca^{2+} channels and opening K^+ channels that tend to keep the cell near its resting potential (Greif et al. 1995; Lin et al. 1996; Cantrell and Catterall 2001; Dong et al. 2004; Hu et al. 2005; Nasif et al. 2005; Perez et al. 2006; Peterson et al. 2006; Surmeier et al. 2007).

D_3, D_4, and D_5 receptors are also expressed in mesoforebrain areas, although at lower levels compared with D_1 and D_2 receptors (Sokoloff et al. 1990; Bergson et al. 1995; Surmeier et al. 1996; Ariano 1997; Ciliax et al. 2000; Diaz et al. 2000; Khan et al. 2000; Oak et al. 2000; Wong and Van Tol 2003). D_3 and D_4 receptors often use the same signaling pathway used by D_2 receptors, leading to similar cellular effects (Surmeier et al. 1996). D_5 receptors induce excitation/depolarization in striatal cholinergic interneurons and fast-spiking GABA interneurons (Centonze et al. 2003) and in the subthalamic neurons (Baufreton et al. 2003).

9.2.3 DA Receptor Modulation of 5-HT Neurons and 5-HT Signaling

Anatomical studies indicate that midbrain nigral and VTA DA neurons provide a descending innervation to the dorsal raphe nucleus (Kalen et al. 1988; Peyron et al. 1995; Kitahama et al. 2000; Lee et al. 2005). Raphe 5-HT neurons express D_2-type receptors (Mansour et al. 1990; Suzuki et al. 1998). DA ligands that activate D_2-type receptors also increase 5-HT release (Ferre and Artigas 1993; Ferre et al. 1994; Matsumoto et al. 1996; Pineyro and Blier 1999; Martin-Ruiz et 2al. 2001). More recent physiological studies show that DA can also directly excite dorsal raphe 5-HT neurons by activating D_2-type receptors coupled to a novel PLC-mediated signaling mechanism that eventually induces a mixed, depolarizing cation current (Haj-Dahmane 2001; Aman et al. 2007).

There also are relatively scattered DA neurons in the hypothalamus and periventricular area that form a dense local DA plexus and innervate the pituitary (Bjorklund and Lindvall 1986). 5-HT terminals also innervate these brain areas, and evidence suggests that DA terminals in these brain areas cross-uptake 5-HT via DATs (Vanhatalo and Soinila 1994, 1995, 1998).

An inherent potential for interaction between DA and 5-HT signaling arises within the intracellular cascades of the G-protein coupled receptors. DA-mediated signaling pathways are diverse and suggest DA and 5-HT may interaction along these intracellular pathways because most 5-HT receptors also are G-protein coupled (Svenningsson et al. 2002b, 2004; Neve et al. 2004).

9.2.4 DA Transporters in the Mesostriatal DA System

Dopamine transporters (DATs) are the primary route for returning extracellular DA molecules to their nerve terminals and for maintaining proper extracellular DA concentration profiles (Povlock and Amara 1997; Gainetdinov et al. 1998; Torres et al. 2003; Cragg and Rice 2004). DATs belong to a family of plasma membrane neurotransmitter transporters that are Na^{+}/Cl^{-} dependent. In rodents and primates, including humans, DAT specific radioligands

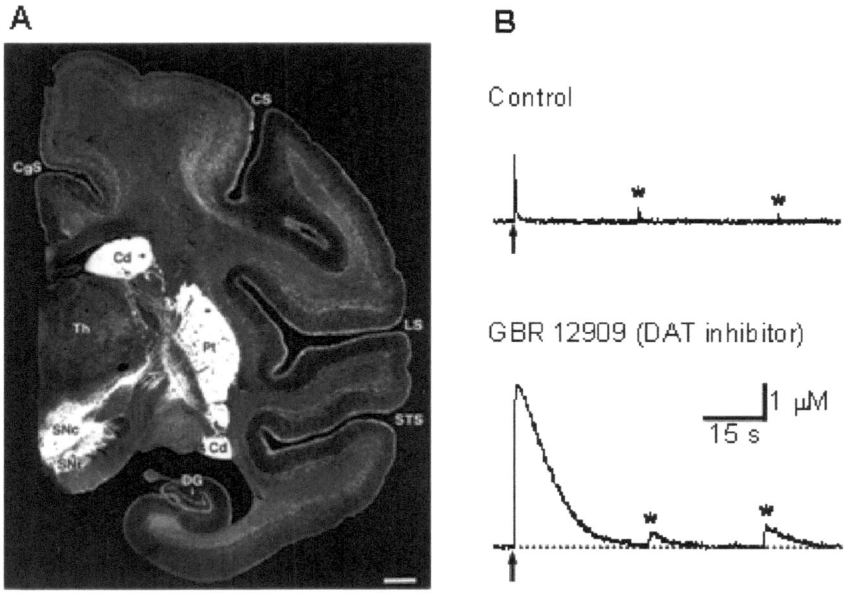

Fig. 9.3 The spatial extent of DA signals is profoundly affected by DAT-mediated reuptake. **(a)** The dense expression of DATs in the nigrostriatal DA system are revealed by DAT antibody immunostaining. From Lewis et al. (2001) with permission. **(b)** Extracellular DA signals measured by cyclic voltammetry. GBR12909 (1 μM, a DAT inhibitor) increases and extends the extracellular DA concentration profile arising from electrically evoked vesicular DA release (arrows), indicating the profound effect of DATs on the profile of the DA signal. Asterisks (*) indicate spontaneous DA release. The recording carbon fiber was 10 μm in diameter and about 60 μm in length. *Unpublished data of Zhou and Dani*

(Richfield 1991; Marcusson et al. 1988; Mennicken et al. 1992; Coulter et al. 1996; Chen et al. 1997) and DAT antibodies (Ciliax et al. 1995; Freed et al. 1995; Sesack et al. 1998; Lewis et al. 2001; Sesack 2002) have revealed dense DAT expression in the mesostriatal DA system (Fig. 9.3a). DATs are expressed on the axons, dendrites, and cell bodies of midbrain DA neurons (Nirenberg et al. 1996; Nirenberg et al. 1997a,b; Ciliax et al. 1999), including the numerous varicosities located along the axons (Hersch et al. 1997; Sesack et al. 1998).

Because of the high packing density of DA axons and varicosities, each point in the striatum is endowed with adequate DAT-mediated capacity to remove DA and to shape the DA signal spatially and temporally. By limiting the spatial and temporal spread of DA from its release sites (Cragg and Rice 2004), DATs help control the extent, intensity, and duration of the DA signal. Pharmacological or genetic inactivation of DATs drastically increases the amplitude and prolongs the duration of the DA signal (Fig. 9.3b) and also decreases DA tissue content (Jones et al. 1998a,b; Torres et al. 2003; Zhou et al. 2005).

9.3 Forebrain Serotonin System

9.3.1 HT Innervation

The majority of 5-HT neurons are located in the raphe nuclei, extending from caudal midbrain to pontine tegmentum in the brainstem (Steinbusch 1981; Wiklund et al. 1981; Leger and Wiklund 1982; Tork 1990; Baumgarten and Grozdanovic 1997; Descarries and Mechawar 2000; Hornung 2003). In human brainstem, there are nearly 300,000 5-HT neurons providing ascending 5-HT fibers that innervate virtually all forebrain structures including midbrain DA cell areas and striatal DA terminal areas (Fig. 9.1b) (Steinbusch 1981; Tork 1990; Baker et al. 1991; Jacobs and Azmitia 1992; Mrini et al. 1995; Smiley and Goldman-Rakic 1996; Baumgarten and Grozdanovic 1997; Charara and Parent 1998). The dorsal raphe nucleus contains the largest 5-HT neuron group, numbering about 165,000 5-HT neurons in humans, that projects to the forebrain. The dorsal raphe nucleus is also the dominant origin of the mesostriatal 5-HT innervation (Azmitia and Segal 1978; Moore et al. 1978; Okumura et al. 2000). In contrast to the mesoforebrain DA system, which more densely innervates the striatum than any other brain area, the 5-HT innervation in the brain is diffuse and relatively more evenly distributed, though variations in density exist among different brain areas. There is also a descending 5-HT projection to the brainstem and spinal cord that regulates spinal neurons and is important for functions such as respiration, motor control, ejaculation, and pain (Garraway et al. 2001; de Jong et al. 2006; Paterson et al. 2006; Lopez-Garcia 2006; Zhong et al. 2006).

Like DA neurons, 5-HT neurons often fire spontaneous action potentials with a tonic frequency of a few Hz (Jacobs and Azmitia 1992). The firing frequency of 5-HT neurons is influenced by 5-HT_{1A} autoreceptors and by synaptic inputs from cortical areas, DA areas, the locus ceruleus, and local GABAergic interneurons (Pineyro and Blier 1999; Stamford et al. 2000).

Quantitative ultrastructural studies have estimated 5-HT terminal density in mesostriatal DA areas in mammalian animals (Mori et al. 1985, 1987; Lavoie and Parent 1990). In the rat, the density of 5-HT axonal varicosities is about $9 \times 10^6/mm^3$ in the SN pars reticulata (SNr) and is about $6 \times 10^6/mm^3$ in the SNc (Herve et al. 1987; Moukhles et al. 1997). In the striatum, the mean density is around 3×10^6 varicosities/mm^3 with the ventro-medial striatum having a higher density than the dorsal striatum (Soghomonian et al. 1987; Descarries and Mechawar 2000; Baumgarten and Grozdanovic 1997). The globus pallidus has 4.5×10^6 5-HT varicosities/mm^3. On average, the ratio of 5-HT terminals to DA terminals in the striatum is about 1:20 (Feuerstein et al. 1986; Baumgarten and Grozdanovic 1997; Descarries and Mechawar 2000; Zhou et al. 2005).

In these DA areas, 5-HT axon terminals form both synaptic and non-synaptic contacts (Van Bockstaele et al. 1994, 1996; Moukhles et al. 1997; Okumura et al. 2000). In SNc, 5-HT axonal terminals make both synaptic and non-synaptic contacts with DA neurons whereas they make mostly synaptic contacts with GABA neurons in SNr (Moukhles et al. 1997). In the VTA, 5-HT axon terminals are in close contact with somata and large dendrites of DA neurons (Van Bockstaele et al. 1994). In the rat nucleus accumbens, 5-HT terminals are in apposition to DA axon terminals (Van Bockstaele and Pickel 1993). 5-HT terminals are also in close contact with the dendrites and axon terminals of striatal GABA neurons, striatal cholinergic neurons, and presumed glutamatergic synapses on striatal neurons (Van Bockstaele et al. 1996; Okumura et al. 2000). These synaptic and non-synaptic contacts between DA and 5-HT terminals provide an anatomical basis for the two systems to interact with each other.

9.3.2 5-HT Transporters

After 5-HT is released from 5-HT axon terminals, serotonin transporters (SERTs) normally return 5-HT from the extracellular space back into their nerve terminals, thereby, regulating the spatial and temporal profile of extracellular 5-HT signaling (Bunin and Wightman 1998; Torres et al. 2003). SERTs reuptake 5-HT with a Km of approximately 100 nM in synaptosomes or brain slices (Shaskan and Snyder 1970; O'Reilly and Reith 1988; Wong et al. 1995). Cloned SERTs take up 5-HT with a Km of approximately 500 nM (Ramamoorthy et al. 1993; Miller et al. 2001). Like the 5-HT innervation in the brain, SERT expression in the striatum is far less dense than DAT expression (Nirenberg et al. 1996, 1997a; Sur et al. 1996; Chen and Reith 1997; Pickel and Chan 1999). Consistent with the 5-HT innervation, SERT expression in the striatum is rather diffuse and evenly distributed (Coulter et al. 1996; Chen et al. 1997; Nirenberg et al. 1996; Nirenberg et al. 1997a;

Pickel and Chan 1999). SERTs are present on 5-HT neuron somata, dendrites, axons, and axon terminals but not on DA neurons (Pickel and Chan 1999).

9.3.3 5-HT Receptors

Over a dozen 5-HT receptor subtypes have been identified (Barnes and Sharp 1999; Hoyer et al. 2002; Bockaert et al. 2006; Alexander et al. 2007). All 5-HT receptors except the 5-HT_3 receptor are G protein-coupled receptors. Activation of 5-HT_1 receptors may decrease the formation of cyclic adenosine monophosphate (cAMP). cAMP, via the protein kinase A (PKA) system, may decrease the activities of ion channels, synaptic vesicle release, and neuronal excitability (Greengard 2001; Hilfiker et al. 2001; Laurent et al. 2002). 5-HT_{1A}, 5-HT_{1B}, 5-HT_{1D}, 5-HT_{1E}, 5-HT_{1F}, and 5-HT_5 are $G_{i/o}$-coupled receptors and generally inhibit neuronal activity and neurotransmitter release. 5-HT_{1A} and 5-HT_{1D} receptors often serve as autoreceptors on 5-HT cell body and terminals while 5-HT_{1B} receptor may presynaptically inhibit glutamate release (Muramatsu et al. 1998; Morikawa et al. 2000; Laurent et al. 2002; Seeburg et al. 2004) probably by inhibiting Ca^{2+} channels on the presynaptic terminals (Mizutani et al. 2006). 5-HT_{1B} receptor activation may also inhibit GABA release (Cameron and Williams 1994; Bramley et al. 2005). 5-HT_4, 5-HT_6, and 5-HT_7 receptors are G_s-coupled receptors and generally enhance neuronal excitability/activity and neurotransmitter release (Barnes and Sharp 1999; Xiang et al. 2005). 5-HT_{2A}, 5-HT_{2B}, and 5-HT_{2C} receptors are coupled to $G_{q/11}$ and activate phospholipase C (PLC), generating inositol phosphates (IP) and diacylglycerol (DAG). These pathways then increase intracellular Ca signaling, neuronal activity and neurotransmitter release (Zhou and Hablitz 1999; Lambe and Aghajanian 2001; Puig et al. 2003; Fink and Göthert 2007).

5-HT_3 receptors (including 5-HT_{3A}, 5-HT_{3B}, and 5-HT_{3C}) are the only 5-HT receptors directly coupled to a cation channel permeable to Na^+, K^+ and Ca^{2+}, causing fast membrane depolarization and excitation (Barnes and Sharp 1999; Zhou and Hablitz 1999; van Hooft and Vijverberg 2000; van Hooft and Yakel 2003). 5-HT_3 receptor expression level is relatively low in the entire forebrain including the mesostriatal DA system, although they are often selectively expressed at high levels in forebrain GABA interneurons (Morales et al. 1998; Zhou and Hablitz 1999; Ferezou et al. 2002; van Hooft and Yakel 2003).

9.3.4 5-HT Receptor Modulation of DA Neurons in the Midbrain

In midbrain DA areas, 5-HT_2 type receptors are the best studied. VTA DA and GABA neurons express 5-HT_{2A} receptors that mediate a modest depolarization (Pessia et al. 1994; Doherty and Pickel 2000; Nocjar et al. 2002) that increases DA neuron activity and DA release. 5-HT_{2A} receptor activation in VTA GABA neurons may indirectly inhibit DA neuron activity (Tepper et al. 1995; Esposito 2006; Invernizzi et al. 2007), creating another layer of complexity within the DA-5-HT interaction.

5-HT_{2C} receptor expression is relatively high in the mesostriatal DA system. 5-HT_{2C} receptors are found in neurons within the midbrain DA centers (Eberle-Wang et al. 1997; Clemett et al. 2000; Lopez-Gimenez et al. 2001; Giorgetti and Tecott 2004; Bubar and Cunningham 2007). Consistent with these anatomical studies, electrophysiological experiments have found that 5-HT_{2C} receptor activation can strongly activate midbrain GABA neurons and consequently inhibit DA neurons (Rick et al. 1995; Stanford and Lacey 1996; Di Matteo et al. 2001; De Deurwaerdere and Spampinato 1999; Di Giovanni et al. 1999, 2001, 2006; Olijslagers et al. 2004). Blocking 5-HT_{2C} receptors may enhance DA neuron activity and increase DA release (Di Matteo et al. 2001; Esposito 2006). Therefore, 5-HT may inhibit DA neurons in the midbrain whereas 5-HT_{2C} antagonists activate the mesoforebrain DA system (Invernizzi et al. 2007). Consequently, selective serotonin reuptake inhibitors (SSRIs) can influence VTA DA neuron activity by increasing extracellular 5-HT levels (Prisco and Esposito 1995; Di Matteo et al. 2001; Esposito 2006).

9.3.5 5-HT Receptor Modulation of the Striatal Neurons

Neuronal populations in the dorsal striatum and nucleus accumbens express multiple types of 5-HT receptors. In rat NAc, 5-HT inhibits synaptic glutamatergic inputs to medium spiny neurons via presynaptic 5-HT_{1B} receptors, located most likely on glutamate afferent terminals from cortical and limbic areas (Muramatsu et al. 1998; Morikawa et al. 2000). Histochemical studies indicate that 5-HT_{2A} receptors are expressed in striatal medium spiny neurons and possibly also on cortical and pallidal afferent terminals in the striatum (Cornea-Hebert et al. 1999; Rodriguez et al. 1999; Bubser et al. 2001; Bockaert et al. 2006). Activation of 5-HT_2 receptors causes a depolarization by decreasing a potassium conductance in rat NAc medium spiny neurons (North and Uchimura 1989). Recent electrophysiological and single cell RT-PCR studies indicate that 5-HT may excite striatal cholinergic interneurons via 5-HT_{2C}, 5-HT_6, and 5-HT_7 receptors (Blomeley and Bracci 2005; Bonsi et al. 2007). Because the striatal cholinergic interneurons tonically affect DA release via presynaptic nAChRs (Zhou et al. 2001; Rice and Cragg 2004), 5-HT excitation of striatal ACh interneurons may also indirectly alter DA release.

Studies using in situ hybridization, immunostaining and radioligand binding all indicated that expression of 5-HT_3 receptors is relatively low in the striatum in rodents and humans (Kilpatrick et al. 1987; Waeber et al. 1988; Laporte et al. 1992; Tecott et al. 1993; Morales et al. 1998; Chameau and van Hooft 2006). Experimental lesion of the nigrostriatal DA pathway does not affect 5-HT_3

receptors in the striatum, indicating that the majority of 5-HT_3 receptors are not on the mesostriatal DA axon terminals (Kidd et al. 1993). In human basal ganglia, 5-HT_3 receptor expression is reduced in Huntington's disease but not by Parkinson's disease, indicating that 5-HT_3 receptors are likely to be expressed on striatal spiny projection neurons (Steward et al. 1993). Functional studies indicate that 5-HT increases intra-terminal Ca^{2+} via 5-HT_3 receptor activation (Nichols and Mollard 1996; Nayak et al. 1999).

Relatively high densities of 5-HT_4 binding sites are found in the striatum, the globus pallidus, and SN in human, mouse, rat and guinea pig brains (Waeber et al. 1993, 1996). Based on ligand binding data combined with mRNA measurement and selective lesion, 5-HT_4 receptors are expressed on the soma and axon terminals of striatal projection neurons, but not on DA neurons or DA terminals (Patel et al. 1995; Vilaro et al. 2005). 5-HT_4 receptors increase intracellular cAMP levels (Barnes and Sharp 1999; Hoyer et al. 2002). Activation of 5-HT_4 receptors and 5-HT_6 receptors contributes to fluoxetine's antidepressant effects by altering the phosphorylation states of brain DARPP-32 (dopamine and cAMP-regulated phosphoprotein, 32 kDa, a protein phosphatase-1 inhibitor), a key player in DA signaling pathways (Greengard 2001; Svenningsson et al. 2002a,b, 2003, 2004, 2007). DARPP-32-mediated mechanisms can regulate membrane excitability, neurotransmitter release, and receptor/ion channel functionality.

9.4 Dopamine and Serotonin Co-Transmission

9.4.1 DA Transporter as a Low Efficiency 5-HT Uptake Route

The reuptake transporters for monoamine neurotransmitters are not perfectly selective. For example, DA and norepinephrine are taken up by both the norepinephrine transporter and the DAT, with the norepinephrine transporter having a higher affinity for DA than for norepinephrine (Masson et al. 1999; Stanford 1999; Torres et al. 2003). DATs also have a low affinity for 5-HT that is about 1/15 of its affinity for DA (Shaskan and Snyder 1970; Giros et al. 1991; Pristupa et al. 1994; Eshleman et al. 1999). Because DATs vastly outnumber SERTs in the striatum, DATs have opportunities to uptake 5-HT. Because 5-HT is stored in synaptic vesicles at concentrations as high as 270 mM (Bruns et al. 2000), 5-HT at release sites can far exceed the concentration necessary for uptake by DATs. Since there is no extracellular enzyme to degrade 5-HT, as 5-HT spreads by diffusion, various transporters uptake 5-HT until the concentration falls sufficiently low to prevent further reuptake by each transport type (Jones et al. 1998; Sora et al. 2001; Torres et al. 2003; Shen et al. 2004). Given the close anatomical apposition of DA terminals and 5-HT terminals in the striatum, the extremely dense striatal DA terminals coupled with the relatively low selectivity of DATs produces a situation that enables crosstalk between the DA and 5-HT neurotransmitter systems.

Experimental or biological circumstances that elevate extracellular 5-HT favor DAT uptake of 5-HT, especially in the striatum where there is an exceptionally high density of DATS. Genetic deletion of SERTs or monoamine oxidase A (a cytoplasmic enzyme that degrades monoamine neurotransmitters such as 5-HT) causes DAT-mediated 5-HT transport into DA neurons and terminals (Cases et al. 1998; Zhou et al. 2002; Mossner et al. 2006). Direct elevation of 5-HT in mouse brain slices also causes 5-HT to be transported into DA terminals (Zhou et al. 2005). In mice and rats, particularly during early postnatal developmental, chronic in vivo SSRI treatment induces slow DAT-mediated 5-HT uptake into midbrain DA neurons and into striatal DA terminals, where 5-HT is then taken into DA vesicles (Cases et al. 1998; Zhou et al. 2002; Zhou et al. 2005). The plasma membrane DAT transports DA and some elevated extracellular 5-HT into the cytoplasm of the striatal DA terminals. Then, the non-selective vesicular monoamine transporter-2 (VMAT2), which has similar affinities for DA and 5-HT and is located in the synaptic vesicle membrane (Yelin and Schuldiner 2002), transports both DA and 5-HT into the DA vesicles. Because DATs are not highly selective and they are at a high density, particular circumstances (such as SSRI treatment) favors some co-localization of 5-HT in DA synaptic vesicles (Zhou et al. 2005).

9.4.2 5-HT Enters DA neurons and DA Terminals

Several studies indicate that in both mice and rats, particularly in their early postnatal developmental stage, chronic SSRI treatment induces DAT-mediated 5-HT uptake into midbrain DA cell bodies (Fig. 9.4a,b) and into striatal DA terminals (Cases et al. 1998; Zhou et al. 2002; Zhou et al. 2005; Mossner et al. 2006). Genetic deletion of SERT results in stronger DAT-mediated ectopic 5-HT localization into DA neurons (Zhou et al. 2002). Elevating the extracellular 5-HT level by inhibiting or genetically deleting monoamine oxidase A (MAOA) also leads to DAT uptake of 5-HT into DA neurons and terminals (Cases et al. 1998). These findings are particularly interesting because disrupted 5-HT signaling after MAOA and/or SERT inhibition leads to significant neurodevelopment abnormalities (Vitalis et al. 1998; Boylan et al. 2000; Gaspar et al. 2003; Xu et al. 2004).

Antidepressants that are selective serotonin reuptake inhibitors (SSRIs) elevate extracellular 5-HT and alter the temporal and spatial relationship between DA and 5-HT signaling in the striatum. Antibodies against tyrosine hydroxylase (TH, an enzyme for DA synthesis in the striatum) and against 5-HT were used to visualize DA and 5-HT co-localization in striatal axon terminals (Fig. 9.4c-e). Under normal conditions, the modest number of 5-HT terminals (Fig. 9.4c, red, control 5-HT label) and the dense DA terminals

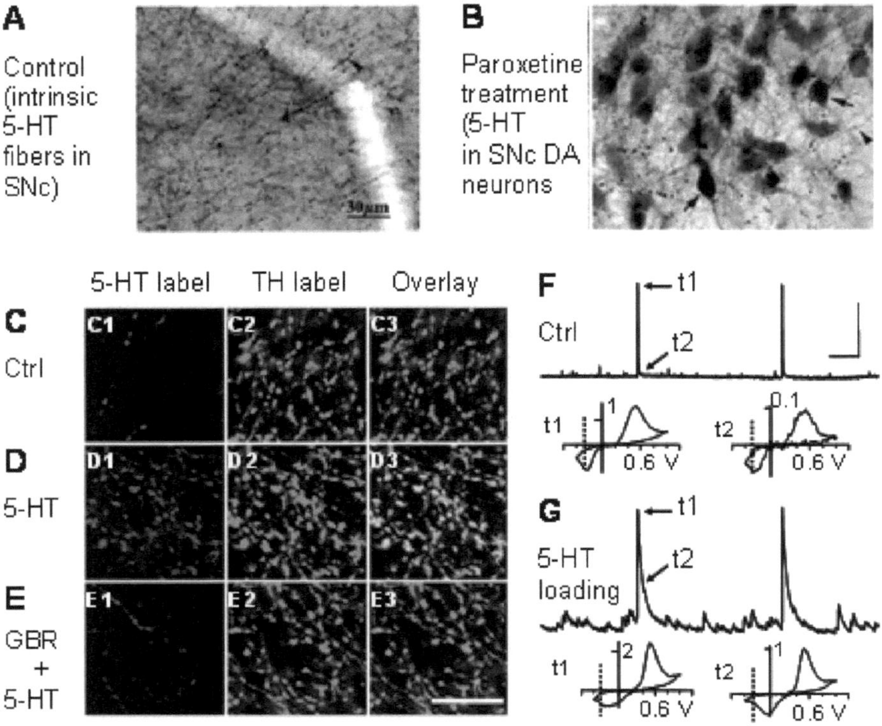

Fig. 9.4 5-HT enters DA neuron somata and axon terminals. **(a)** Under normal condition, 5-HT-immunostained fibers but not cell bodies are seen in the SNc. **(b)** Treatment of animals with paroxetine, a commonly used SSRI antidepressant, induces 5-HT accumulation in DA neurons in SNc in mice. From Zhou FC et al. (2002) with permission. **(c)** In striatal brain slices under control conditions, 5-HT axon terminals as detected by anti-5-HT antibody are relatively sparse (red, C1), and DA terminals as detected by anti-tyrosine hydroxylase (TH) antibody are dense (green, C2). The DA and 5-HT terminals generally do not overlap (absence of yellow, C3). **(d)** Incubation of striatal brain slices with 5-HT (2 μM) increases the number of 5-HT-positive terminals (red, D1), and DA terminals are relatively unchanged (green, D2). Most of the 5-HT-positive terminals coincide with DA terminals (yellow, D3), suggesting that DA terminals in this case also contain 5-HT. **(e)** Pretreatment with the DAT inhibitor, GBR12909 (1 μM), blocks the effect of 5-HT (2 μM). 5-HT terminals remain sparse (red, E1), DA terminals are dense and unchanged (green, E2), and the 5-HT-positive terminals rarely coincide with TH-positive terminals (yellow, E3). Scale bar = 10 μm. **(f)** A control trace showing small spontaneous voltammetric release events and two large electrically stimulated events. The scale bars are 1 μM DA and 30 s. Time t1 indicates the peak of the signal. At time t2, 4 s after the peak, the signal is small but still detectable. The two voltammograms (below) from t1 and t2 are typical DA voltammograms with the oxidation peak near 500 mV and the reduction peak near –230 mV (dotted vertical lines). The y-axis of the voltammograms varies, and is labeled for comparison in nA. **(g)** In the presence of 5-HT (2 μM) after subtracting off the 5-HT background, both the spontaneous and evoked events are larger and longer lasting than in the control case. The voltammogram from t1 is a mixed DA/5-HT voltammogram with a broad reduction peak from –230 mV (DA, dotted vertical line) to –30 mV (near 5-HT’s reduction peak), whereas the voltammogram from t2 is a predominantly 5-HT voltammogram with a reduction peak at –30 mV. (1) and (b) are from *Zhou FC et al.* 2002 *with permission.* (c-g) are from *Zhou et al.* 2005 *with permission.* (*See* Color Plate 9)

(Fig. 9.4c, green, control TH label) are next to each other, but they do not generally overlap even in 2-dimensional images (Fig. 9.4c, overlay) (Descarries and Mechawar 2000; Zhou et al. 2005). When the external 5-HT concentration is increased, the number of 5-HT positive terminals increase (Fig. 9.4d, red, 5-HT, left), but the number of TH-positive (i.e., DA terminals) remains relatively constant (Fig. 9.4d, green, middle). When 5-HT is elevated, the majority (~80%) of the 5-HT and DA terminals overlap with each other (Fig. 9.4d, yellow, right). With the DATs inhibited (by GBR12909), elevated 5-HT does not increase the number of 5-HT-positive terminals (Fig. 9.4e). The result suggests that when extracellular 5-HT is elevated, then 5-HT enters striatal DA terminals, filling DA vesicles with both DA and 5-HT. Because DA terminals do not express 5-HT transporters (Pickel and Chan 1999), these results indicate that DATs uptake 5-HT into striatal DA terminals when extracellular 5-HT is elevated.

The imaging results are supported by quantitative high performance liquid chromatography (HPLC) results using tissue from striatal brain slices. Elevating the extracellular 5-HT level increases 5-HT content but decreases DA content in a DAT-dependent manner (Zhou et al. 2005). The content of the DA metabolite, DOPAC, is also increased. These results independently indicate that elevated 5-HT is taken up into DA terminals, thereby, displacing DA slightly from DA vesicles. This process increases DA outside of DA vesicles, which leads to more breakdown by monoamine oxidases and increases the concentration of DOPAC. HPLC analysis also detects a small GBR12909-sensitive DAT-dependent 5-HT component in the striatum after prolonged chronic treatment with fluoxetine in vivo, indicating an ectopic 5-HT in DA terminals (Zhou et al. 2005).

9.4.3 Vesicular Co-release of DA and 5-HT from Striatal DA Terminals

Recent evidence indicates that after chronic SSRI treatment, a small proportion of 5-HT is released together with DA from DA terminals. Fast-scan cyclic voltammetry was used to identify DA and 5-HT based on their respective oxidation-reduction profiles (or voltammograms) (Stamford et al. 1990; Jackson and Wightman 1995; Bunin and Wightman 1998; Zhou et al. 2001, 2005). Measured with a carbon-fiber microelectrode, the distinct voltammograms of DA and 5-HT are particularly different near their reduction minima, which is near –250 mV (–20 to –280 mV) for DA and is near –30 mV (0 to –50 mV) for 5-HT.

Under control conditions, both spontaneous and electrically evoked DA release events are detected by voltammetry (Fig. 9.4f). A 5-HT component is undetectable under normal conditions because the 5-HT content is very low, and the physiologically released 5-HT is below the level of detection. When extracellular 5-HT level is elevated, spontaneous and electrically evoked voltammetric signals contained both a DA and 5-HT component as indicated by the voltammograms (Fig. 9.4 g). Furthermore, in elevated 5-HT, the DA component is reduced, and the decay of the signals becomes much prolonged. At the peak of the response (t1), the signal is a mixture of DA and 5-HT, showing an extended reduction peak (from –30 to –250) in the voltammogram. The DA component is significantly decreased compared to the control, but the 5-HT concentration increases drastically. Four seconds after the maximum (t2), a 5-HT signal is practically the only component. This finding is consistent with the more rapid re-uptake of DA by the DATs and slower clearance of 5-HT from the carbon fiber when compared to DA. The 5-HT component is also completely prevented by blocking DATs with GBR12909, further confirming that DAT-mediated uptake of 5-HT into DA terminals is required for the vesicular release of 5-HT with DA.

In addition to the stimulation-evoked vesicular DA release, spontaneous vesicular DA release also contains a 5-HT component after the extracellular 5-HT level is elevated (Fig. 9.4f,g). Under control condition, there are small, spontaneous, fast DA release spikes indicated by the characteristic DA voltammogram (Fig. 9.4f). After extracellular 5-HT level is elevated, there are more detectable small release events. In all of the events analyzed, the rising phase and the maximum of these spontaneous events are mixed signals of both DA and 5-HT (Fig. 9.4 g). The later falling phase of the events, however, is predominantly 5-HT with only a small DA component. Four seconds after the maximum response, only the 5-HT component is detectable. These results indicate that DA and 5-HT are being released together, and that there is more rapid re-uptake of DA by DATs and more rapid clearance of DA than 5-HT from the carbon-fiber microelectrode. The evidence indicates 5-HT and DA are being released from the same terminals because all the spontaneous events have both DA and 5-HT components.

Antidepressants such as fluoxetine are selective serotonin re-uptake inhibitors (SSRIs) that prolong and elevate the extracellular 5-HT (Wong et al. 1995, 2005; Knobelman et al. 2000; Smith et al. 2000). Increasing the extracellular 5-HT level in vivo with chronic SSRI treatment also induces a small 5-HT component in DA release events that is sensitive to DAT inhibition (Zhou et al. 2005), indicating that during SSRI treatment a portion of released 5-HT is taken up into DA terminals and then released together with DA (Fig. 9.5).

Based on above discussion, an interesting scenario emerges. Normally 5-HT is efficiently recycled into 5-HT terminals by 5-HT transporters (Fig. 9.5a). When patients take fluoxetine (Prozac) and inhibit 5-HT re-uptake into 5-HT terminals (Wong et al. 1995, 2005), DATs have a greater opportunity to uptake 5-HT into DA cell bodies and terminals (especially in the striatum when there is a high density of DATs) (Cases et al. 1998; Zhou et al. 2002; Zhou et al. 2005). When SSRIs block 5-HT transporters, the extracellular 5-HT level is elevated. It remains in the extracellular space longer and diffuses farther. Under those conditions, the striatal DATs are able to re-uptake significant amounts of 5-HT

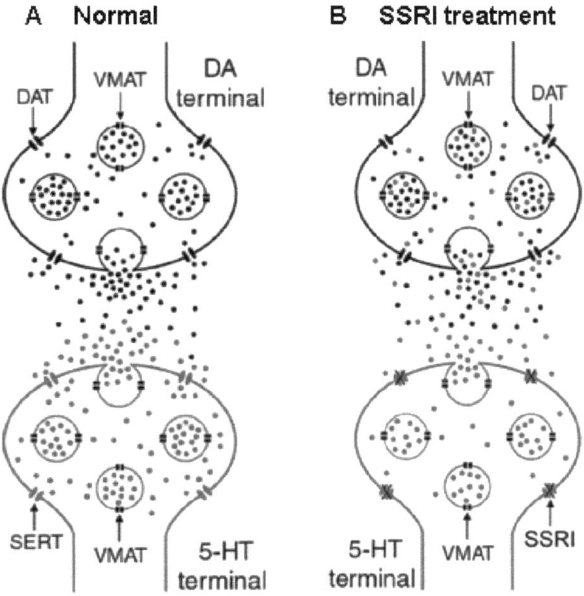

Fig. 9.5 Schematic, didactic representation of the hypothesized DAT-dependent 5-HT accumulation into striatal DA terminals after chronic SSRI antidepressant treatment. **(a)** Under normal conditions when the 5-HT transporters (SERTs) are functional, the released 5-HT molecules (red dots) are efficiently taken back into 5-HT terminals. Consequently, the extracellular 5-HT concentration remains low. DATs on DA terminals have little chance to uptake 5-HT. **(b)** DATs may uptake a portion of endogenously released 5-HT into DA terminals when the extracellular 5-HT concentration is elevated after the SERTs are inhibited by SSRIs. Once inside the DA terminals, 5-HT is further sequestered into DA terminals through the vesicular monoamine transporter (VMAT), which has similar affinity for DA and 5-HT. Subsequently, 5-HT partially displaces DA (black dots) from DA vesicles, and the two neurotransmitters are co-released. *From Zhou et al.* 2005 *with permission*. (*See* Color Plate 10)

into DA terminals (Fig. 9.5b). This process likely proceeds slowly during SSRI treatment of patients, possibly contributing to the many days of SSRI treatment often required before achieving anxiolytic benefits (Nemeroff and Owens 2002; Baldessarini 2006; Zhou and Dani 2006).

9.4.4 DA in 5-HT Neurons and Terminals during L-dopa Treatment

While some circumstances favor 5-HT uptake into DA neurons and terminals, other conditions favor accumulation of DA in 5-HT terminals. Parkinson's disease (PD) is commonly treated with the DA precursor, L-3,4-dihydroxyphenylalanine (L-dopa) that commonly reaches high blood concentrations (10–20 μM) (Olanow

2004; Olanow et al. 2006). During PD, however, many DA neurons and their terminals have degenerated, limiting the potential for L-dope uptake. Under those conditions, L-dopa becomes available to other aromatic amino acid transporters, such as those on 5-HT neurons and terminals (Arai et al. 1996; Miller and Abercrombie 1999; Mura et al. 2000; Lopez-Real et al. 2003; Maeda et al. 2005).

Once inside the 5-HT neurons and terminals, aromatic L-amino acid decarboxylase (AADC) catalyzes L-dopa to DA (Fig. 9.6) (Arai et al. 1995, 1996; Mura et al. 2000; Lopez-Real et al. 2003; Yamada et al. 2007). AADC, which is normally present in DA and 5-HT neurons, converts L-dopa and 5-hydroxytryptophan to DA and 5-HT, respectively. In a 6-hydroxydopamine (6-OHDA, a DA neuron toxin)-lesioned rats, L-dopa administration induced DA-immunoreactivity in 5-HT axon terminals of the striatum and SN reticulata (Yamada et al. 2007). The relatively high density of 5-HT terminals in the SN reticulata, a key basal ganglia output nucleus, makes L-dopa conversion to DA in 5-HT terminals particularly significant. This process is also enhanced by the 5-HT hyper-innervation that occurs in the striatum after DA denervation (Kostrzewa et al. 1998; Maeda et al. 2003, 2005; Brown and Gerfen 2006).

DA synthesized from L-dopa in 5-HT neurons is probably sequestered into 5-HT vesicles and co-released with 5-HT. Evidence indicates that after release

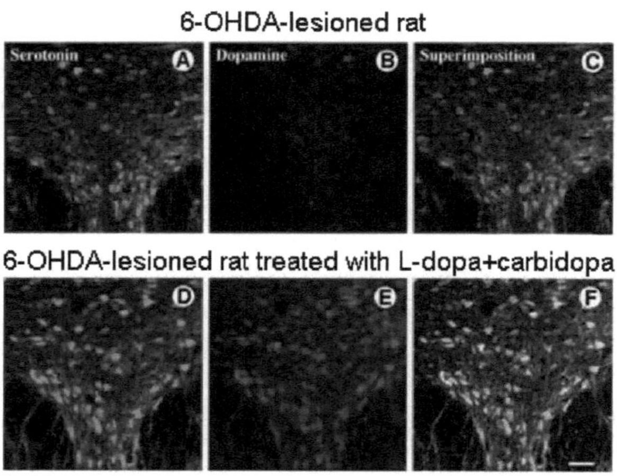

Fig. 9.6 DA in 5-HT neurons of the raphe. **(a)** Dorsal raphe 5-HT neurons immunostained with anti-5-HT antibody in a 6-hydroxydopamine (6-OHDA, a DA neuron toxin) lesioned rat. **(b)** Only a few DA terminals are revealed by an anti-DA antibody in the raphe area in the same section as in (A). **(c)** Superimposition of (A) and (B) shows no co-localization of 5-HT and DA-containing processes. **(d)** Dorsal raphe 5-HT neurons immunostained with anti-5-HT antibody in a 6-OHDA-lesioned rat treated with L-Dopa and carbidopa. **(e)** DA-containing neurons in the same section as in (D) revealed by an anti-DA antibody. **(f)** Superimposition of (D) and (E) shows that in addition to intrinsic DA terminals, DA co-localizes within raphe 5-HT neurons. The results indicate that L-dopa enters 5-HT neurons and, then, is converted to DA in 5-HT neurons. *From Yamada et al.* (2007) *with permission.* Scale bar = 10 μm. (*See* Color Plate 11)

into the extracellular space, this ectopically synthesized DA may also be uptaken by SERTs into 5-HT terminals. Indeed, fluoxetine, a selective SERT inhibitor, increases the L-dopa-derived extracellular DA level in the striatum after DA denervation, presumably by inhibiting SERT-mediated DA uptake (Tanaka et al. 1999; Kannari et al. 2006).

9.4.5 DA and 5-HT Co-signaling and Parkinson's Disease

The key pathology of Parkinson's disease (PD) is progressive degeneration of the midbrain DA neurons, particularly those in the substantia nigra pars compacta (SNc). The DA neuron loss disrupts DA signaling in the basal ganglia and alters the normal operation of the motor control circuit (Albin et al. 1989; Alexander and Crutcher 1990; Damier et al. 1999; Hornykiewicz 2001; Chan et al. 2007; DeLong and Wichmann 2007). As described above, DA replacement therapy supplies L-dopa that is then converted to DA mainly in DA neurons, but 5-HT neurons also uptake L-dopa and convert it to DA (Fig. 9.6) (Arai et al. 1995, 1996; Mura et al. 2000; Lopez-Real et al. 2003; Yamada et al. 2007). In addition, under conditions of elevated DA, 5-HT neurons may also uptake DA via SERTs (Tanaka et al. 1999). Functionally, ectopic DA release from 5-HT terminals may compensate partially for the loss of DA terminals during Parkinson's disease (Brotchie 2005; Cenci and Lundblad 2006; Cenci 2007). This scenario is made more plausible because of the development of supersensitive DA receptors after DA denervation (Miller and Abercrombie 1999; Gerfen 2003). However, recent studies also suggest that DA release from 5-HT terminals after L-dopa treatment in a rat model of PD may contribute to L-dopa-induced dyskinesia (Carlsson et al. 2007; Carta et al. 2007). It is likely that release patterns and release regulation are different at DA terminals and 5-HT terminals.

Other changes occur during PD within these interacting neurotransmitter systems. Degeneration of the DA system alters the expression and function of 5-HT receptor subtypes differently in different brain areas. For example, DA neuron lesions increased 5-HT_{2A} receptor expression but decreased 5-HT_{2C} receptor expression in the striatum (Chu et al. 2004; Zhang et al. 2007). L-dopa treatment prevented the 5-HT_{2A} over-expression. These and other forms of crosstalk within the DA and 5-HT systems may be significant in the pathophysiology and treatment of Parkinson's disease.

9.4.6 DA and 5-HT Co-signaling at the Single Cell Level

Many neuronal types in the mesostriatal DA system express both DA and 5-HT receptors (Bonsi et al. 2007; Pisani et al. 2007; Surmeier et al. 2007). Thus, when there is DA and 5-HT co-released, the potential exists for interactions between DA and 5-HT signaling pathways. Some of these DA and 5-HT receptors share common intracellular signaling components and couple to common effectors, such as ion channels and intracellular Ca^{2+} (Hoyer et al. 2002; Svenningsson et al. 2002b, 2004; Neve et al. 2004; Bockaert et al. 2006). DA and 5-HT co-localization and consequent co-release alter the timing, location, intensity, and interaction of 5-HT and DA signaling cascades that influence neuronal circuits in ways not achieved through DA or 5-HT signaling alone. In summary, additive, synergistic, or opposing interactions may occur via DA and 5-HT intracellular signaling components and the final effectors.

9.4.7 DA and 5-HT Co-signaling and Serotonin Syndrome

When an SSRI, momoamine oxidase inhiditor (MAOI), 5-HT agonist, or dietary 5-HT precursor tryptophan or a tricyclic antidepressant are used in combinations or the gap between their use is not long enough, an abnormally high brain 5-HT level or hyper-serotonergic state may be induced, leading to the so called serotonin syndrome (Sternbach 1991; Boyer and Shannon 2005; Gillman 2006). Symptoms include hyperthermia, diaphoresis, excitement or confusion, hyper-reflex, hypotension, and tremor and muscle rigidity. In severe cases, loss of consciousness and death may occur.

With increased extracellular 5-HT levels and SERTs inhibited by SSRIs, DAT uptake of 5-HT into DA neurons is likely, particularly in the striatum, substantia nigra pars reticulata (SNr), and hypothalamus where there are dense DA terminals juxtaposed with 5-HT terminals. When excess 5-HT precursor, 5-hydroxytryptophan (5-HTP), is available, DA neurons may also uptake 5-HTP and convert it to 5-HT. Once inside the DA neuron, 5-HT, obtained via uptake or 5-HTP conversion, is sequestered in DA vesicles. This process may lead to increased tissue 5-HT content and decreased tissue DA content (Bedard and Pycock 1977; Shioda et al. 2004; Izumi et al. 2006). Indeed, spontaneous and evoked DA and 5-HT vesicular co-release from DA terminals also occurs in the striatum and possibly in other brain areas with converging DA and 5–HT innervation after 5-HTP loading (Stamford et al. 1990; Jackson and Wightman 1995). The ectopic 5-HT accumulation in DA neurons and terminals and subsequent alterations in DA and 5-HT signaling may contribute to the clinical symptoms.

9.5 Summary

Although DA and 5-HT co-transmission is likely to be minimal in general, circumstances can favor co-release from either DA or 5-HT terminals. During SSRI antidepressant treatment, 5-HT reuptake by SERTs is inhibited, elevating and prolonging extracellular 5-HT concentrations. The elevated extracellular 5-HT becomes more readily available to the relatively non-selective DATs, which then transport some 5-HT into DA terminals. This process is strongest in the striatum, where DATs are at a very high density and in close proximity to 5-HT release sites. 5-HT then competes with DA for DAT, VMAT2, and intravesicular space that may lead to functional changes in the DA system.

During L-dopa treatment of Parkinson's disease, 5-HT neurons may uptake L-dopa and convert it to DA, leading to co-release of DA with intrinsic 5-HT. The cross uptake and co-release may partly compensate for the loss of DA terminals and contribute to the therapeutic mechanisms and/or side effects of L-dopa treatment.

These occasions of DA and 5-HT co-release influence inherent on-going DA and 5-HT signaling interactions that arise from innervation and receptor localization within these systems. For example, midbrain DA neurons innervate and influence raphe 5-HT neuron activity via DA receptors expressed in 5-HT neurons. In addition, DA and 5-HT influence afferent innervation onto their own and the others neurons. Co-release consequently alters the timing, location, intensity, and interaction of 5-HT and DA signaling cascades. The synergistic and/or opposing interactions occur at multiple levels within intracellular signaling cascades and within neuronal circuits to influence behavior, disease progression, and therapeutic interventions.

Acknowledgments The authors are supported by grants from the American Parkinson Disease Association, the National Alliance for Research on Schizophrenia and Depression, and the National Institutes of Health (NIDA, NIMH, and NINDS).

References

Aghajanian GK, Marek GJ (1999) Serotonin and hallucinogens. Neuropsychopharmacology 21(2 Suppl):16S–23S

Aizman O, Brismar H, Uhlen P, Zettergren E, Levey AI, Forssberg H, et al (2000) Anatomical and physiological evidence for D_1 and D_2 dopamine receptor colocalization in neostriatal neurons. Nat Neurosci 3:226–230

Albin RL, Young AB, Penney JB (1989) The functional anatomy of basal ganglia disorders. Trends Neurosci 12:366–375

Alexander GE, Crutcher MD (1990) Functional architecture of basal ganglia circuits: neural substrates of parallel processing. Trends Neurosci 13:266–271

Alexander SP, Mathie A, Peters JA (2007) 7TM Receptors. Br J Pharmacol 150 Suppl 1:S4–S81

Aman TK, Shen RY, Haj-Dahmane S (2007) D2-like dopamine receptors depolarize dorsal raphe serotonin neurons through the activation of nonselective cationic conductance. J Pharmacol Exp Ther 320:376–385

Anden NE, Dahlstrom A, Fuxe K, Larsson K (1965) Mapping out of catecholamine and 5-hydroxytryptamine neurons innervating the telencephalon and diencephalon. Life Sci 4:1275–1279

Arai R, Karasawa N, Geffard M, Nagatsu I (1995) L-DOPA is converted to dopamine in serotonergic fibers of the striatum of the rat: a double-labeling immunofluorescence study. Neurosci Lett 195:195–198

Arai R, Karasawa N, Nagatsu I (1996) Aromatic L-amino acid decarboxylase is present in serotonergic fibers of the striatum of the rat. A double-labeling immunofluorescence study. Brain Res 706:177–179

Arbuthnott GW, Wickens J (2007) Space, time and dopamine. Trends Neurosci 30:62–69

Ariano MA (1997) Distribution of dopamine receptors. In: Neve KA, RL Neve RL (ed) The dopamine receptors. Humana Press, Totowa, New Jersey. pp. 77–103

Azmitia EC, Segal M (1978) An autoradiographic analysis of the differential ascending projections of the dorsal and median raphe nuclei in the rat. J Comp Neurol 179:641–667

Baker KG, Halliday GM, Hornung JP, Geffen LB, Cotton RG, Tork I (1991) Distribution, morphology and number of monoamine-synthesizing and substance P-containing neurons in the human dorsal raphe nucleus. Neuroscience 42:757–775

Baldessarini RJ (2006) Drug Therapy of Depression and Anxiety Disorders. In: Brunton LL, Lazo JS, Parker KL (ed) Goodman and Gilman's the Pharmacological Basis of Therapeutics. McGraw-Hill, New York. pp. 429–459

Bamford NS, Zhang H, Schmitz Y, Wu NP, Cepeda C, Levine MS, et al (2004) Heterosynaptic dopamine neurotransmission selects sets of corticostriatal terminals. Neuron 42:653–663

Barnes NM, Sharp T (1999) A review of central 5-HT receptors and their function. Neuropharmacology 38:1083–1152

Baufreton J, Garret M, Rivera A, de la Calle A, Gonon F, Dufy B, Bioulac B, Taupignon A (2003) D5 (not D1) dopamine receptors potentiate burst-firing in neurons of the subthalamic nucleus by modulating an L-type calcium conductance. J Neurosci 23:816–825

Baumgarten HG, Grozdanovic Z (1997) Anatomy of the central serotoninergic projection systems. In: Baumgarten HG, Gothert M (ed) Serotoninergic neurons and 5-HT receptors in the CNS. Springer-Verlag, Berlin. pp. 41–89

Bedard P, Pycock CJ (1977) "Wet-dog" shake behaviour in the rat: a possible quantitative model of central 5-hydroxytryptamine activity. Neuropharmacology 16:663–670

Bentivoglio M, Morelli M (2005) The organisation and circuits of mesencephalic dopaminergic neurons and the distribution of dopamine receptors in the brain. In: Dunnett SB, Bentivoglio M, Bjorklund A, Hokfelt T (ed) Handbook of Chemical Neuroanatomy: Dopamine Vol. 21, Elsevier, Amsterdam. pp. 1–107.

Bergson C, Mrzljak L, Smiley JF, Pappy M, Levenson R, Goldman-Rakic PS (1995) Regional, cellular, and subcellular variations in the distribution of D_1 and D_5 dopamine receptors in primate brain. J Neurosci 15:7821–7836

Bertler A, Rosengren E (1959) Occurrence and distribution of dopamine in brain and other tissues. Experientia 15:10–11

Bjorklund A, Dunnett SB (2007) Dopamine neuron systems in the brain: an update. Trends Neurosci 30:194–202

Bjorklund A, Lindvall O (1984) Dopamine-containing systems in the CNS. In: Björklund A, Hökfelt T Ed, Classical transmitters in the CNS, Part I. Elsevier, Amsterdam. pp. 55–122

Bjorklund A, Lindvall O (1986) Cathecholaminergic brain stem regulator systems. In: Bloom FE (ed) Handbook of Physiology. Section 1: The Nervous System. vol IV. Intrinsic regulatory system of the brain. Waverly Press, Baltimore, Maryland. pp 155–235.

Blomeley C, Bracci E (2005) Excitatory effects of serotonin on rat striatal cholinergic interneurones. J Physiol 569:715–721

Bockaert J, Claeysen S, Becamel C, Dumuis A, Marin P (2006) Neuronal 5-HT metabotropic receptors: fine-tuning of their structure, signaling, and roles in synaptic modulation. Cell Tissue Res 326:553–572

Bonsi P, Cuomo D, Ding J, Sciamanna G, Ulrich S, Tscherter A, Bernardi G, Surmeier DJ, Pisani A (2007) Endogenous serotonin excites striatal cholinergic interneurons via the activation of 5-HT_{2C}, 5-HT_6, and 5-HT_7 serotonin receptors: implications for extrapyramidal side effects of serotonin reuptake inhibitors. Neuropsychopharmacology 32:1840–1854

Boyer EW, Shannon M (2005) The serotonin syndrome. N Engl J Med 352:1112–1120

Boylan CB, Bennett-Clarke CA, Crissman RS, Mooney RD, Rhoades RW (2000) Clorgyline treatment elevates cortical serotonin and temporarily disrupts the vibrissae-related pattern in rat somatosensory cortex. J Comp Neurol 427:139–149

Bramley JR, Sollars PJ, Pickard GE, Dudek FE (2005) 5-HT_{1B} receptor-mediated presynaptic inhibition of GABA release in the suprachiasmatic nucleus. J Neurophysiol 93:3157–3164

Brodie BB, Shore PA, Silver SL (1955) Potentiating action of chlorpromazine and reserpine. Nature 175:1133–1134

Brotchie JM (2005) Nondopaminergic mechanisms in levodopa-induced dyskinesia. Movement Disorders 20:919–931

Brown P, Gerfen CR (2006) Plasticity within striatal direct pathway neurons after neonatal dopamine depletion is mediated through a novel functional coupling of serotonin 5-HT2 receptors to the ERK 1/2 map kinase pathway. J Comp Neurol 498:415–430

Bruns D, Riedel D, Klingauf J, Jahn R (2000) Quantal release of serotonin. Neuron 28:205–220

Bubar MJ, Cunningham KA (2007) Distribution of serotonin 5-HT_{2C} receptors in the ventral tegmental area. Neuroscience 146:286–297

Bubser M, Backstrom JR, Sanders-Bush E, Roth BL, Deutch AY (2001) Distribution of serotonin 5-HT_{2A} receptors in afferents of the rat striatum. Synapse 39:297–304

Bunin MA, Wightman RM (1998) Quantitative evaluation of 5-hydroxytryptamine (serotonin) neuronal release and uptake: an investigation of extrasynaptic transmission. J Neurosci 18:4854–4860

Cameron DL, Williams JT (1993) Dopamine D_1 receptors facilitate transmitter release. Nature 366:344–347

Cameron DL, Williams JT (1994) Cocaine inhibits GABA release in the VTA through endogenous 5-HT. J Neurosci 14:6763–6767

Cantrell AR, Catterall WA (2001) Neuromodulation of Na^+ channels: an unexpected form of cellular plasticity. Nat Rev Neurosci 2:397–407

Carlsson A (1959) The occurrence, distribution and physiological role of catecholamines in the nervous system. Pharmacol Rev 11:490–493

Carlsson A (2001) A half-century of neurotransmitter research: impact on neurology and psychiatry. Nobel lecture. Nobel Foundation, Stockholm, Sweden.

Carlsson A (2002a) Treatment of Parkinson's with L-DOPA. The early discovery phase, and a comment on current problems. J Neural Transm 109:777–787

Carlsson A (2002b) Birth of dopamine: A Cinderella saga. In: Di Chiara G (ed) Dopamine in CNS I. Springer-Verlag, Berlin. pp 23–41

Carlsson T, Carta M, Winkler C, Bjorklund A, Kirik D (2007) Serotonin neuron transplants exacerbate L-DOPA-induced dyskinesias in a rat model of Parkinson's disease. J Neurosci 27:8011–8022

Carlsson A, Falck B, Hillarp N-A (1962) Cellular localization of brain monoamines. Acta Physiol Scand 56 Suppl 196:1–28

Carlsson A, Lindqvist M (1963) Effect of Chlorpromazine or haloperidol on the formation of 3-methoxytyramine and normetanephrine in mouse brain. Acta Pharmacol Toxicol 20:140–144

Carlsson A, Lindqvist M, Magnusson T, Waldeck B (1958) On the presence of 3-hydroxytyramine in brain. Science 127:471

Carta M, Carlsson T, Kirik D, Bjorklund A (2007) Dopamine released from 5-HT terminals is the cause of L-DOPA-induced dyskinesia in parkinsonian rats. Brain 130:1819–1833

Cases O, Lebrand C, Giros B, Vitalis T, De Maeyer E, Caron MG, et al (1998) Plasma membrane transporters of serotonin, dopamine, and norepinephrine mediate serotonin accumulation in atypical locations in the developing brain of monoamine oxidase A knock-outs. J Neurosci 18:6914–6927

Cenci MA (2007) Dopamine dysregulation of movement control in L-DOPA-induced dyskinesia. Trends Neurosci 30:236–243

Cenci MA, Lundblad M (2006) Post- versus presynaptic plasticity in L-DOPA-induced dyskinesia. J Neurochem 99:381–392
Centonze D, Grande C, Usiello A, Gubellini P, Erbs E, Martin AB, Pisani A, Tognazzi N, Bernardi G, Moratalla R, Borrelli E, Calabresi P (2003) Receptor subtypes involved in the presynaptic and postsynaptic actions of dopamine on striatal interneurons. J Neurosci. 23:6245–6254
Chameau P, van Hooft JA (2006) Serotonin 5-HT_3 receptors in the central nervous system. Cell Tissue Res 326:573–581
Chan CS, Guzman JN, Ilijic E, Mercer JN, Rick C, Tkatch T, Meredith GE, Surmeier DJ (2007) 'Rejuvenation' protects neurons in mouse models of Parkinson's disease. Nature 447:1081–1086
Charara A, Parent A (1998) Chemoarchitecture of the primate dorsal raphe nucleus. J Chem Neuroanat 15: 111–127
Chen N-H, Reith MEA (1997) Role of axonal and somatodendritic monoamine transporters in action of uptake blockers. In: Reith MEA (ed) Neurotransmitter transporters: structure, function, and regulation. Humana Press, Totowa, New Jersey. pp. 345–391
Chu Y, Kompoliti K, Cochran EJ, Mufson EJ, Kordower JH (2002) Age-related decreases in Nurr1 immunoreactivity in the human substantia nigra. J Comp Neurol 450:203–214
Chu YX, Liu J, Feng J, Wang Y, Zhang QJ, Li Q (2004) Changes of discharge rate and pattern of 5-hydroxytrypamine neurons of dorsal raphe nucleus in a rat model of Parkinson's disease. Sheng Li Xue Bao 56:597–602
Ciliax BJ, Drash GW, Staley JK, Haber S, Mobley CJ, Miller GW, et al (1999) Immunocytochemical localization of the dopamine transporter in human brain. J Comp Neurol 409:38–56
Ciliax BJ, Heilman C, Demchyshyn LL, Pristupa ZB, Ince E, Hersch SM, et al (1995) The dopamine transporter: immunochemical characterization and localization in brain. J Neurosci 15:1714–1723
Ciliax BJ, Nash N, Heilman C, Sunahara R, Hartney A, Tiberi M, Rye DB, Caron MG, Niznik HB, Levey AI (2000) Dopamine D_5 receptor immunolocalization in rat and monkey brain. Synapse 37:125–145
Clemett DA, Punhani T, Duxon MS, Blackburn TP, Fone KC (2000) Immunohistochemical localisation of the 5-HT_{2C} receptor protein in the rat CNS. Neuropharmacology 39:123–132
Cornea-Hebert V, Riad M, Wu C, Singh SK, Descarries L (1999) Cellular and subcellular distribution of the serotonin 5-HT_{2A} receptor in the central nervous system of adult rat. J Comp Neurol 409:187–209
Coulter CL, Happe HK, Murrin LC (1996) Postnatal development of the dopamine transporter: a quantitative autoradiographic study. Brain Res Dev Brain Res 92:172–181
Cragg SJ, Rice ME (2004) DAncing past the DAT at a DA synapse. Trends Neurosci 27:270–277
Creese I, Burt, DR, Snyder SH (1976) Dopamine receptor binding predicts clinical and pharmacological potencies of anti-schizophrenic drugs. Science 19:481–483
Dahlstrom A, Fuxe K (1964) Evidence for the existence of monoamine-containing neurons in the central nervous system. I. Demonstration of monoamines in the cell bodies of brain stem neurons. Acta Physiol Scand 62:1–55
Dahlstrom A, Fuxe K, Olson L, Ungerstedt U (1964) Ascending systems of catecholamine neurons from the lower brain stem. Acta Physiol Scand 62:485–486
Damier P, Hirsch EC, Agid Y, Graybiel AM (1999) The substantia nigra of the human brain. II. Patterns of loss of dopamine-containing neurons in Parkinson's disease. Brain 122:1437–1448
Day M, Wang Z, Ding J, An X, Ingham CA, Shering AF, et al. (2006) Selective elimination of glutamatergic synapses on striatopallidal neurons in Parkinson disease models. Nat Neurosci 9:251–259

De Deurwaerdere P, Bonhomme N, Lucas G, Le Moal M, Spampinato U (1996) Serotonin enhances striatal dopamine outflow in vivo through dopamine uptake sites. J Neurochem 66:210–215

de Jong TR, Veening JG, Waldinger MD, Cools AR, Olivier B (2006) Serotonin and the neurobiology of the ejaculatory threshold. Neurosci Biobehav Rev 30:893–907

DeLong MR, Wichmann T (2007) Circuits and circuit disorders of the basal ganglia. Arch Neurol 64:20–24

Descarries L, Mechawar N (2000) Ultrastructural evidence for diffuse transmission by monoamine and acetylcholine neurons of the central nervous system. Prog Brain Res 125:27–47

Di Chiara G, Bassareo V (2007) Reward system and addiction: what dopamine does and doesn't do. Curr Opin Pharmacol 7:69–76

Di Giovanni G, De Deurwaerdére P, Di Mascio M, Di Matteo V, Esposito E, Spampinato U (1999) Selective blockade of serotonin-2C/2B receptors enhances mesolimbic and mesostriatal dopaminergic function: a combined in vivo electrophysiological and microdialysis study. Neuroscience 91:587–597

Di Giovanni G, Di Matteo V, La Grutta V, Esposito E (2001) m-Chlorophenylpiperazine excites non-dopaminergic neurons in the rat substantia nigra and ventral tegmental area by activating serotonin-2C receptors. Neuroscience 103:111–116

Di Giovanni G, Di Matteo V, Pierucci M, Benigno A, Esposito E (2006) Central serotonin2C receptor: from physiology to pathology. Curr Top Med Chem 6:1909–1925

Di Matteo V, De Blasi A, Di Giulio C, Esposito E (2001) Role of 5-HT_{2C} receptors in the control of central dopamine function. Trends Pharmacol Sci 22:229–232

Diana M, Tepper JM (2002) Electrophysiological pharmacology of mesencephalic dopaminergic neurons. In: Di Chiara G (ed) Dopamine in the CNS II. Springer-Verlag, Berlin. pp 1–62

Diaz J, Pilon C, Le Foll B, Gros C, Triller A, Schwartz JC, Sokoloff P (2000) Dopamine D_3 receptors expressed by all mesencephalic dopamine neurons. J Neurosci 20:8677–8684

Doherty MD, Pickel VM (2000) Ultrastructural localization of the serotonin2A receptor in dopaminergic neurons in the ventral tegmental area. Brain Res 864:176–185

Dong Y, Cooper D, Nasif F, Hu XT, White FJ (2004) Dopamine modulates inwardly rectifying potassium currents in medial prefrontal cortex pyramidal neurons. J Neurosci 24:3077–3085

Doucet G, Descarries L, Garcia S (1986) Quantification of the dopamine innervation in adult rat neostriatum. Neuroscience 19:427–445

Dunah AW, Standaert DG (2001) Dopamine D1 receptor-dependent trafficking of striatal NMDA glutamate receptors to the postsynaptic membrane. J Neurosci 21:5546–5558

Emborg ME, Ma SY, Mufson EJ, Levey AI, Taylor MD, Brown WD, Holden JE, Kordower JH (1998) Age-related declines in nigral neuronal function correlate with motor impairments in rhesus monkeys. J Comp Neurol 401:253–265

Eberle-Wang K, Mikeladze Z, Uryu K, Chesselet MF (1997) Pattern of expression of the serotonin2C receptor messenger RNA in the basal ganglia of adult rats. J Comp Neurol 384:233–247

Eshleman AJ, Carmolli M, Cumbay M, Martens CR, Neve KA, Janowsky A (1999) Characteristics of drug interactions with recombinant biogenic amine transporters expressed in the same cell type. J Pharmacol Exp Ther 289:877–885

Esposito E (2006) Serotonin-dopamine interaction as a focus of novel antidepressant drugs. Curr Drug Targets 7:177–185

Falck B (1962) Observations on the possibilities of the cellular localization of monoamines by a fluorescence method. Acta Physiol Scand 56 Suppl 197:1–25

Ferezou I, Cauli B, Hill EL, Rossier J, Hamel E, Lambolez B (2002) 5-HT_3 receptors mediate serotonergic fast synaptic excitation of neocortical vasoactive intestinal peptide/cholecystokinin interneurons. J Neurosci 22:7389–7397

Ferre S, Artigas F (1993) Dopamine D_2 receptor-mediated regulation of serotonin extracellular concentration in the dorsal raphe nucleus of freely moving rats. J Neurochem 61:772–775

Ferre S, Cortés R, Artigas F (1994) Dopaminergic regulation of the serotonin raphe-striatal pathway: Microdialysis studies in freely moving rats. J Neurosci 14:4839–4846
Feuerstein TJ, Hertting G, Lupp A, Neufang B (1986) False labelling of dopaminergic terminals in the rabbit caudate nucleus: uptake and release of [3H]-5-hydroxytryptamine. Br J Pharmacol 88:677–684
Fink KB, Göthert M (2007) 5-HT receptor regulation of neurotransmitter release. Pharmacol Rev 59:360–417
Freed C, Revay R, Vaughan RA, Kriek E, Grant S, Uhl GR, et al (1995) Dopamine transporter immunoreactivity in rat brain. J Comp Neurol 359:340–349
Gainetdinov RR, Jones SR, Fumagalli F, Wightman RM, Caron MG (1998) Re-evaluation of the role of the dopamine transporter in dopamine system homeostasis. Brain Res Rev 26:148–153
Gao C, Sun X, Wolf ME (2006) Activation of D_1 dopamine receptors increases surface expression of AMPA receptors and facilitates their synaptic incorporation in cultured hippocampal neurons. J Neurochem 98:1664–1677
Garraway SM, Hochman S (2001) Modulatory actions of serotonin, norepinephrine, dopamine, and acetylcholine in spinal cord deep dorsal horn neurons. J Neurophysiol 86:2183–2194
Gaspar P, Cases O, Maroteaux L (2003) The developmental role of serotonin: news from mouse molecular genetics. Nat Rev Neurosci 4:1002–1012
Gerfen CR (1992) The neostriatal mosaic: multiple levels of compartmental organization in the basal ganglia. Annu Rev Neurosci 15:285–320
Gerfen CR (2000) Molecular effects of dopamine on striatal-projection pathways. Trends Neurosci 23(10 Suppl):S64–70
Gerfen CR (2003) D1 dopamine receptor supersensitivity in the dopamine-depleted striatum animal model of Parkinson's disease. Neuroscientist 9:455–462
Gerfen CR (2006) Indirect-pathway neurons lose their spines in Parkinson disease. Nat Neurosci 9:157–158
German DC, Manaye KF (1993) Midbrain dopaminergic-neurons (Nuclei A8, A9, and A10): 3-dimensional reconstruction in the rat. J Comp Neurol 331:297–309
Gillman PK (2006) A review of serotonin toxicity data: implications for the mechanisms of antidepressant drug action. Biol Psychiatry 59:1046–1051
Giorgetti M, Tecott LH (2004) Contributions of 5-HT_{2C} receptors to multiple actions of central serotonin systems. Eur J Pharmacol 488:1–9
Giros B, el Mestikawy S, Bertrand L, Caron MG (1991) Cloning and functional characterization of a cocaine-sensitive dopamine transporter. FEBS Lett 295:149–1454
Gong S, Zheng C, Doughty ML, Losos K, Didkovsky N, Schambra UB, et al (2003) A gene expression atlas of the central nervous system based on bacterial artificial chromosomes. Nature 425:917–925
Gonon F (1997) Prolonged and extrasynaptic excitatory action of dopamine mediated by D_1 receptors in the rat striatum in vivo. J Neurosci 17:5972–5978
Goto Y, Grace AA (2007) The Dopamine system and the pathophysiology of schizophrenia: A basic science perspective. Int Rev Neurobiol 78:41–68
Grace AA (1991) Phasic versus tonic dopamine release and the modulation of dopamine system responsivity: a hypothesis for the etiology of schizophrenia. Neuroscience 41:1–24
Greengard P (2001) The neurobiology of slow synaptic transmission. Science 294:1024–1030
Greif GJ, Lin YJ, Liu JC, Freedman JE (1995) Dopamine-modulated potassium channels on rat striatal neurons: specific activation and cellular expression. J Neurosci 15:4533–4544
Groves PM, Linder JC, Young SJ (1994) 5-hydroxydopamine-labeled dopaminergic axons: three-dimensional reconstructions of axons, synapses and postsynaptic targets in rat neostriatum. Neuroscience 58:593–604
Guillin O, Abi-Dargham A, Laruelle M (2007) Neurobiology of dopamine in schizophrenia. Int Rev Neurobiol 78:1–39

Haber SN, Fudge JL, McFarland NR (2000) Striatonigrostriatal pathways in primates form an ascending spiral from the shell to the dorsolateral striatum. J Neurosci 20:2369–2382
Haber SN (2003) The primate basal ganglia: parallel and integrative networks. J Chem Neuroanat 26:317–330
Haj-Dahmane S (2001) D2-like dopamine receptor activation excites rat dorsal raphe 5-HT neurons in vitro. Eur J Neurosci 14:125–134
Hallett PJ, Spoelgen R, Hyman BT, Standaert DG, Dunah AW (2006) Dopamine D1 activation potentiates striatal NMDA receptors by tyrosine phosphorylation-dependent subunit trafficking. J Neurosci 26:4690–4700
Hernandez-Lopez S, Tkatch T, Perez-Garci E, Galarraga E, Bargas J, Hamm H, Surmeier DJ (2000) D2 dopamine receptors in striatal medium spiny neurons reduce L-type Ca^{2+} currents and excitability via a novel PLCβ1-IP3-calcineurin-signaling cascade. J Neurosci 20:8987–8995
Hersch SM, Ciliax BJ, Gutekunst CA, Rees HD, Heilman CJ, Yung KK, Bolam JP,Ince E, Yi H, Levey AI (1995)Electron microscopic analysis of D1 and D2 dopamine receptor proteins in the dorsal striatum and their synaptic relationships with motor corticostriatal afferents. J Neurosci 15:5222–5237
Hersch SM, Yi H, Heilman CJ, Edwards RH, Levey AI (1997) Subcellular localization and molecular topology of the dopamine transporter in the striatum and substantia nigra. J Comp Neurol 388:211–227
Herve D, Pickel VM, Joh TH, Beaudet A (1987) Serotonin axon terminals in the ventral tegmental area of the rat: fine structure and synaptic input to dopaminergic neurons. Brain Res 435:71–83
Herve D, Girault A (2005) Signal transduction of dopamine receptors. In: Dunnett SB, Bentivoglio M, Bjorklund A, Hokfelt T (ed) Handbook of Chemical Neuroanatomy: Dopamine Vol. 21, Elsevier, Amsterdam. pp. 109–151
Hilfiker S, Czernik AJ, Greengard P, Augustine GJ (2001) Tonically active protein kinase A regulates neurotransmitter release at the squid giant synapse. J Physiol 531:141–146
Hopf FW, Cascini MG, Gordon AS, Diamond I, Bonci A (2003) Cooperative activation of dopamine D_1 and D_2 receptors increases spike firing of nucleus accumbens neurons via G-Protein βγ subunits. J Neurosci 23:5079–5087
Hornung JP (2003) The human raphe nuclei and the serotonergic system. J Chem Neuroanat 26:331–343
Hornykiewicz O (2001) Chemical neuroanatomy of the basal ganglia — normal and in Parkinson's disease. J Chem Neuroanat 22:3–12
Hornykiewicz O (2002) Brain dopamine: A historical perspective. In: Di Chiara G (ed) Dopamine in CNS I. Springer-Verlag, Berlin. pp. 1–22
Hoyer D, Hannon JP, Martin GR (2002) Molecular, pharmacological and functional diversity of 5-HT receptors. Pharmacol Biochem Behav 71:533–554
Hu XT, Dong Y, Zhang XF, White FJ (2005) Dopamine D_2 receptor-activated Ca^{2+} signaling modulates voltage-sensitive sodium currents in rat nucleus accumbens neurons. J Neurophysiol 93:1406–1417
Hyman SE, Malenka RC, Nestler EJ (2006) Neural mechanisms of addiction: the role of reward-related learning and memory. Annu Rev Neurosci 29:565–598
Ibanez-Sandoval O, Hernandez A, Floran B, Galarraga E, Tapia D, Valdiosera R, Erlij D, Aceves J, Bargas J (2006) Control of the subthalamic innervation of substantia nigra pars reticulata by D_1 and D_2 dopamine receptors. J Neurophysiol 95:1800–1811
Inoue Y, Yao L, Hopf FW, Fan P, Jiang Z, Bonci A, Diamond I (2007) Nicotine and ethanol activate protein kinase A synergistically via Gi βγ ubunits in nucleus accumbens/ventral tegmental cocultures: the role of dopamine D_1/D_2 and adenosine A_{2A} receptors. J Pharmacol Exp Ther 322:23–29
Invernizzi RW, Pierucci M, Calcagno E, Di Giovanni G, Di Matteo V, Benigno A, Esposito E (2007) Selective activation of $5\text{-}HT_{2C}$ receptors stimulates GABAergic function in the rat

substantia nigra pars reticulata: A combined in vivo electrophysiological and neurochemical study. Neuroscience 144:1523–1535
Izumi T, Iwamoto N, Kitaichi Y, Kato A, Inoue T, Koyama T (2006) Effects of co-administration of a selective serotonin reuptake inhibitor and monoamine oxidase inhibitors on 5-HT-related behavior in rats. Eur J Pharmacol 532:258–264
Jackson BP, Wightman RM (1995) Dynamics of 5-hydroxytryptamine released from dopamine neurons in the caudate putamen of the rat. Brain Res 674:163–166
Jacobs BL, Azmitia EC (1992) Structure and function of the brain serotonin system. Physiol Rev 72:165–229
Jin LQ, Wang HY, Friedman E (2001) Stimulated D_1 dopamine receptors couple to multiple Galpha proteins in different brain regions. J Neurochem 78:981–990
Jones SR, Gainetdinov RR, Jaber M, Giros B, Wightman RM, Caron MG (1998a) Profound neuronal plasticity in response to inactivation of the dopamine transporter. Proc Natl Acad Sci USA 95:4029–4034
Jones SR, Gainetdinov RR, Wightman RM, Caron MG (1998b) Mechanisms of amphetamine action revealed in mice lacking the dopamine transporter. J Neurosci 18:1979–1986
Kalen P, Skagerberg G, Lindvall O (1988) Projections from the ventral tegmental area and mesencephalic raphe to the dorsal raphe nucleus in the rat. Exp Brain Res 73:69–77
Kannari K, Shen H, Arai A, Tomiyama M, Baba M (2006) Reuptake of L-DOPA-derived extracellular dopamine in the striatum with dopaminergic denervation via serotonin transporters. Neurosci Lett 402:62–65
Kapur S, Mamo D (2003) Half a century of antipsychotics and still a central role for dopamine D_2 receptors. Prog Neuropsychopharmacol Biol Psychiatry 27:1081–1090
Kebabian JW, Calne DB (1979) Multiple receptors for dopamine. Nature 277:93–96
Khan ZU, Gutierrez A, Martin R, Penafiel A, Rivera A, de la Calle A (2000) Dopamine D5 receptors of rat and human brain. Neuroscience 100:689–699
Khan ZU, Mrzljak L, Gutierrez A, de la Calle A, Goldman-Rakic PS (1998) Prominence of the dopamine D2 short isoform in dopaminergic pathways. Proc Natl Acad Sci USA 95:7731–7736
Kidd EJ, Laporte AM, Langlois X, Fattaccini CM, Doyen C, Lombard MC, Gozlan H, Hamon M (1993) 5-HT_3 receptors in the rat central nervous system are mainly located on nerve fibres and terminals. Brain Res 612:289–298
Kilpatrick GJ, Jones BJ, Tyers MB (1987) Identification and distribution of 5-HT_3 receptors in rat brain using radioligand binding. Nature 330:746–748
Kitahama K, Nagatsu I, Geffard M, Maeda T (2000) Distribution of dopamine-immunoreactive fibers in the rat brainstem. J Chem Neuroanat 18:1–9
Knobelman DA, Kung HF, Lucki I (2000) Regulation of extracellular concentrations of 5-hydroxytryptamine (5-HT) in mouse striatum by 5-HT_{1A} and 5-HT_{1B} receptors. J Pharmacol Exp Ther 292:1111–1117
Kostrzewa RM, Reader TA, Descarries L (1998) Serotonin neural adaptations to ontogenetic loss of dopamine neurons in rat brain. J Neurochem 70:889–898
Lacey MG (1993) Neurotransmitter receptors and ionic conductances regulating the activity of neurones in substantia nigra pars compacta and ventral tegmental area. Prog Brain Res 99:251–276
Lacey MG, Mercuri NB, North RA (1987) Dopamine acts on D2 receptors to increase potassium conductance in neurones of the rat substantia nigra zona compacta. J Physiol 392:397–416
Lacey MG, Mercuri NB, North RA (1988) On the potassium conductance increase activated by GABAB and dopamine D_2 receptors in rat substantia nigra neurones. J Physiol 401:437–453
Lambe EK, Aghajanian GK (2001) The role of Kv1.2-containing potassium channels in serotonin-induced glutamate release from thalamocortical terminals in rat frontal cortex. J Neurosci 21:9955–9963

Laporte AM, Koscielniak T, Ponchant M, Verge D, Hamon M, Gozlan H. (1992) Quantitative autoradiographic mapping of 5-HT3 receptors in the rat CNS using [125I]iodozacopride and [3H]zacopride as radioligands. Synapse 10:271–281

Laurent A, Goaillard JM, Cases O, Lebrand C, Gaspar P, Ropert N (2002) Activity-dependent presynaptic effect of serotonin 1B receptors on the somatosensory thalamocortical transmission in neonatal mice. J Neurosci 22:886–900

Lavoie B, Parent A (1990) Immunohistochemical study of the serotoninergic innervation of the basal ganglia in the squirrel monkey. J Comp Neurol 299: 1–16

Lee SP, So CH, Rashid AJ, Varghese G, Cheng R, Lança AJ, O'Dowd BF, George SR (2004) Dopamine D1 and D2 receptor Co-activation generates a novel phospholipase C-mediated calcium signal. J Biol Chem 279:35671–35678

Lee HS, Kim MA, Waterhouse BD (2005) Retrograde double-labeling study of common afferent projections to the dorsal raphe and the nuclear core of the locus coeruleus in the rat. J Comp Neurol 481:179–193

Leger L, Wiklund L (1982) Distribution and numbers of indoleamine cell bodies in the cat brainstem determined with Falck-Hillarp fluorescence histochemistry. Brain Res Bull 9:245–251

Levey AI, Hersch SM, Rye DB, Sunahara RK, Niznik HB, Kitt CA, Price DL, Maggio R, Brann MR, Ciliax BJ, et al. (1993) Localization of D_1 and D_2 dopamine receptors in brain with subtype-specific antibodies. Proc Natl Acad Sci USA 90:8861–8865

Lewis DA, Melchitzky DS, Sesack SR, Whitehead RE, Auh S, Sampson A (2001) Dopamine transporter immunoreactivity in monkey cerebral cortex: regional, laminar, and ultrastructural localization. J Comp Neurol 432:119–136

Lin YJ, Greif GJ, Freedman JE (1996) Permeation and block of dopamine-modulated potassium channels on rat striatal neurons by cesium and barium ions. J Neurophysiol 76:1413–1422

Liu XY, Chu XP, Mao LM, Wang M, Lan HX, Li MH, Zhang GC, Parelkar NK, Fibuch EE, Haines M, Neve KA, Liu F, Xiong ZG, Wang JQ (2006) Modulation of D2R-NR2B interactions in response to cocaine. Neuron 52:897–909

Lopez-Garcia JA (2006) Serotonergic modulation of spinal sensory circuits. Curr Top Med Chem 6:1987–1996

Lopez-Real A, Rodriguez-Pallares J, Guerra MJ, Labandeira-Garcia JL (2003) Localization and functional significance of striatal neurons immunoreactive to aromatic L-amino acid decarboxylase or tyrosine hydroxylase in rat Parkinsonian models. Brain Res 969:135–146

Lopez-Gimenez JF, Mengod G, Palacios JM, Vilaro MT (2001) Regional distribution and cellular localization of 5-HT2C receptor mRNA in monkey brain: comparison with [3H]mesulergine binding sites and choline acetyltransferase mRNA. Synapse 42:12–26

Maeda T, Kannari K, Shen H, Arai A, Tomiyama M, Matsunaga M, Suda T (2003) Rapid induction of serotonergic hyperinnervation in the adult rat striatum with extensive dopaminergic denervation. Neurosci Lett 343:17–20

Maeda T, Nagata K, Yoshida Y, Kannari K (2005) Serotonergic hyperinnervation into the dopaminergic denervated striatum compensates for dopamine conversion from exogenously administered l-DOPA. Brain Res 1046:230–233

Mangiavacchi S, Wolf ME (2004) D1 dopamine receptor stimulation increases the rate of AMPA receptor insertion onto the surface of cultured nucleus accumbens neurons through a pathway dependent on protein kinase A. J Neurochem 88:1261–1271

Mannoury la Cour C, Vidal S, Pasteau V, Cussac D, Millan MJ (2007) Dopamine D1 receptor coupling to $G_{s/olf}$ and G_q in rat striatum and cortex: a scintillation proximity assay (SPA)/antibody-capture characterization of benzazepine agonists. Neuropharmacology 52:1003–1014

Mansour A, Meador-Woodruff JH, Zhou Q, Civelli O, Akil H, Watson SJ (1992) A comparison of D_1 receptor binding and mRNA in rat brain using receptor autoradiographic and in situ hybridization techniques. Neuroscience 46:959–971

Marcusson J, Eriksson K (1988) [3H]GBR-12935 binding to dopamine uptake sites in the human brain. Brain Res 457:122–129

Martin-Ruiz R, Ugedo L, Honrubia MA, Mengod G, Artigas F (2001) Control of serotonergic neurons in rat brain by dopaminergic receptors outside the dorsal raphe nucleus. J Neurochem 77:762–775

Masson J, Sagne C, Hamon M, El Mestikawy S (1999) Neurotransmitter transporters in the central nervous system. Pharmacol Rev 51:439–464

Matsumoto M, Yoshioka M, Togashi H, Ikeda T, and Saito H (1996) Functional regulation by dopamine receptors of serotonin release from the rat hippocampus: In vivo microdialysis study. Naunyn-Schmiedeberg's Arch Pharmacol 353:621– 629.

Meador-Woodruff JH, Mansour A, Healy DJ, Kuehn R, Zhou QY, Bunzow JR, Akil H, Civelli O, Watson SJ, Jr (1991) Comparison of the distributions of D_1 and D_2 dopamine receptor mRNAs in rat brain. Neuropsychopharmacology 5:231–242

Mennicken F, Savasta M, Peretti-Renucci R, Feuerstein C (1992) Autoradiographic localization of dopamine uptake sites in the rat brain with 3H-GBR 12935. J Neural Transm Gen Sect 87:1–14

Mercuri NB, Saiardi A, Bonci A, Picetti R, Calabresi P, Bernardi G, et al (1997) Loss of autoreceptor function in dopaminergic neurons from dopamine D_2 receptor deficient mice. Neuroscience 79:323–327

Miller DW, Abercrombie ED (1999) Role of high-affinity dopamine uptake and impulse activity in the appearance of extracellular dopamine in striatum after administration of exogenous L-DOPA: studies in intact and 6-hydroxydopamine-treated rats. J Neurochem 724:1516–1522

Miller GM, Yatin SM, De La Garza R 2nd, Goulet M, Madras BK (2001) Cloning of dopamine, norepinephrine and serotonin transporters from monkey brain: relevance to cocaine sensitivity. Brain Res Mol Brain Res 87:124–143

Missale C, Nash SR, Robinson SW, Jaber M, Caron MG (1998) Dopamine receptors: from structure to function. Physiol Rev 78:189–225

Miyazaki T, Lacey MG (1998) Presynaptic inhibition by dopamine of a discrete component of GABA release in rat substantia nigra pars reticulata. J Physiol 513:805–817

Mizutani H, Hori T, Takahashi T (2006) 5-HT_{1B} receptor-mediated presynaptic inhibition at the calyx of Held of immature rats. Eur J Neurosci 24:1946–1954.

Moukhles H, Bosler O, Bolam JP, Vallee A, Umbriaco D, Geffard M, et al (1997) Quantitative and morphometric data indicate precise cellular interactions between serotonin terminals and postsynaptic targets in rat substantia nigra. Neuroscience 76:1159–1171

Moore RY, Halaris AE, Jones BE (1978) Serotonin neurons of the midbrain raphe: ascending projections. J Comp Neurol 180:417–438.

Morales M, Battenberg E, Bloom FE (1998) Distribution of neurons expressing immunoreactivity for the 5HT3 receptor subtype in the rat brain and spinal cord. *J Comp Neurol* 402:385–401

Mori S, Ueda S, Yamada H, Takino T, Sano Y (1985) Immunohistochemical demonstration of serotonin nerve fibers in the corpus striatum of the rat, cat and monkey. Anat Embryol (Berl) 173:1–5

Mori S, Matsuura T, Takino T, Sano Y (1987) Light and electron microscopic immunohistochemical studies of serotonin nerve fibers in the substantia nigra of the rat, cat and monkey. Anat Embryol (Berl) 176:13–18

Morikawa H, Manzoni OJ, Crabbe JC, Williams JT (2000) Regulation of central synaptic transmission by 5-HT_{1B} auto- and heteroreceptors. Mol Pharmacol 58:1271–1278

Mossner R, Simantov R, Marx A, Lesch KP, Seif I (2006) Aberrant accumulation of serotonin in dopaminergic neurons. Neurosci Lett 401:49–54

Mrini A, Soucy JP, Lafaille F, Lemoine P, Descarries L (1995) Quantification of the serotonin hyperinnervation in adult rat neostriatum after neonatal 6-hydroxydopamine lesion of nigral dopamine neurons. Brain Res 669:303–308

Mura A, Linder JC, Young SJ, Groves PM (2000) Striatal cells containing aromatic L-amino acid decarboxylase: an immunohistochemical comparison with other classes of striatal neurons. Neuroscience 98:501–511

Muramatsu M, Lapiz MD, Tanaka E, Grenhoff J (1998) Serotonin inhibits synaptic glutamate currents in rat nucleus accumbens neurons via presynaptic 5-HT_{1B} receptors. Eur J Neurosci 10:2371–2379

Nasif FJ, Hu XT, White FJ (2005) Repeated cocaine administration increases voltage-sensitive calcium currents in response to membrane depolarization in medial prefrontal cortex pyramidal neurons. J Neurosci 25:3674–3679

Nayak SV, Ronde P, Spier AD, Lummis SC, Nichols RA (1999) Calcium changes induced by presynaptic 5-HT_3 serotonin receptors on isolated terminals from various regions of the rat brain. Neuroscience 91:107–117

Nelson EL, Liang CL, Sinton CM, German DC (1996) Midbrain dopaminergic neurons in the mouse: computer-assisted mapping. J Comp Neurol 369:361–371

Nemeroff CB, Owens MJ (2002) Treatment of mood disorders. Nat Neurosci 5 Suppl:1068–1070

Neve KA, Neve RL (1997) Molecular biology of dopamine receptors. In: Neve KA and Neve RL (ed) The dopamine receptors. Humana Press, Totowa, New Jersey. pp. 27–76.

Neve KA, Seamans JK, Trantham-Davidson H (2004) Dopamine receptor signaling. J Recept Signal Transduct Res 24:165–205

Nichols RA, Mollard P (1996) Direct observation of serotonin 5-HT_3 receptor-induced increases in calcium levels in individual brain nerve terminals. J Neurochem 67:581–592

Nicola SM, Surmeier J, Malenka RC (2000) Dopaminergic modulation of neuronal excitability in the striatum and nucleus accumbens. Annu Rev Neurosci 23:185–215

Nirenberg MJ, Vaughan RA, Uhl GR, Kuhar MJ, Pickel VM (1996) The dopamine transporter is localized to dendritic and axonal plasma membranes of nigrostriatal dopaminergic neurons. J Neurosci 16:436–447

Nirenberg MJ, Chan J, Pohorille A, Vaughan RA, Uhl GR, Kuhar MJ, Pickel VM (1997a) The dopamine transporter: comparative ultrastructure of dopaminergic axons in limbic and motor compartments of the nucleus accumbens. J Neurosci 17:6899–6907

Nirenberg MJ, Chan J, Vaughan RA, Uhl GR, Kuhar MJ, Pickel VM (1997b) Immunogold localization of the dopamine transporter: an ultrastructural study of the rat ventral tegmental area. J Neurosci 17:5255–5262

Nocjar C, Roth BL, Pehek EA (2002) Localization of 5-HT_{2A} receptors on dopamine cells in subnuclei of the midbrain A10 cell group. Neuroscience 111:163–176

North RA, Uchimura N (1989) 5-Hydroxytryptamine acts at 5-HT_2 receptors to decrease potassium conductance in rat nucleus accumbens neurones. J Physiol 417:1–12

Oak JN, Oldenhof J, Van Tol HH (2000) The dopamine D_4 receptor: one decade of research. Eur J Pharmacol 405:303–327

Obeso JA, Olanow CW, Nutt JG (2000) Levodopa motor complications in Parkinson's disease. Trends Neurosci 23:S2–7

Okumura T, Dobolyi A, Matsuyama K, Mori F, Mori S (2000) The cat neostriatum: relative distribution of cholinergic neurons versus serotonergic fibers. Brain Dev 22:S27–37

Olanow CW (2004) The scientific basis for the current treatment of Parkinson's disease. Annu Rev Med 55:41–60

Olanow CW, Obeso JA, Stocchi F (2006) Continuous dopamine-receptor treatment of Parkinson's disease: scientific rationale and clinical implications. Lancet Neurol 5:677–687

Olijslagers JE, Werkman TR, McCreary AC, Siarey R, Kruse CG, Wadman WJ (2004) 5-HT2 receptors differentially modulate dopamine-mediated auto-inhibition in A9 and A10 midbrain areas of the rat. Neuropharmacology 46:504–510

Oorschot DE (1996) Total number of neurons in the neostriatal, pallidal, subthalamic, and substantia nigral nuclei of the rat basal ganglia: a stereological study using the cavalieri and optical disector methods. J Comp Neurol 366:580–599

O'Reilly CA, Reith ME (1988) Uptake of [3H]serotonin into plasma membrane vesicles from mouse cerebral cortex. J Biol Chem 263:6115–6121
Patel S, Roberts J, Moorman J, Reavill C (1995) Localization of serotonin-4 receptors in the striatonigral pathway in rat brain. Neuroscience 69:1159–1167
Paterson DS, Trachtenberg FL, Thompson EG, Belliveau RA, Beggs AH, Darnall R, Chadwick AE, Krous HF, Kinney HC (2006) Multiple serotonergic brainstem abnormalities in sudden infant death syndrome. JAMA 296:2124–2132
Perez MF, White FJ, Hu XT (2006) Dopamine D_2 receptor modulation of K^+ channel activity regulates excitability of nucleus accumbens neurons at different membrane potentials. J Neurophysiol 96:2217–2228
Pessia M, Jiang ZG, North RA, Johnson SW (1994) Actions of 5-hydroxytryptamine on ventral tegmental area neurons of the rat in vitro. Brain Res 654:324–330
Peters JL, Michael AC (2000) Changes in the kinetics of dopamine release and uptake have differential effects on the spatial distribution of extracellular dopamine concentration in rat striatum. J Neurochem 74:1563–1673
Peterson JD, Wolf ME, White FJ (2006) Repeated amphetamine administration decreases D1 dopamine receptor-mediated inhibition of voltage-gated sodium currents in the prefrontal cortex. J Neurosci 26:3164–3168
Peyron C, Luppi PH, Kitahama K, Fort P, Hermann DM, Jouvet M (1995) Origin of the dopaminergic innervation of the rat dorsal raphe nucleus. Neuroreport 6:2527–2531
Pickel VM, Chan J (1999) Ultrastructural localization of the serotonin transporter in limbic and motor compartments of the nucleus accumbens. J Neurosci 19:7356–7366
Pineyro G, Blier P (1999) Autoregulation of serotonin neurons: role in antidepressant drug action. Pharmacol Rev 51:533–591
Pisani A, Bernardi G, Ding J, Surmeier DJ (2007) Re-emergence of striatal cholinergic interneurons in movement disorders. Trends Neurosci 30:545–553
Pletscher A, Shore PA, Brodie BB (1956) Serotonin as a mediator of reserpine action in brain. J Pharmacol Exp Ther 116:84–89
Povlock SL, Amara SG (1997) The structure and function of norepinephrine, dopamine, and serotonin transporters. In: Reith MEA (ed) Neurotransmitter transporters: structure, function, and regulation. Humana Press, Totowa, New Jersey. pp. 1–28
Prensa L, Parent A (2001) The nigrostriatal pathway in the rat: A single-axon study of the relationship between dorsal and ventral tier nigral neurons and the striosome/matrix striatal compartments. J Neurosci 21:7247–7260
Prisco S, Esposito E (1995) Differential effects of acute and chronic fluoxetine administration on the spontaneous activity of dopaminergic neurones in the ventral tegmental area. Br J Pharmacol 116:1923–1931
Pristupa ZB, Wilson JM, Hoffman BJ, Kish SJ, Niznik HB (1994) Pharmacological heterogeneity of the cloned and native human dopamine transporter: disassociation of [3H]WIN 35,428 and [3H]GBR 12,935 binding. Mol Pharmacol 45:125–135
Puig MV, Celada P, Diaz-Mataix L, Artigas F (2003) In vivo modulation of the activity of pyramidal neurons in the rat medial prefrontal cortex by 5-HT_{2A} receptors: relationship to thalamocortical afferents. Cereb Cortex 13:870–882
Ramamoorthy S, Bauman AL, Moore KR, Han H, Yang-Feng T, Chang AS, Ganapathy V, Blakely RD (1993) Antidepressant- and cocaine-sensitive human serotonin transporter: molecular cloning, expression, and chromosomal localization. Proc Natl Acad Sci USA 90:2542–2546
Rashid AJ, O'Dowd BF, Verma V, George SR (2007a) Neuronal $G_{q/11}$-coupled dopamine receptors: an uncharted role for dopamine. Trends Pharmacol Sci 28:551–555
Rashid AJ, So CH, Kong MM, Furtak T, El-Ghundi M, Cheng R, O'Dowd BF, George SR (2007b) D_1-D_2 dopamine receptor heterooligomers with unique pharmacology are coupled to rapid activation of $G_{q/11}$ in the striatum. Proc Natl Acad Sci USA 104:654–659
Richfield EK (1991) Quantitative autoradiography of the dopamine uptake complex in rat brain using [3H]GBR 12935: binding characteristics. Brain Res 540:1–13

Rick CE, Stanford IM, Lacey MG (1995) Excitation of rat substantia nigra pars reticulata neurons by 5-HT in vitro: evidence for a direct action mediated by 5-HT_{2C} receptors. Neuroscience 69:903–913
Rodríguez JJ, Garcia DR, Pickel VM (1999) Subcellular distribution of 5-hydroxytryptamine2A and N-methyl-D-aspartate receptors within single neurons in rat motor and limbic striatum. J Comp Neurol 413:219–231
Schmitz Y, Benoit-Marand M, Gonon F, Sulzer D (2003) Presynaptic regulation of dopaminergic neurotransmission. J Neurochem 87:273–289
Schultz W (2002) Getting formal with dopamine and reward. Neuron 36:241–263
Schultz W (2007) Multiple dopamine functions at different time courses. Annu Rev Neurosci 30:259–288
Seeburg DP, Liu X, Chen C (2004) Frequency-dependent modulation of retinogeniculate transmission by serotonin. J Neurosci 24:10950–10962
Seeman P, Lee T (1975) Antipsychotic drugs: Direct correlation between clinical potency and presynaptic action on dopamine neurons. Science 188:1217–1219
Seeman P (1980) Brain dopamine receptors. Pharmacol Rev 32:229–313
Sesack SR, Hawrylak VA, Matus C, Guido MA, Levey AI (1998) Dopamine axon varicosities in the prelimbic division of the rat prefrontal cortex exhibit sparse immunoreactivity for the dopamine transporter. J Neurosci 18:2697–2708
Sesack SR (2002) Synaptology of dopamine neurons. In: Di Chiara G (ed) Dopamine in the CNS I. Springer-Verlag, Berlin. pp 63–119
Shaskan EG, Snyder SH (1970) Kinetics of serotonin accumulation into slices from rat brain: relationship to catecholamine uptake. J Pharmacol Exp Ther 175:404–418
Shen HW, Hagino Y, Kobayashi H, Shinohara-Tanaka K, Ikeda K, Yamamoto H, et al. (2004) Regional differences in extracellular dopamine and serotonin assessed by in vivo microdialysis in mice lacking dopamine and/or serotonin transporters. Neuropsychopharmacology 29:1790–1799
Shioda K, Nisijima K, Yoshino T, Kato S (2004) Extracellular serotonin, dopamine and glutamate levels are elevated in the hypothalamus in a serotonin syndrome animal model induced by tranylcypromine and fluoxetine. Prog Neuropsychopharmacol Biol Psychiatry 28:633–640
Shore PA, Silver SL, Brodie BB (1955) Interaction of reserpine, serotonin, and lysergic acid diethylamide in brain. Science 122:284–285.
Smiley JF, Goldman-Rakic PS (1996) Serotonergic axons in monkey prefrontal cerebral cortex synapse predominantly on interneurons as demonstrated by serial section electron microscopy. J Comp Neurol 367:431–443
Smith TD, Kuczenski R, George-Friedman K, Malley JD, Foote SL (2000) In vivo microdialysis assessment of extracellular serotonin and dopamine levels in awake monkeys during sustained fluoxetine administration. Synapse 38:460–470
Snyder GL, Allen PB, Fienberg AA, Valle CG, Huganir RL, Nairn AC, Greengard P (2000) Regulation of phosphorylation of the GluR1 AMPA receptor in the neostriatum by dopamine and psychostimulants in vivo. J Neurosci 20:4480–4488
Soghomonian JJ, Doucet G, Descarries L (1987) Serotonin innervation in adult rat neostriatum. I. Quantified regional distribution. Brain Res 425:85–100
Sokoloff P, Giros B, Martres MP, Bouthenet ML, Schwartz JC (1990) Molecular cloning and characterization of a novel dopamine receptor (D3) as a target for neuroleptics. Nature 347:146–151
Sora I, Hall FS, Andrews AM, Itokawa M, Li XF, Wei HB, et al (2001) Molecular mechanisms of cocaine reward: combined dopamine and serotonin transporter knockouts eliminate cocaine place preference. Proc Natl Acad Sci USA 98:5300–5305
Stanford IM, Lacey MG (1996) Differential actions of serotonin, mediated by 5-HT_{1B} and 5-HT_{2C} receptors, on GABA-mediated synaptic input to rat substantia nigra pars reticulata neurons in vitro. J Neurosci 16:7566–7573

Stamford JA, Kruk ZL, Millar J (1990) Striatal dopamine terminals release serotonin after 5-HTP pretreatment: in vivo voltammetric data. Brain Res 515:173–180
Stamford JA, Davidson C, McLaughlin DP, Hopwood SE (2000) Control of dorsal raphe 5-HT function by multiple 5-HT_1 autoreceptors: parallel purposes or pointless plurality Trends Neurosci 23:459–465
Stanford SC (1999) SSRI-induced changes in catecholaminergic transmission. In: Stanford SC (ed) Selective serotonin reuptake inhibitors (SSRIs): Past, present and future. RG Landes Company, Austin, Texas. pp. 147–169
Steinbusch HW, Verhofstad AA, Joosten HW (1978) Localization of serotonin in the central nervous system by immunohistochemistry: description of a specific and sensitive technique and some applications. Neuroscience 3:811–819
Steinbusch HW (1981) Distribution of serotonin-immunoreactivity in the central nervous system of the rat-cell bodies and terminals. Neuroscience 6:557–618
Sternbach H (1991) The serotonin syndrome. Am J Psychiatry 148:705–713
Steward LJ, Bufton KE, Hopkins PC, Davies WE, Barnes NM (1993) Reduced levels of 5-HT_3 receptor recognition sites in the putamen of patients with Huntington's disease. Eur J Pharmacol 242:137–143
Sulzer D (2007) Multiple hit hypotheses for dopamine neuron loss in Parkinson's disease. Trends Neurosci 30:244–250
Sun X, Zhao Y, Wolf M (2005) Dopamine receptor stimulation modulates AMPA receptor synaptic insertion in prefrontal cortex neurons. J Neurosci 25:7342–7351
Sur C, Betz H, Schloss P (1996) Immunocytochemical detection of the serotonin transporter in rat brain. Neuroscience *73*:217–231
Surmeier DJ, Song WJ, Yan Z (1996) Coordinated expression of dopamine receptors in neostriatal medium spiny neurons. J Neurosci 16:6579–6591
Surmeier DJ, Ding J, Day M, Wang Z, Shen W (2007) D_1 and D_2 dopamine-receptor modulation of striatal glutamatergic signaling in striatal medium spiny neurons. Trends Neurosci 30:228–235
Suzuki M, Hurd YL, Sokoloff P, Schwartz JC, Sedvall G (1998) D_3 dopamine receptor mRNA is widely expressed in the human brain. Brain Res 779:58–74
Svenningsson P, Nishi A, Fisone G, Girault JA, Nairn AC, Greengard P (2004) DARPP-32: an integrator of neurotransmission. Annu Rev Pharmacol Toxicol. 44:269–296
Svenningsson P, Tzavara ET, Liu F, Fienberg AA, Nomikos GG, Greengard P (2002a) DARPP-32 mediates serotonergic neurotransmission in the forebrain. Proc Natl Acad Sci USA 99:3188–3193
Svenningsson P, Tzavara ET, Witkin JM, Fienberg AA, Nomikos GG, Greengard P (2002b) Involvement of striatal and extrastriatal DARPP-32 in biochemical and behavioral effects of fluoxetine (Prozac). Proc Natl Acad Sci USA 99:3182–3187
Svenningsson P, Tzavara ET, Carruthers R, Rachleff I, Wattler S, Nehls M, McKinzie DL, Fienberg AA, Nomikos GG, Greengard P (2003) Diverse psychotomimetics act through a common signaling pathway. Science 302:1412–1415
Svenningsson P, Tzavara ET, Qi H, Carruthers R, Witkin JM, Nomikos GG, Greengard P (2007) Biochemical and behavioral evidence for antidepressant-like effects of 5-HT_6 receptor stimulation. J Neurosci 27:4201–4209
Tanaka H, Kannari K, Maeda T, Tomiyama M, Suda T and Matsunaga M (1999) Role of serotonergic neurons in L-DOPA-derived extracellular dopamine in the striatum of 6-OHDA-lesioned rats. NeuroReport 10:631–634
Tanaka S (2006) Dopaminergic control of working memory and its relevance to schizophrenia: a circuit dynamics perspective. Neuroscience 139:153–171
Tecott LH, Maricq AV, Julius D (1993) Nervous system distribution of the serotonin 5-HT_3 receptor mRNA. Proc Natl Acad Sci USA 90:1430–1434
Tepper JM, Martin LP, Anderson DR (1995) GABAA receptor-mediated inhibition of rat substantia nigra dopaminergic neurons by pars reticulata projection neurons. J Neurosci 15:3092–3103

Tork I (1990) Anatomy of the serotonergic system. Ann N Y Acad Sci 600:9–34
Torres GE, Gainetdinov RR, Caron MG (2003) Plasma membrane monoamine transporters: structure, regulation and function. Nat Rev Neurosci 4:13–25
Twarog BM, Page IH (1953) Serotonin content of some mammalian tissues and urine and a method for its determination. Am J Physiol 175:157–161
Usiello A, Baik JH, Rouge-Pont F, Picetti R, Dierich A, LeMeur M, et al (2000) Distinct functions of the two isoforms of dopamine D2 receptors. Nature 408:199–203
Van Bockstaele EJ, Cestari DM, Pickel VM (1994) Synaptic structure and connectivity of serotonin terminals in the ventral tegmental area: potential sites for modulation of mesolimbic dopamine neurons. Brain Res 647:307–322
Van Bockstaele EJ, Chan J, Pickel VM (1996) Pre- and postsynaptic sites for serotonin modulation of GABA-containing neurons in the shell region of the rat nucleus accumbens. J Comp Neurol 371:116–128
Van Bockstaele EJ, Pickel VM (1993) Ultrastructure of serotonin-immunoreactive terminals in the core and shell of the rat nucleus accumbens: cellular substrates for interactions with catecholamine afferents. J Comp Neurol 334:603–617
van Hooft JA, Vijverberg HP (2000) 5-HT_3 receptors and neurotransmitter release in the CNS: a nerve ending story? Trends Neurosci 23:605–610
van Hooft JA, Yakel JL (2003) 5-HT_3 receptors in the CNS: 3B or not 3B? Trends Pharmacol Sci 24:157–160
van Rossum JM (1966) The significance of dopamine-receptor blockade for the mechanism of action of neuroleptic drugs. Arch Int Pharmacodyn Ther 160:492–494
Vanhatalo S, Soinila S (1994) Pharmacological characterization of serotonin synthesis and uptake suggest a false neurotransmitter role for serotonin in the pituitary intermediate lobe. Neurosci Res 21:143–149
Vanhatalo S, Soinila S (1995) Release of false transmitter serotonin from the dopaminergic nerve terminals of the rat pituitary intermediate lobe. Neurosci Res 22:367–374
Vanhatalo S, Soinila S (1998) Serotonin is not synthesized, but specifically transported in the neurons of the hypothalamic dorsomedial nucleus. Eur J Neurosci 10:1930–1935
Vilaro MT, Cortes R, Mengod G (2005) Serotonin 5-HT_4 receptors and their mRNAs in rat and guinea pig brain: distribution and effects of neurotoxic lesions. J Comp Neurol 484:418–439
Vitalis T, Cases O, Callebert J, Launay JM, Price DJ, Seif I, Gaspar P (1998) Effects of monoamine oxidase A inhibition on barrel formation in the mouse somatosensory cortex: determination of a sensitive developmental period. J Comp Neurol 393:169–184
Waeber C, Dixon K, Hoyer D, Palacios JM (1988)Localisation by autoradiography of neuronal 5-HT_3 receptors in the mouse CNS. Eur J Pharmacol 151:351–362
Waeber C, Sebben M, Grossman C, Javoy-Agid F, Bockaert J, Dumuis A (1993) [3H]-GR113808 labels 5-HT_4 receptors in the human and guinea-pig brain. Neuroreport 4:1239–1242
Waeber C, Sebben M, Bockaert J, Dumuis A (1996) Regional distribution and ontogeny of 5-HT_4 binding sites in rat brain. Behav Brain Res 73:259–262
Wang HY, Undie AS, Friedman E (1995) Evidence for the coupling of G_q protein to D1-like dopamine sites in rat striatum: possible role in dopamine-mediated inositol phosphate formation. Mol Pharmacol 48:988–994
West AR, Grace AA (2002) Opposite influences of endogenous dopamine D1 and D2 receptor activation on activity states and electrophysiological properties of striatal neurons: studies combining in vivo intracellular recordings and reverse microdialysis. J Neurosci 22:294–304
Whitaker-Azmitia PM (1999) The discovery of serotonin and its role in neuroscience. Neuropsychopharmacology 21(2 Suppl):2S–8S
White FJ (1996) Synaptic regulation of mesocorticolimbic dopamine neurons. Annu Rev Neurosci 19:405–436
Wiklund L, Leger L, Persson M (1981) Monoamine cell distribution in the cat brain stem. A fluorescence histochemical study with quantification of indolaminergic and locus coeruleus cell groups. J Comp Neurol 203:613–647

Wong AH, Van Tol HH (2003) The dopamine D_4 receptors and mechanisms of antipsychotic atypicality. Prog Neuropsychopharmacol Biol Psychiatry 27:1091–1099
Wong DT, Bymaster FP, Engleman EA (1995) Prozac (fluoxetine, Lilly 110140), the first selective serotonin uptake inhibitor and an antidepressant drug: twenty years since its first publication. Life Sci 57:411–441
Wong DT, Perry KW, Bymaster FP (2005) Case history: the discovery of fluoxetine hydrochloride (Prozac). Nat Rev Drug Discov 4:764–774
Woolley DW, Shaw W (1954) A biochemical and pharmacological suggestion about certain mental disorders. Proc Natl Acad Sci USA 40:228–231
Wu JP, Hablitz JJ (2005) Cooperative activation of D_1 and D_2 dopamine receptors enhances a hyperpolarization-activated inward current in layer I interneurons. J Neurosci 25:6322–6328
Wu J, Dougherty JJ, Nichols RA (2006) Dopamine receptor regulation of Ca^{2+} levels in individual isolated nerve terminals from rat striatum: comparison of presynaptic D1-like and D2-like receptors. J Neurochem 98:481–494
Xiang Z, Wang L, Kitai ST (2005) Modulation of spontaneous firing in rat subthalamic neurons by 5-HT receptor subtypes. J Neurophysiol 93:1145–1157
Xu Y, Sari Y, Zhou FC (2004) Selective serotonin reuptake inhibitor disrupts organization of thalamocortical somatosensory barrels during development. Dev Brain Res 150:151–161
Yamada H, Aimi Y, Nagatsu I, Taki K, Kudo M, Arai R (2007) Immunohistochemical detection of l-DOPA-derived dopamine within serotonergic fibers in the striatum and the substantia nigra pars reticulata in Parkinsonian model rats. Neurosci Res 59:1–7
Yelin R, Schuldiner S (2002) Vesicular neurotransmitter transporters: pharmacology, biochemistry, and molecular analysis. In: Reith MEA (ed) Neurotransmitter transporters: structure, function, and regulation. Humana Press, Totowa, New Jersey. pp 313–354
Yung KK, Bolam JP, Smith AD, Hersch SM, Ciliax BJ, Levey AI (1995) Immunocytochemical localization of D_1 and D_2 dopamine receptors in the basal ganglia of the rat: light and electron microscopy. Neuroscience 65:709–730
Zhang X, Andren PE, Svenningsson P (2007) Changes on 5-HT_2 receptor mRNAs in striatum and subthalamic nucleus in Parkinson disease model. Physiol Behav 92:29–33
Zhang H, Sulzer D (2004) Frequency-dependent modulation of dopamine release by nicotine. Nat Neurosci 7:581–582
Zhong G, Diaz-Rios M, Harris-Warrick RM (2006) Intrinsic and functional differences among commissural interneurons during fictive locomotion and serotonergic modulation in the neonatal mouse. J Neurosci 26:6509–6517
Zhou FC, Lesch KP, Murphy DL (2002) Serotonin uptake into dopamine neurons via dopamine transporters: a compensatory alternative. Brain Res 942:109–119
Zhou F-M, Dani JA (2006) Antidepressants alter mesostriatal dopamine interactions with serotonin signaling. Current Psychiatry Reviews 2:453–461
Zhou F-M, Hablitz JJ (1999) Activation of serotonin receptors modulates synaptic transmission in rat cerebral cortex. J Neurophysiol 82:2989–2999
Zhou F-M, Liang Y, Dani JA (2001) Endogenous nicotinic cholinergic activity regulates dopamine release in the striatum. Nat Neurosci 4:1224–1249
Zhou F-M, Liang Y, Salas R, Zhang L, De Biasi M, Dani JA (2005) Corelease of dopamine and serotonin from striatal dopamine terminals. Neuron 46:65–74

Chapter 10
The Dual Glutamatergic/GABAergic Phenotype of Hippocampal Granule Cells

R. Gutiérrez

Abstract Not only are markers of the glutamatergic and GABAergic phenotype both found in developing hippocampal granule cells but, the activation of these cells simultaneously produces responses mediated by glutamate and GABA receptors in their postsynaptic cells. In the adult, the markers of the GABAergic phenotype and as a consequence, GABAergic transmission disappears. However, these elements can still be transiently expressed in an activity-dependent manner. Indeed, the induction of hyperexcitability in granule cells by LTP-like stimulation in vitro, or by seizures in vivo, up-regulates the GABAergic machinery such that stimulation again provokes monosynaptic GABA receptor mediated responses in their target cells. The putative release of GABA from the mossy fibers (MFs) has been shown to differ according to the postsynaptic cell type in the CA3 area and such release can even activate GABA-A receptors located on the MFs themselves. Accordingly, and despite the emergence of this aberrant GABAergic influx on the CA3 area, the effective inhibitory control on CA3 activity exerted by the dentate gyrus is preserved.

The co-release of glutamate and GABA from single cells gives the central nervous system a powerful computational tool and provides a valuable element to help understand the plasticity inherent in neuronal communication.

10.1 Introduction

The dentate gyrus (DG) is the main target of cortical inputs to the hippocampus. From there, its principal cells, the granule cells, project through the mossy fiber (MF) pathway to the CA3 area. It is believed that the MFs form glutamatergic excitatory synapses with pyramidal cells and local inhibitory interneurons of the CA3 (Crawford and Connor, 1973; Acsády et al., 1998). Indeed,

R. Gutiérrez (✉)
Department of Physiology, Biophysics and Neurosciences, Center for Research and Advanced Studies of the National Polytechnic Institute, Post Box 14–740, México D.F. 07000
e-mail: grafael@fisio.cinvestav.mx

R. Gutierrez (ed.), *Co-Existence and Co-Release of Classical Neurotransmitters*,
DOI 10.1007/978-0-387-09622-3_10, © Springer Science+Business Media, LLC 2009

these synapses were originally identified as both asymmetric and symmetric (Hamlyn, 1961). However, besides releasing glutamate, granule cells also contain and release dynorphin, enkephalin, Zn^{2+}, brain derived neurotrophic factor (BDNF), and other peptides like somatostatin, neuropeptide Y, neurokinin B and cholecystokinin (reviewed in Jaffe and Gutiérrez, 2007). The latest addition to this list of substances present in the granule cells is the amino acid GABA.

Several reviews have addressed different aspects of the activity of granule cells (Henze et al., 2000; Urban et al., 2001; Bischofberger and Jonas, 2002; Nicoll and Schmitz, 2005; Jaffe and Gutiérrez, 2007). Hence, here we shall focus on the evidence that the granule cells express GABAergic markers, as well as those of the glutamatergic phenotype. In particular, we will focus on the activity-dependent plasticity of their expression that supports the hypothesis that MF synapses use GABA as a fast neurotransmitter.

The mechanistic definition of a transmitter, which identifies a specific neuron as glutamatergic or GABAergic, fully applies to the glutamatergic/GABAergic granule cells. Hence, the transmitters that these cells use are synthesized, stored and released during nerve activity. Subsequently, they interact with specific receptors on the postsynaptic membrane to produce changes in postsynaptic activity. Thus, besides releasing glutamate, granule cells have the necessary machinery for the synthesis, vesiculation and release of GABA. The transient expression of their GABAergic phenotype is constitutive during development (see also Safiulina et al., in this volume) and it is stimulated in the adult by enhanced activity.

10.1.1 Granule Cells Transiently Express Markers of the GABAergic Phenotype

10.1.1.1 GABA and Glutamic Acid Decarboxylase (GAD)

The presence of GABA in the *stratum lucidum* was initially described by Ottersen and Storm Mathisen (1984), who concluded that GABA was present within the MFs. Using electron microscopy, Sandler and Smith (1991) later reported that GABA existed in MF terminals of the monkey and human hippocampus which made asymmetric synaptic contacts with spines on the large dendrites of CA3 pyramidal cells. They also showed that GABA colocalized with glutamate in the same terminals, although no GABA immunoreactivity can be detected in granule cell bodies. Since it was not known that granule cells also contained GAD, it was concluded that GABA must be obtained from either the extracellular space or that it originated from an alternative route of synthesis, from γ-hydroxybutyrate. Moreover, irrespective of the origin of GABA in the MF, it was suggested that at least one component of the inhibitory synaptic potentials in pyramidal cells provoked by DG stimulation had to be of MF origin. In this way, GABA released by MFs could modulate the normal glutamatergic responses.

Immunohistological methods were used to conclusively demonstrate the presence of GABA, and of its synthetic enzyme, GAD_{67}, in rat, monkey and human MFs (Sloviter et al., 1996). Interestingly, seizures upregulate the GAD_{65}, GAD_{67}, and GABA content (Sloviter et al., 1996), and GAD_{67} and its mRNA (but not GAD_{65}) is also transiently upregulated after seizures provoked by kainic acid or by the kindling method in rats (Schwarzer and Sperk, 1995; Lehmann et al., 1996; Ramírez and Gutiérrez, 2001; Maqueda et al., 2003). Therefore, if granule cells have the enzymes necessary for GABA synthesis and they indeed contain GABA itself, the granule cells that synthesize GABA could use it as a neurotransmitter. Hence, GABA synthesis in the MFs may be a mechanism to counteract the enhanced excitability caused by epileptic activity. Complementing and extending the aforementioned findings, immunogold cytochemistry confirmed the coexistence of glutamate and GABA in MF synapses, which also contained the corresponding receptors in apposition to the presynaptic terminal (Bergersen et al., 2003). Both amino acids coexist in all the MF terminals examined, spatially associated with synaptic vesicles. Indeed, immunogold particles binding to both GABA and glutamate were shown to be located at a distance that suggests their presence inside vesicles in the release zones. However, GABA was estimated to be at much lower concentration than glutamate within the MF terminals and even lower than that in identified inhibitory terminals (Bergersen et al., 2003).

On the other hand, the upregulation of GAD_{67} and its mRNA in granule cells was further confirmed in other seizure models and after the induction of epilepsy by chemical convulsants or electrical stimulation (Ding et al., 1998; Makiura et al., 1999; Szabó et al., 2000; Ramírez and Gutiérrez, 2001), as well as in a genetic epilepsy model (Sirvanci et al., 2003). Moreover, it was also established that GAD_{67} could be upregulated in an activity-dependent manner, in the absence of epileptiform activity (Ramírez and Gutiérrez, 2001; Gutiérrez, 2002). Thus, GAD_{67} expression can be induced by activity in a specific temporal and spatial pattern within the somata and axons of granule cells. A single a seizure induces the rapid upregulation of the enzyme in MFs but not in the somata of granule cells, probably to rapidly replenish the GABA pool. Thereafter, long-lasting periods of stimulation (repeated seizures or long-lasting synaptic potentiation in vitro) upregulate GAD_{67}, such that it can also be detected in the somata of granule cells (Fig. 10.1; Ramírez and Gutiérrez, 2001; Sloviter et al., 1996).

When the expression of GAD_{67} is studied at different ages, this isoform but not GAD_{65}, is also developmentally regulated in MFs (Maqueda et al., 2003). Indeed, GAD_{67} is expressed in the MFs early in life and it is then downregulated after 23–24 days, once development is completed (Gutiérrez et al., 2003; Maqueda et al., 2003). Despite the reported downregulation of GAD_{67} at the end of development, this enzyme does not disappear from the MFs in the adult rodent (Sloviter et al., 1996; Uchigashima et al., 2007). By contrast, GABA-immunoreactive neurons characteristic of granule cells are found in the *stratum granulare* of the DG in developing rats but not in adults (Gutiérrez et al., 2003). Only this GAD_{67} isoform coincides in granule cells and the MF with the expression of

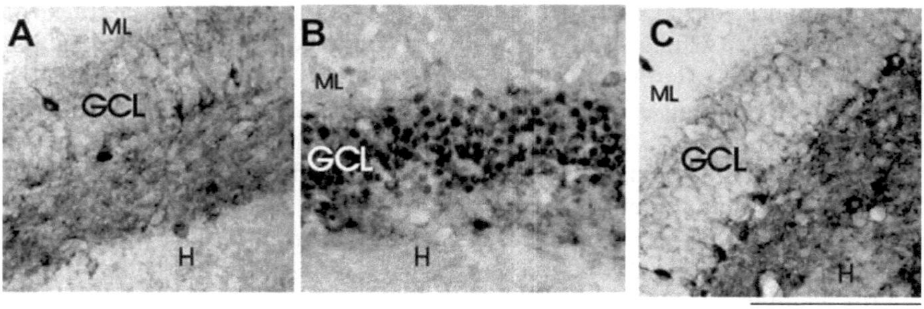

Fig. 10.1 Immunoreactivity for GAD67 in the DG of a control rat **(a)** and of a kindled rat 24 h after the last of several seizures **(b)**, as well as in a kindled rat a month after the last of several seizures **(c)**. The DG of control slices (a) lacks immunoreactivity in the granule cell layer although some putative interneurons and granule cells are stained. Repeated kindling seizures induce overexpression of GAD_{67} in the granule cell layer (b), that returns to control conditions a month later (c). Abbreviations: ML, molecular layer; GCL, granule cell layer; H, hilar region. The bar represents 50 μm. *Modified from Ramírez and Gutiérrez*, 2001, *with permission*

GABA. Neurochemical analysis has shown that stimulation of the perforant pathway for 24 hours, that culminated in seizures (Sloviter et al., 1996) and kindled seizures (Gómez-Lira et al., 2002) significantly increased the concentrations of GAD_{67} and GABA, as well as those of the GABA precursors glutamate and glutamine in the DG. Moreover, up-regulation of such GABAergic markers depends on protein synthesis since they are not expressed in hippocampal slices of rats subjected to seizures or a LTP-like protocol in the presence of cycloheximide. Likewise, calcium entry is also required to up-regulate these GABergic markers (Romo-Parra et al., 2003; Gutiérrez, 2002).

Finally, as well as epileptic activity and LTP, the exposure of normal adult tissue (hippocampal slices or dissociated cells) to Brain Derived Neurotrophic Factor (BDNF) can also induce the GABAergic phenotype in granule cells. Moreover, when dissociated and exposed to kainic acid or BDNF, the levels of GAD_{67}, GABA and VGAT mRNA were strongly up-regulated in glutamatergic granule cells from adult rats (Fig. 10.2). This up-regulation of GABAergic markers following exposure of dissociated granule cells to kainate or BDNF is also prevented in the presence of cycloheximide or a calcium channel blocker (Gómez-Lira et al., 2005). Thus, the determination of the neurotransmitter phenotype of granule cells is neither limited by a critical developmental period nor is it restricted by their insertion in their natural network.

10.1.1.2 The GABA Membrane Transporter (GAT-1)

The GABA membrane transporter GAT-1 is only found in neurons that synthesize and release GABA, and in glial cells (Iversen and Kelly, 1975; Radian et al., 1990; Ribak et al., 1996a). Some GABAergic cells, characterized by the presence of GAD_{67} or GABA, do not contain GAT-1 or they only

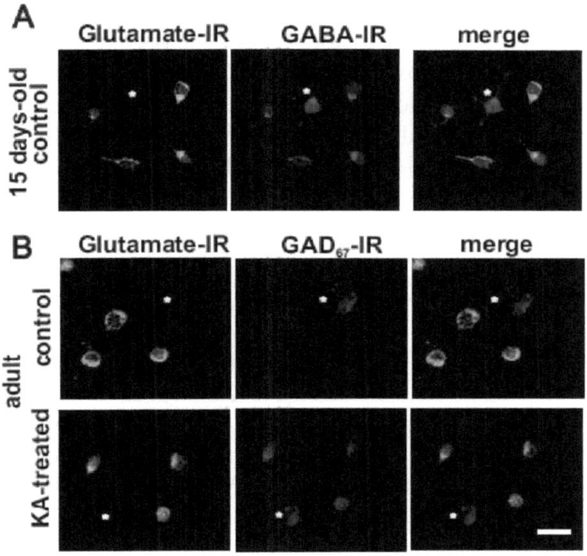

Fig. 10.2 Immunocytochemistry for glutamate, GAD_{67}, and GABA in cultured GCs. **(a)** Confocal images of short-term GC and interneuron cultures (asterisks) from a 15 day-old rat showing the co-expression of glutamate and GABA in GCs. **(b)** In cultures from adult rats, GCs express glutamate but not GABAergic markers unless they are exposed to KA or BDNF, whereupon GAD_{67} and glutamate-IR colocalize. Scale bar in B (15 μm) also applies to panel A. *From Gómez-Lira et al.*, 2005, *with permission*. (*See* Color Plate 12)

contain traces of this transporter, but not vice versa (Rattray and Priestley, 1993; Ribak et al., 1996b). Interestingly, immunohistological studies have shown that GAT-1 is present in granule cells (Frahm et al., 2000). Moreover, MF terminals from control animals capture GABA, a process that can be blocked by nipecotic acid (Gómez-Lira et al., 2002) and electrophysiological evidence supports these immunohistological and neurochemical data (Vivar and Gutiérrez, 2005).

10.1.1.3 The Vesicular GABA Transporter (VGAT)

The identification of vesicular transporter proteins has proven to be an invaluable tool to correctly assess the neurotransmitter phenotype of neurons. In recent years, vesicular transporters of a given neurotransmitter have been shown to be present in neurons thought to use a neurotransmitter other than that carried by the transporter, evidence that more than one neurotransmitter is released by a single neuron. In particular, glutamate and GABA seem to share the load (Seal and Edwards, 2006).

Evidence that VGAT is present in glutamatergic granule cells was first provided by the detection of its mRNA in a homogenate of micro-dissected DG and in MF synaptosomes (Lamas et al., 2001). This study showed that

VGAT mRNA could be barely detected in preparations obtained from control rats but transcripts were more readily found in preparations obtained from kindled epileptic rats. Moreover, antidromic stimulation of the granule cells in the presence of glutamatergic blockers up-regulated the expression of VGAT mRNA in an activity-dependent manner. These initial results were confirmed and extended using single-cell PCR analysis (Gómez-Lira et al., 2005; Fig. 10.3). Indeed, electrophysiologically identified granule cells, but not interneurons or pyramidal cells dissociated from developing or adult epileptic rats, co-expressed glutamate transporter VGLUT-1 and VGAT mRNA. By contrast, the VGAT transcript was not detected or barely detected in the majority of granule cells obtained from healthy adult rats. Interestingly, when dissociated granule cells from control rats were exposed to kainic acid or BDNF, the levels of VGAT mRNA were strongly up-regulated (Gómez-Lira et al., 2005).

VGAT mRNA is normally expressed during development, and like GABA and GAD_{67}, it is downregulated once development is completed. However, the protein has not been identified in either developing or in adult epileptic rats

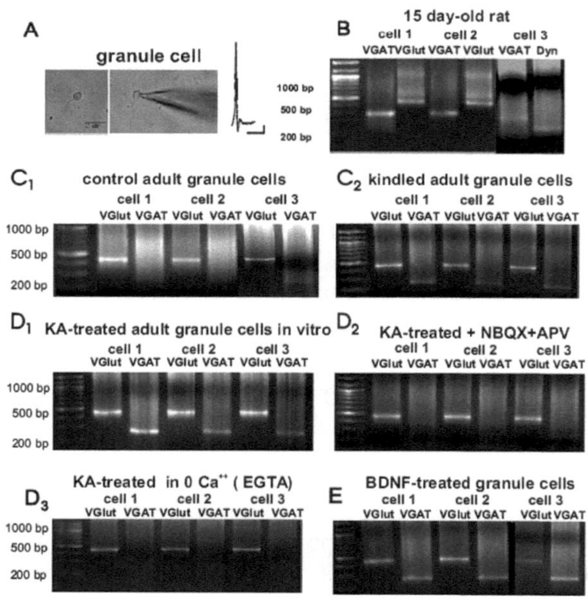

Fig. 10.3 **(a)** Single-cell RT-PCR of electrophysiologically identified granule cells (GCs). **(b)** Developing GCs co-express VGAT and VGlut-1 transcripts. Dynorphin mRNA was used as a marker of GCs. (**c1**) Adult DG only express VGlut and Dyn. (**c2**) By contrast, GCs from kindled rats express VGlut and VGAT. (**d1**) GCs from adult rats exposed to KA express the transcripts of both vesicular transporters and such expression was prevented with glutamatergic blockers (**d2**) or in the absence of calcium (**d3**). E) GCs from adult rats exposed to BDNF also express transcripts for both vesicular transporters. *From Gómez-Lira et al.* (2005), *with permission*

(Chaudhry et al., 1998; Sperk et al., 2003; Uchigashima et al., 2007; but see Safiulina et al., 2006). Therefore, it was suggested that an unidentified GABA vesicular transporter might exist, possibly generated through some post-transcriptional mechanism. Indeed, GABA can be loaded into vesicles of non-GABAergic cells in the absence of VGAT, suggesting the presence of ectopic synthetic enzymes and transporters for GABA (Bekkers et al., 2005). However, this important issue has still to be confirmed.

10.2 Activation of Granule Cells can Transiently Evoke Monosynaptic $GABA_A$-R Mediated Intracellular Responses and Population Responses in the CA3

The intracellular responses of CA3 pyramidal neurons to granule cell activation were initially described by Yamamoto (1972). These consist of a biphasic response involving a small excitatory postsynaptic potential (EPSP) followed by a larger, overlapping compound inhibitory postsynaptic potential (IPSP). The latter was presumably mediated by either feed-forward/feed-back inhibition, or by the direct stimulation of inhibitory interneurons. Thus, the MFs form glutamatergic excitatory synapses on pyramidal cells and local inhibitory interneurons (Crawford and Connor, 1973; Acsády et al., 1998), which in turn inhibit pyramidal cells (Yamamoto, 1972; Brown and Johnston, 1983; Buzsáki, 1984). All synaptic responses can be blocked by perfusion of the NMDA type glutamate receptor antagonist 5APV and the AMPA-Kainate type receptor antagonist NBQX (Fig. 10.4), demonstrating that inhibitory transmission in the MF-to-CA3 projection is mediated disynaptically.

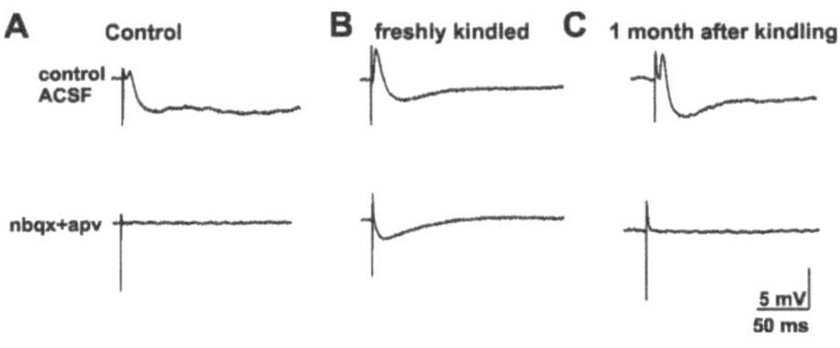

Fig. 10.4 In healthy adult rats (control), granule cell activation provokes EPSP/IPSP sequences in CA3 pyramidal cells that can be blocked by the perfusion of glutamate receptor antagonists. However, 24 hours after the last kindled seizure, a fast GABA-R-mediated IPSP is observed during the blockage of glutamate receptors (freshly kindled). This response is transient because granule cell activation does not evoke GABA-mediated synaptic responses when the recording is made one month after provoking the seizure (1 month after kindling). The latencies of the control EPSP and the pharmacologically isolated IPSP do not differ in any case. All traces are an average of 6–10 responses. (*Adapted from Gutiérrez and Heinemann*, 2001, *with permission*)

Although MFs are clearly glutamatergic (Crawford and Connor, 1973), the evidence of the presence of GABA in MF terminals suggested that they may contribute to CA3 inhibition (Sandler and Smith, 1991; Sloviter et al., 1996; Vogt and Nicoll, 1999). In particular, the evidence that seizures or repeated activation of the dentate gyrus up-regulated GAD_{67} in granule cells (Schwarzer and Sperk, 1995; Sloviter et al., 1996; Ramírez and Gutiérrez, 2001) suggested that MFs could synthesize and use GABA as a neurotransmitter under special circumstances, especially after seizures.

Evidence of monosynaptic GABA-mediated transmission from the dentate gyrus to the CA3 was initially obtained from kindled epileptic but not from control healthy rats (Gutiérrez and Heinemann, 1997, 2001; Gutiérrez, 2000). As mentioned, the MFs form glutamatergic excitatory synapses on pyramidal cells and local inhibitory interneurons, which in turn inhibit pyramidal cells. Thus, while perfusion of the NMDA type glutamate receptor antagonist 5APV and the AMPA-Kainate type receptors' antagonist NBQX blocks all synaptic responses in control rats, in kindled epileptic animals, the blockage of glutamatergic transmission isolated a bicuculline-sensitive IPSP (Fig. 10.4). This IPSP had the same latency as the control EPSP and could be inhibited by activation of group III metabotropic glutamate receptors (mGluR) with L-AP4 (see Fig. 10.6). Furthermore, this response was transient because it was observed 24–48 hours after the last kindled seizure but not a month after the last seizure (Fig. 10.4; see correspondence of immunodetection of GAD_{67} in Fig. 10.1).

Nevertheless, if seizures are again produced these monosynaptic GABAergic responses reappear (Gutiérrez, 2000; Gutiérrez and Heinemann, 2001) and moreover, the kindled epileptic state is not necessary for this monosynaptic IPSP to be expressed, as it could be provoked by a single seizure (Gutiérrez, 2000). Indeed, the IPSP produced in CA3 pyramidal cells by the activation of the DG persisted in low Ca^{2+} condition or in the presence of the $GABA_B$ agonist, baclofen, and glutamatergic blockers. These manipulations depressed the amplitude of the DG-evoked IPSP without altering its onset latency or slope. This unequivocally establishes that the IPSP is a monosynaptic response since under these conditions it is unlikely that local interneurons that might evoke the IPSP are synaptically recruited.

Due to the difficulty in unequivocally identifying responses of MF origin, an experimental design was developed to record from the same cell before and after the induction of MF GABAergic transmission in vitro (Gutiérrez, 2002). This stimulation protocol produced hyperexcitability in the absence of epileptic activity by stimulating the perforant path in a kindling-like manner (three 1-sec trains of 0.1 ms pulses at 100 Hz, 1 min apart from each other every 15 min for 3 hours). After the protocol was completed, the perfusion of glutamatergic antagonists blocked the EPSP and isolated a fast bicuculline-sensitive IPSP (Fig. 10.5). Expression of the GABAergic potential was prevented if the stimulation was provided for one hour alone, or if it was completed in the presence of the protein synthesis inhibitor cycloheximide. This establishes

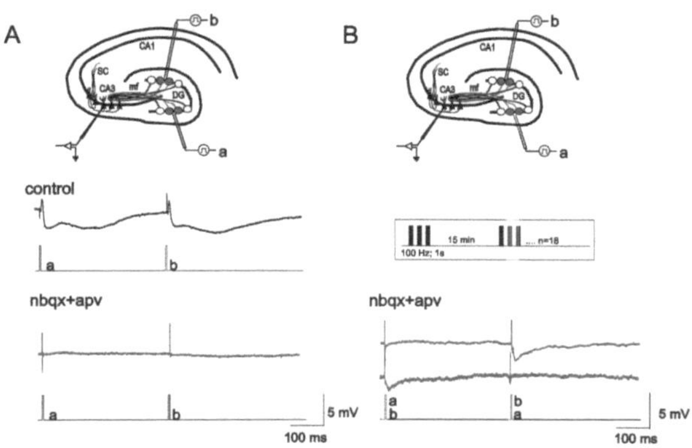

Fig. 10.5 Spatial specificity of the kindling-induced inhibitory responses in vitro. **(a)** Schematic representation of a hippocampus, showing two sites of stimulation in the DG (a,b); actual separation during the experiments 200–300 μm) and the recording electrode in CA3. Initially, test pulse stimulation at both sites of stimulation evoked EPSP/IPSP sequences in the pyramidal cell at a RMP of –64 mV (control), which were blocked by GluRAs (blue trace). **(b)** After applying 17 ± 1 kindling trials to site b (in red), test pulse stimulation of the kindled site (b) but not of the non-kindled site (a, in blue), provoked a fast IPSP in the pyramidal cell. Changing the order of stimulation inverted the order of the responses. All traces are an average of 10 responses. *From Gutiérrez* (2002), *with permission*. (*See* Color Plate 13)

that a certain level of excitation is required for granule cells to express their GABAergic phenotype (Gutiérrez, 2002). Alternatively, when the control synaptic responses (EPSP/IPSP sequences) of a given cell were blocked first, the direct kindling stimulation of the same site during perfusion of glutamatergic antagonists induced fast GABAergic potentials after 3 hours of stimulation that was provided every 15–20 min. Furthermore, this phenomenon was very specific in spatial terms as a pulse stimulation applied to an alternative non-kindled parallel MF input did not evoke GABAergic responses.

Further confirmation that the induction of GABAergic markers and MF GABAergic transmission is dependent on activity and protein synthesis was presented by Romo-Parra et al. (2003). Slices obtained from rats immediately after a pentylenetetrazol-induced seizure were incubated in the presence of protein synthesis blockers. This prevented the MF but not interneuronal GABAergic transmission to CA3 pyramidal cells. Moreover, granule cells dissociated from naïve rats express all GABAergic markers 3 hours after their exposure (30 min) to kainate or BDNF. However, this expression can be prevented by cycloheximide or by blocking calcium entry (Gómez-Lira et al., 2005).

The aforementioned results were obtained by recording the responses of pyramidal cells to activation of the granule cells. However, interneurons rather than the pyramidal cells seem to be the main targets of MFs (Acsády et al., 1998). MF activation of EPSP/IPSP sequences can be totally blocked in the presence of glutamatergic antagonists and hence, inhibitory transmission onto interneurons, like that onto pyramidal cells, is mediated disynaptically. However, as previously described in pyramidal cells, MF activation in the presence of glutamatergic antagonists provokes monosynaptic IPSPs with the same latency as the control EPSP (Romo-Parra et al., 2003; Treviño et al., 2007; Fig. 10.6).

Although MF GABAergic transmission to interneurons occurs, the effect of MF glutamatergic transmission surpasses the inhibitory influence. In animals subjected to seizures or in preparations stimulated with the kindling-like protocol in vitro, stimulation of the MF pathway provokes a long-lasting EPSP of large amplitude that totally obscures the underlying IPSP in interneurons. By contrast, pyramidal cells display an enhanced IPSP that partially shunts the EPSP. Thus, interneurons are more readily excited than pyramidal cells, which continue to receive strong inhibition from the interneurons within the CA3, besides the MF GABAergic signal (Romo-Parra et al., 2003). Therefore, the activation of MFs at high frequencies provokes more pronounced IPSPs in

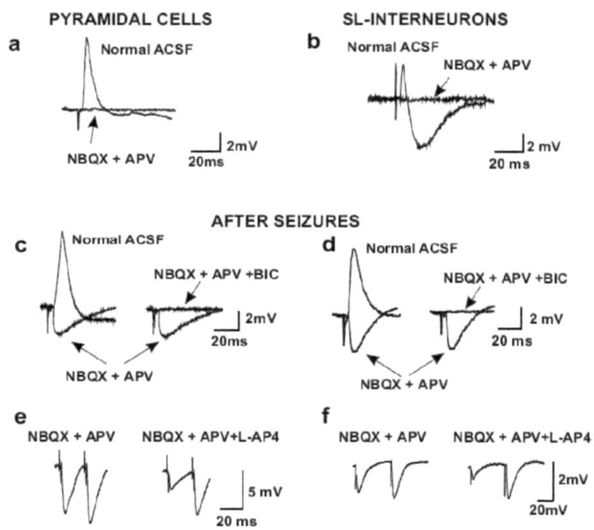

Fig. 10.6 Emergence of MF-evoked $GABA_A$-R-mediated monosynaptic responses in PyrC and SL-Int after seizures. In electrophysiologically and anatomically identified PyrC **(a)** and SL-Int **(b)** from control preparations, MF stimulation produced EPSP/IPSP sequences that were blocked by iGluR antagonists. In preparations from rats that were subjected to seizures, the same stimulation produced a monosynaptic $GABA_A$-R-mediated IPSP in the presence of NBQX and APV. This IPSP was blocked by addition of bicuculline. **(c,d)**. A paired pulse stimulation paradigm that produced potentiation of the second response was used to reveal the presynaptic inhibition of the $GABA_A$-R-mediated responses by activation of mGluR-III with L-AP4. The first response to the pair of pulses was depressed by L-AP4, while the $IPSP_2$/$IPSP_1$ ratio was enhanced, both in pyramidal cells **(e)** and in interneurons **(f)**. *From Treviño et al.* (2007), *with permission*

pyramidal cells, which can be potentiated to hyperpolarize them, while the same kind of stimulation produces summation of EPSPs in interneurons (Gutiérrez and Heinemann, 2001; Treviño et al., 2007). These data show that although pyramidal cells and interneurons receive the same dual MF input, the ratio of the glutamate/GABA released onto the target cells is probably different. Indeed, the concentration of GABA was estimated to be much lower than that of glutamate within the MF terminals impinging onto hilar neurons (Bergersen et al., 2003), highlighting an aspect of MF compartmentalization. These data also confirm that all the targets of the MFs receive both glutamatergic and GABAergic neurotransmission. Moreover, all MF terminals analyzed have been shown to coexpress both glutamate and GABA (Sloviter et al., 1996; Bergersen et al., 2003). However, these transmitters are not contained within single vesicles but rather they seem to be packaged separately. Physiological evidence for this was provided by Walker et al. (2001) who showed that minimal stimulation of MFs provoked glutamate-mediated, GABA-mediated and compound currents mediated by both amino acids in pyramidal cells. This indicated that both signals have a common origin and that both neurotransmitters can be released synchronously or asynchronously, ruling out the possibility that they are packaged in single vesicles (Fig. 10.7; see also Safiulina et al., 2006).

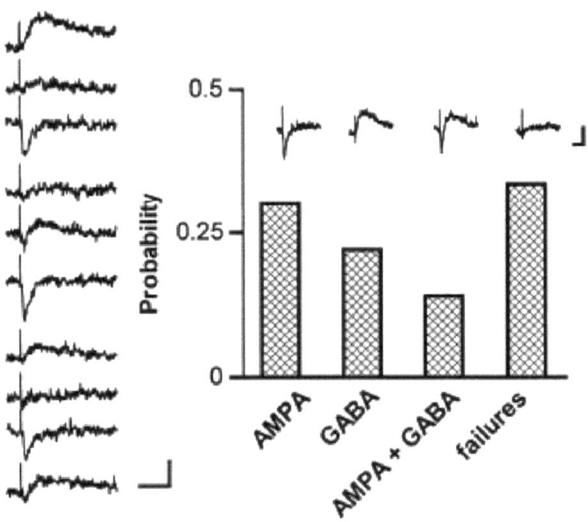

Fig. 10.7 Mossy fiber stimulation in the presence of DL-APV (100 mM) at a holding potential of 240 mV elicits failures, inward PSCs (glutamatergic), outward PSCs (GABAergic), and biphasic PSCs (dual component). Single traces (left) are shown together with a histogram depicting the relative frequency of each type of response, classified by eye (right, the traces are derived from averages of ten responses of each type). Calibration bars in 5 pA, 50 ms. *From Walker et al. (2003), with permission*

10.2.1 Glutamate and GABA Receptors Appose Mossy Fiber Terminals

If pyramidal cells and interneurons in the CA3 area respond to MF-GABAergic signaling, $GABA_A$ receptors must appose the MF terminals. It has been shown that $GABA_A$ receptors cluster with glutamatergic receptors in pyramidal cells apposing both glutamatergic and GABAergic terminals (Rao et al., 2000). Accordingly, while $GABA_A$ receptors were clustered opposite to some terminals, NMDA receptors were clustered opposite to other terminals of a single axon. Moreover, $GABA_A$ receptors form clusters even in the absence of GABAergic innervation. These studies suggest that the mismatching of glutamatergic and GABAergic elements may reflect a common signal involved in the alignment of presynaptic and postsynaptic components during the formation of excitatory and inhibitory synapses. In fact, the mismatched GABA receptors apposing the glutamatergic terminal may be functional if the presynaptic terminal is loaded with GABA (Bekkers, 2005). This alignment of heterogeneous populations of receptors or the clustering of different receptors in the same subsynaptic site also takes place in other neurotransmitter systems (Tsen et al., 2000). In vivo, such findings are mirrored by the colocalization of AMPA and $GABA_A$ receptors at MF synapses (Fig. 10.8, Bergersen et al., 2003). The fact that $GABA_A$ receptors are present in the MF synapse is in accordance with physiological studies showing that MF-GABAergic transmission is normally detected in CA3 targets in developing rodents (Walker et al., 2001; Gutiérrez et al., 2003; Safiulina et al., 2006). Indeed, MF-GABAergic transmission can be induced in the absence of postsynaptic glutamate receptor activation in the adult (Gutiérrez, 2002), which favors presynaptic changes and not the postsynaptic relocation of receptors, even though this might occur.

10.2.2 $GABA_A$ Receptors in Mossy Fiber Terminals

Like the post-synaptic sites apposing MF terminals (Bergersen et al., 2003), MFs themselves also posses $GABA_A$ receptors (Ruiz et al., 2003). Since the MFs also release GABA, this GABA could provoke presynaptic inhibition of a collateral MF. Activity-dependent autoinhibition of the MF pathway is primarily exerted by glutamate, acting through its metabotropic receptors. High frequency stimulation provokes its spill over, which acts on presynaptic mGluR and KAR (Min et al., 1998; Vogt and Nicoll, 1999; Schmitz et al., 2001). Furthermore, in preparations from epileptic but not from healthy rats, when a pulse applied in the DG is preceded by a pulse applied to another site of the DG, the response to the second pulse is depressed in a bicuculline sensitive manner in the presence of iGluR antagonists (Treviño and Gutiérrez, 2005). This phenomenon is consistent with $GABA_A$-receptor mediated collateral inhibition. Thus, GABA release from the MF may also exert strong control over

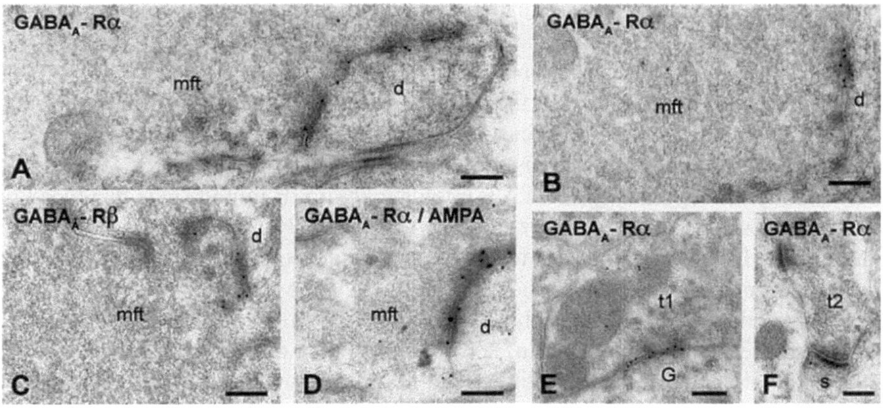

Fig. 10.8 $GABA_A$ receptors are present in MF synapses. (**a** and **b**) Localization of gold particles indicative of the $GABA_A$ receptor α_1 in synapses between large MF terminals (mft) and dendritic spines (d) in the hilus of fascia dentanta (a) and the CA3 of hippocampus (**b**). Bars 150 nm. **(c)** Gold particles indicative of the $GABA_A$ receptor $\beta_{2/3}$ receptors are present in the synapse between a large MF and a dendritic spine (d) in the hilus of fascia dentata. Bar, 150 nm. **(d)** Colocalization of $GABA_A$ receptors α_1 (small gold particles) and AMPA receptors (large gold particles) in a synapse formed by a large MF terminal (mft) and a dendritic spine (d). Bar, 150 nm. **(e)** $GABA_A$ receptors α_1 are enriched in a synapse between a terminal (tl) that forms a symmetric synapse with a granule cell body (G). Bar, 150 nm. **(f)** $GABA_A$ receptor α_1 are not expressed in an asymmetric synapse formed by a small terminal and a dendritic spine (s) in the inner third of the dentate molecular layer. Bar, 150 nm. *From Bergersen et al.* (2003), *with permission*

any subsequent release by fast activation of presynaptic $GABA_A$ receptors. Moreover, field responses of the DG to antidromic stimulation of the MF are potentiated by the blockage of $GABA_A$-R. Interestingly, these results and those of Ruiz et al. (2003) show that $GABA_A$-R are tonically active and since the MF-GABA transmission that emerges can still exert collateral inhibition, there appears to be no saturation of the receptors. Thus, postsynaptic dendrites and presynaptic axons alike seem to integrate the input activity in CA3 (Ruiz et al., 2003).

10.3 Function: The DG as an Inhibitory Structure

The ratio of the synaptic contacts that MFs make with interneurons is approximately four-fold higher than those with pyramidal neurons. Assuming that: (i) a single granule cell contacts 15 pyramidal neurons; (ii) each large expansion contacts up to 3 interneurons, either directly or via its filopodia; and (iii) 15 interneurons are innervated by the smaller boutons, then a single granule cell will contact approximately 60 interneurons but only 15 pyramidal neurons (a 4:1 ratio). This emphasizes the extremely efficient control of CA3 excitability by feed-forward inhibition. Thus, the firing of granule cells would have a net inhibitory effect on pyramidal neurons in the CA3 under certain conditions (Acsády et al., 1998; see Jaffe and Gutiérrez, 2007).

Because of this arrangement, the emergence of GABAergic transmission in the MF synapse can inhibit interneurons and, in principle, produce disinhibition. Although after seizures the activation of the MF pathway evokes $GABA_A$-R mediated field responses in the CA3, interestingly no disinhibition is produced in this area (Treviño and Gutiérrez, 2005). As previously noted, this is due to the glutamate/GABA ratio of MF terminals onto interneurons when compared to pyramidal cells, whereby the former are readily excited by the excitatory drive of the MF more efficiently than the inhibition provided by the concurrent GABAergic signal (Gutiérrez and Heinemann, 2001; Romo-Parra et al., 2003; Treviño et al., 2007). In this way, monosynaptic MF inhibition of pyramidal cells adds to the inhibition provided by the interneurons, which continue to be effectively excited (Treviño and Gutiérrez, 2007). In fact, inhibition is built-up in the CA3 area by MF activation as the frequency of this activation augments (Mori et al., 2004).

Another effect that GABA released from the MF may exert on the CA3 area is the activation of presynaptic $GABA_A$ receptors (Treviño and Gutiérrez, 2005) present in the MFs themselves (Ruiz et al., 2003). As mentioned, activation of a site in the dentate gyrus preceded by activation of another site in the presence of iGluR antagonists produces bicuculline-sensitive inhibition of the second afferent volley (Treviño and Gutiérrez, 2005). In this way, $GABA_A$-R mediated presynaptic inhibition produces rapid modulation of neurotransmission, indicative of surprisingly complex signal integration in the MFs. Indeed, MFs detect the release of both glutamate and GABA via ionotropic receptors, and they may rapidly integrate the activity of surrounding neurons, a function normally associated with dendrites (Ruiz et al., 2003).

Finally, spontaneous release of GABA from the MFs after seizure can modulate oscillatory activity in the CA3 area. Such oscillations originate from recurrent excitation of pyramidal cells entrained by the synchronous rhythmic inhibition of local interneurons, although β/γ (20–24 Hz) oscillations originate in the interneuron network (Treviño et al., 2007). Following seizures, the dentate gyrus tonically inhibits these oscillations through GABA-mediated signaling. Indeed, the activation of mGluR with L-AP4, which inhibits the release of GABA from the MFs, prevents this inhibition, and this effect of L-AP4 requires intact DG-to-CA3 connections. Moreover, the influence of mGluR activation is reflected in the spontaneous subthreshold membrane oscillations of CA3 interneurons (Fig. 10.9). It is significant that coincident stimulation of the DG at θ and β/γ frequencies produced frequency-dependent excitation of interneurons and the inhibition of pyramidal cells. These effects are maximal at the frequency that matched the mGluR-sensitive spontaneous field oscillations, suggesting a resonance phenomenon.

In summary, the emergence of GABA transmission from glutamatergic MFs after enhanced excitability, especially following seizures, may be a protective mechanism against further excitation of the system (Gutiérrez, 2003, 2005;

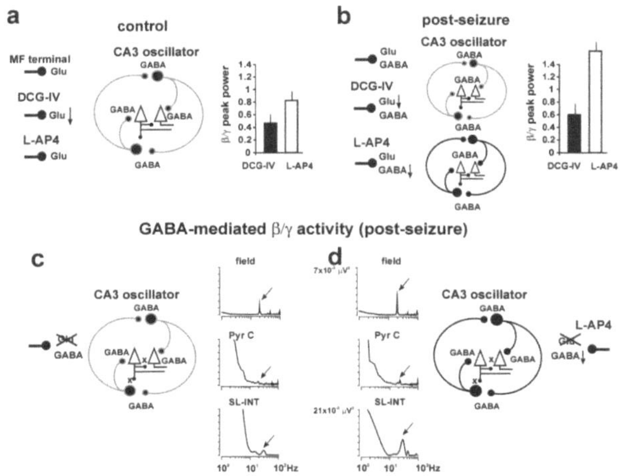

Fig. 10.9 β/γ activity in the CA3 region is inhibited by mGluR-sensitive, GABA-mediated aberrant transmission from the MF. **(a)** The CA3 area is an endogenous oscillator that receives glutamatergic input from the MF. Activation of group II mGlu presynaptic receptors inhibits glutamate release from the MF and strongly depresses oscillatory activity, while activation of group III mGluR produces modest inhibition. In rats subjected to seizures **(b)** the activation of group III mGlu presynaptic receptors with L-AP4, which inhibits GABA release from the MF, produces a noticeable enhancement in theβ/γ power. This reveals the tonic inhibitory control of the MF on oscillatory activity. This control is exerted on GABA-mediated oscillations because it is observed in the presence of iGluR blockers. **(c)** Indeed, in the absence of a glutamatergic input, GABA-mediated oscillatory β/γ activity (arrows) can be observed in field recordings and in spontaneous subthreshold membrane oscillations (SSMO) of SL-Int in the CA3 area after seizures. d) Activation of group III mGluR relieves the inhibition of GABA-mediated β/γ oscillations (arrows) in the CA3 field, which is primarily reflected in the SSMO of SL-Int

Gutiérrez and Heinemann, 2006). The potential protection afforded by enhanced GABA transmission within the CA3 area may underlie the deficits in memory and cognition observed after epileptic seizures (Gutiérrez and Heinemann, 2001; Treviño et al., 2007).

10.3.1 Development

One of the most important questions related to the possible function of GABA in the mossy fiber synapse is whether it is present under physiological conditions, (i.e., not only after seizures). While monosynaptic GABA transmission from mossy fibers is normally present in developing guinea pigs (Walker et al., 2001), it was subsequently shown to be transient in the rat as it disappears upon the completion of development around postnatal days 22–23. The markers of the GABAergic phenotype GABA, GAD_{67} and VGAT mRNA are present during the developmental period and they are then down regulated (Gutiérrez et al., 2003), such that $GABA_A$-R mediated responses can be no longer evoked by MF stimulation. Moreover, GABA was proposed to act as the primary neurotransmitter in the MFs during the first days of life (Kasyanov et al., 2004; Safiulina et al., 2006, but see Uchigashima et al., 2007). This issue is thoroughly discussed by Safiulina et al., in this volume.

MF-GABAergic transmission accompanies the morphological maturation of pyramidal cells, possibly until the development of dendrites and spines is completed, which more or less occurs concomitantly (Ben-Ari, 2001; Romo-Parra et al., 2008). Thus, the simultaneous release of glutamate and GABA from the MF could fulfill a trophic function during development. At early ages, the depolarization provided by the GABAergic input is needed for the activation of NMDA receptors, suggesting that the co-expression of the glutamatergic and GABAergic phenotypes in a single pathway provides efficient and rapid synergism on their target cells during development (Leinekugel et al., 1997). This activity may contribute to the refinement of neuronal connectivity until adult DG-to-CA3 neurotransmission is established (Romo-Parra et al., 2008). Moreover, it would be inconsistent with the sequential expression of functional GABA and glutamatergic synapses in the hippocampus at early developmental stages (Henou et al., 2002). Furthermore, the developmental expression of the dual glutamatergic–GABAergic phenotype occurs in the immature GABA/glycinergic projections of the medial nucleus of the trapezoid body, which transiently express a glutamatergic phenotype (Gillespie et al., 2005). This has again been proposed to reflect the developmental reorganization of this inhibitory circuit (see Gillespie and Kandler, in this volume).

10.4 Indirect Evidence, Direct Questions

The co-release of two fast-acting neurotransmitters should be reflected by the activation of synaptic currents/potentials by two different receptors in a postsynaptic cell by each action potential in the presynaptic cell. The first clear electrophysiological evidence of the co-release of amino acids was provided by simultaneously recording pairs of interneurons and motor neurons from the spinal cord of neonatal rats (Jonas et al., 1998). In this way it was shown that GABA and glycine are used as fast-acting neurotransmitters in this synapse. So far it has not been possible to apply this paired recording technique to the MF synapse because of the very low probability of recording coupled granule cells and CA3 neurons. Nevertheless, throughout this chapter we have described indirect electrophysiological approaches to study the hypothesis that glutamate and GABA are co-released from the MF, and the resulting data offer compelling support for this possibility. Moreover, there is direct evidence that mRNAs encoding glutamate, its vesicular transporter (VGlut-1), GABA, its synthetic enzyme (GAD_{67}) and the vesicular transporter (VGAT) coexist in developing granule cells, and that they can be up-regulated by activity in the adult. It therefore seems that granule cells do have the machinery to synthesize and release GABA. Moreover, there is indirect but compelling electrophysiological evidence indicating that under certain circumstances, they release GABA to activate post-synaptic $GABA_A$ receptors in pyramidal cells and interneurons, as well as pre-synaptic receptors in the MFs themselves.

There are three alternative explanations for the origin of the monosynaptic $GABA_A$-R mediated responses in the CA3 upon MF stimulation:

(1) Rather than through its synthesis, packaging and action-potential-dependent release, MFs release GABA that is actively transported from the extracellular milieu and that is then released by reversing the direction of its membrane transporter (Sperk et al., 2003). Thus, MF containing GAT-1 could release GABA from their terminals in this manner. This possibility was ruled out as blocking GAT-1 not only prevents GABA release but it also potentiates the GABA-receptor mediated responses in CA3 (Vivar and Gutiérrez, 2005).
(2) Rather than the MFs releasing GABA, the stimulation of some interneurons in either the dentate gyrus or along the MF tract could be responsible for the monosynaptic $GABA_A$-R mediated responses of cells within the CA3 area (Sperk et al., 2003; Mori et al., 2004; Uchigashima et al., 2007). Some evidence suggests that this is unlikely (see Gutiérrez, 2002, 2003). However, it is fair to say that this has not been completely ruled out (Walker et al., 2001; Gutiérrez, 2003; see Uchigashima et al., 2007) and needs direct clarification.
(3) Activation of unknown GABA containing cells projecting from the dentate gyrus to the CA3, with the same physiological and pharmacological characteristics as MF neurotransmission.

Another question that remains unanswered is whether MF terminals contain a GABA transporter other than that already described (McIntire et al., 1997). This would explain why VGAT mRNA is present and modulated by activity, although the protein responsible has not been identified. Moreover, this would help to explain GABA release from other GABA and GAD-containing cells where no VGAT has been detected (Chaudhry et al., 1998).

If MFs do co-release glutamate and GABA, is their release modulated in the same way? To date there is evidence of differential modulation by mGluRs (Gutiérrez, 2005), but what about adenosine and kainate receptors?

Does glutamate and GABA co-release occur from MFs under physiological conditions in the adult rodent? It has been shown that this phenomenon occurs normally in the developing rodent and in different neuronal systems (see also Gillespie and Kandler in this volume), but in the adult MFs co-release has only been detected after seizures. However, there is data consistent with the co-release of glutamate and GABA in adult rodents, whereby the hormonal state controls the expression of neurons presenting a dual glutamatergic/GABAergic phenotype (Ottem et al., 2004). Therefore, is the activity-dependent up-regulation of the GABAergic phenotype of the glutamatergic MFs a graded phenomenon?

Can we detect it? If the phenotype of neurons is plastic (Spitzer et al., 2005), what are the molecular determinants that turn either phenotype on or off? Finally, how do the post-synaptic neurons interpret the simultaneous input of two neurotransmitters of opposing actions?

Many more questions can be formulated, including those addressing the possible effects of simultaneous glutamate and GABA signaling (Mody, 2002). As the co-release of glutamate and GABA is observed in several systems, we begin to envisage several functional consequences and, above all, the value of granule cells and neurons in other parts of the central nervous system releasing both glutamate and GABA. The possibility of a neuron speaking two languages, which is at first glance contrasting but also complementary, provides the central nervous system with a very powerful communication tool.

Acknowledgments This work was supported by grants from the Consejo Nacional de Ciencia y Tecnología and Fundación Miguel Alemán, Mexico. The author thanks the collaborators whose data contributed to this review.

References

Acsády L, Kamondi A, Sik A, Freund T and Buzsáki G (1998) GABAergic cells are the major postsynaptic targets of mossy fibers in the hippocampus. J Neurosci 18:3386–3403

Bekkers JM (2005) Presynaptically silent GABA synapses in hippocampus. J Neurosci 25:4031–4039

Ben-Ari Y (2001) Developing networks play a similar melody. Trends Neurosci 24:353–360

Bergersen L, Ruiz, A, Bjaalie JG, Kullmann DM, Gundersen V (2003) GABA and GABAA receptors at hippocampal mossy fibre synapses. Eur J Neurosci 18:931–941

Bischofberger J, Jonas P (2002) TwoB or not twoB: differential transmission at glutamatergic mossy fiber-interneuron synapses in the hippocampus. Trends Neurosci 25:600–603

Brown TH, Johnston D (1983) Voltage-clamp analysis of mossy fiber synaptic input to hippocampal neurons. J Neurophysiol 50:487–507

Buzsáki G (1984) Feed-forward inhibition in the hippocampal formation. Prog Neurobiol 22:131–153

Chaudhry FA, Reimer R, Bellocchio EE, Danbolt NC, Osen KK, Edwards RH, Storm-Mathisen J (1998) The vesicular GABA transporter, VGAT, localizes to synaptic vesicles in sets of glycinergic as well as GABAergic neurons. J Neurosci 18:9733–9750

Crawford IL, Connor JD (1973) Localization and release of glutamic acid in relation to the hippocampal mossy fibre pathway. Nature 244:442–443

Ding R, Asada H, Obata K (1998) Changes in extracellular glutamate and GABA levels in the hippocampal CA3 and CA1 areas and the induction of glutamic acid decarboxylase-67 in dentate granule cells of rats treated with kainic acid. Brain Res 800:105–113

Frahm C, Engel D, Piechotta A, Heinemann U, Draguhn A (2000) Presence of γ-aminobutyric acid transporter mRNA in interneurons and principal cells of rat hippocampus. Neurosci Lett 388:175–178

Gillespie DC, Kim G, Kandler K (2005) Inhibitory synapses in the developing auditory system are glutamatergic. Nat Neurosci 8:332–338

Gómez-Lira G, Lamas M, Romo-Parra H, Gutiérrez R (2005) Programmed and induced phenotype of the hippocampal granule cells. J Neurosci 25:6939–6946

Gómez-Lira G, Trillo E, Ramírez M, Asai M, Sitges M, Gutiérrez R (2002) The expression of GABA in mossy fiber synaptosomes coincides with the seizure-induced expression of GABAergic transmission in the mossy fiber synapse. Exp Neurol 177:276–283
Gutiérrez R (2000) Seizures induce simultaneous GABAergic and glutamatergic neurotransmission in the dentate gyrus – CA3 system. J Neurophysiol 84:3088–3090
Gutiérrez R (2002) Activity-dependent expression of simultaneous glutamatergic and GABAergic neurotransmission from the mossy fibers in vitro. J Neurphysiol 87:2562–2570
Gutiérrez R (2003) The GABAergic phenotype of the "glutamatergic" granule cells of the dentate gyrus. Prog Neurobiol 71:337–358
Gutierrez R (2005) The dual glutamatergic-GABAergic phenotype of hippocampal granule cells. Trends Neurosci 28:297–303
Gutiérrez R, Heinemann U (1997) Simultaneous release of glutamate and GABA might be induced in mossy fibers after kindling. Neurosci Lett (Suppl. 48), S23
Gutiérrez R, Heinemann U (2001) Kindling induces transient fast inhibition in the dentate gyrus – CA3 projection. Eur J Neuroscience 13:1371–1379
Gutiérrez R, Heinemann U (2006) Co-existence of glutamate and GABA in the hippocampal granule cells: implications for epilepsy. Curr Med Chem 6:975–978
Gutiérrez R, Romo-Parra H, Maqueda J, Vivar C, Ramírez M, Morales MA, Lamas M (2003) Plasticity of the GABAergic phenotype of the "glutamatergic" granule cells of the rat dentate gyrus. J Neurosci 23:5594–5598
Hamlyn LH (1961) Electron microscopy of mossy fibre endings in Ammon's horn. Nature 190:645–646
Hennou S, Khalilov I, Diabira D, Ben-Ari Y, Gozlan H (2002) Early sequential formation of functional GABAA and glutamatergic synapses on CA1 interneurons of the rat foetal hippocampus. Eur J Neurosci 6:197–208
Henze DA, Wittner L, Buzsáki G (2002) Single granule cells reliably discharge targets in the hippocampal CA3 network in vivo. Nat Neurosci 5:790–795
Henze DA, Urban NN, Barrionuevo G (2000) The multifarious hippocampal mossy fiber pathway: a review. Neuroscience 98:407–427
Iversen LL, Kelly JS (1975) Uptake and metabolism of γ-aminobutyric acid by neurones and glial cells. Biochem Pharmacol 24:933–938
Jaffe DB, Gutiérrez R (2007) Mossy fiber synaptic transmission: communication from the dentate gyrus to area CA3. Prog Brain Res 163:109–132
Jonas P, Bischofberger J, Sandkühler J (1998) Corelease of two fast neurotransmitters at a central synapse. Science 281:419–424
Kasyanov AM, Safiulina VF, Voronin LL, Cherubini E (2004) GABA-mediated giant depolarizing potentials as coincidence detectors for enhancing synaptic efficacy in the developing hippocampus. Proc Natl Acad Sci U S A 101:3967–3972
Lamas M, Gómez-Lira G, Gutiérrez R (2001) Vesicular GABA transporter mRNA expression in the dentate gyrus and in mossy fiber synaptosomes. Mol Brain Res 93:209–214
Lehmann H, Ebert U, Löscher W (1996) Immunocytochemical localization of GABA immunoreactivity in dentate granule cells of normal and kindled rats. Neurosci Lett 212:41–44
Leinekugel X, Medina I, Khalilov I, Ben-Ari Y, Khazipov R (1997) $Ca2+$ oscillations mediated by the synergistic excitatory actions of GABA(A) and NMDA receptors in the neonatal hippocampus. Neuron 18:243–255
Makiura Y, Suzuki F, Chevalier E, Onténiente B (1999) Excitatory granule cells of the dentate gyrus exhibit a double inhibitory neurochemical content after intrahippocampal administration of kainate in adult mice. Exp Neurol 159:73–83
Maqueda J, Ramírez M, Lamas M, Gutiérrez R (2003) Glutamic acid decarboxylase $(GAD)_{67}$, but not GAD_{65}, is constitutively expressed during development and transiently overexpressed by activity in the mossy fibers. Neurosci Lett 353:69–71

McIntire SL, Reimer RJ, Schuske K, Edwards RH, Jorgensen EM (1997) Identification and characterization of the vesicular GABA transporter. Nature 389:870–876
Min MY, Rusakov DA, Kullmann DM (1998) Activation of AMPA, kainate, and metabotropic receptors at hippocampal mossy fiber synapses: role of glutamate diffusion. Neuron 21:561–570
Mody I (2002) The GAD-given right of dentate gyrus granule cells to become GABAergic. Epilepsy Curr 2:143–145
Mori M, Abegg MH, Gähwiler BH, Gerber U (2004) A frequency-dependent switch from inhibition to excitation in a hippocampal unitary circuit. Nature 431:453–456
Nicoll RA, Schmitz D (2005) Synaptic plasticity at hippocampal mossy fibre synapses. Nat Rev Neurosci 6:863–876
Ottem EN, Godwin JG, Krishnan S, Petersen SL (2004) Dual-phenotype GABA/glutamate neurons in adult preoptic area: sexual dimorphism and function. J Neurosci 24:8097–8105
Ottersen OP, Storm-Mathisen J (1984) Neurons containing or accumulating transmitter amino acids. In: Bjorklund A, Hokfelt T, Kuhar MJ (eds) Handbook of Chemical Neuroanatomy, Amsterdam, Elsevier
Radian R, Ottersen OP, Sorm-Mathisen J, Castel M, Kanner BI (1990) Immunocytochemical localization of the GABA transporter in rat brain. J Neurosci 10:1319–1330
Ramírez M, Gutiérrez R (2001) Activity-dependent expression of GAD_{67} in the granule cells of the rat hippocampus. Brain Res 917:139–146
Rao A, Cha EM, Craig AM (2000) Mismatched appositions of presynaptic and postsynaptic components in isolated hippocampal neurons. J Neurosci 20:8344–8353
Rattray M, Priestley JV (1993) Differential expression of GABA transporter-1 messenger RNA in subpopulations of GABA neurons. Neurosci Lett 156:163–166
Ribak CE, Tong WMY, Brecha NC (1996a) GABA plasma membrane transporters, GAT-1 and GAT-3, display different distributions in the hippocampus. J Comp Neurol 367:595–606
Ribak CE, Tong WMY, Brecha NC (1996b) Astrocytic processes compensate for the apparent lack of GABA transporters in the axon terminals of cerebellar Purkinje cells. Anat Embryol (Berl) 194:379–390
Romo-Parra H, Treviño M, Heinemann U, Gutiérrez R (2008) GABA actions in hippocampal area CA3 during postnatal development: differential shift from depolarizing to hyperpolarizing in somatic and dendritic compartments. J Neurophysiol, 99:1523–1534
Romo-Parra H, Vivar C, Maqueda J, Morales MA, Gutiérrez R (2003) Activity-dependent induction of multitransmitter signaling onto pyramidal cells and interneurons of area CA3 of the rat hippocampus. J Neurophysiol 89:3155–3167
Ruiz A, Fabian-Fine R, Scott R, Walker MC, Rusakov DA, Kullmann DM (2003) GABAA receptors at hippocampal mossy fibers. Neuron 39:961–973
Safiulina VF, Fattorini G, Conti F, Cherubini E (2006) GABAergic signaling at mossy fiber synapses in neonatal rat hippocampus. J Neurosci 26:597–608
Sandler R, Smith AD (1991) Coexistence of GABA and glutamate in mossy fiber terminals of the primate hippocampus: an ultrastructural study. J Comp Neurol 303:177–192
Schmitz D, Mellor J, Nicoll RA (2001) Presynaptic kainate receptor mediation of frequency facilitation at hippocampal mossy fiber synapses. Science 291:1972–1976
Schwarzer C, Sperk G (1995) Hippocampal granule cells express glutamic acid decarboxylase-67 after limbic seizures in the rat. Neuroscience 69:705–709
Seal RP, Edwards RH (2006) Functional implications of neurotransmitter corelease: glutamate and GABA share the load. Curr Opinion Pharmacol 6:114–119
Sirvanci S, Meshul CK, Onat F, San T (2003) Immunocytochemical analysis of glutamate and GABA in hippocampus of genetic absence epilepsy rats (GAERS). Brain Res 988:180–188
Sloviter RS, Dichter MA, Rachinsky TL, Dean E, Goodman JH, Sollas AL, Martin DL (1996) Basal expression and induction of glutamate decarboxylase and GABA in excitatory granule cells of the rat and monkey hippocampal dentate gyrus. J Comp Neurol 373:593–618

Sperk G, Schwarzer C, Heilman J, Furtinger S, Reimer RJ, Edwards RH, Nelson N (2003) Expression of plasma membrane GABA transporters but not of the vesicular GABA transporter in dentate granule cells after kainic acid seizures. Hippocampus 13:806–815

Spitzer NC, Borodinsky LN, Root CM (2005) Homeostatic activity-dependent paradigm for neurotransmitter specification. Cell Calcium 37:417–423

Szabó G, Kartarova Z, Hoertnagl B, Somogyi R, Sperk G (2000) Differential regulation of adult and embryonic glutamate decarboxylases in rat dentate granule cells after kainate-induced limbic seizures. Neuroscience 100:287–295

Treviño M, Gutierrez R (2005) The GABAergic projection of the dentate gyrus to hippocampal area CA3 of the rat: pre- and postsynaptic actions after seizures. J Physiol 567:939–949

Treviño M, Vivar C, Gutiérrez R (2007) β/γ Oscillatory activity in the CA3 hippocampal area is depressed by aberrant GABAergic transmission from the dentate gyrus after seizures. J Neurosci 27:251–259

Tsen G, Williams B, Allaire P, Zhou YD, Ikonomov O, Kondova I, Jacob MH (2000) Receptors with opposing functions are in postsynaptic microdomains under one presynaptic terminal. Nat Neurosci 3:126–132

Uchigashima M, Fukaya M, Watanabe M, Kamiya H (2007) Evidence against GABA release from glutamatergic mossy fiber terminals in the developing hippocampus. J Neurosci 27:8088–8100

Urban NN, Henze DA, Barrionuevo G (2001) Revisiting the role of the hippocampal mossy fiber synapse. Hippocampus 11:408–417

Vivar C, Gutierrez R (2005) Blockade of the membranal GABA transporter potentiates GABAergic responses evoked in pyramidal cells by mossy fiber activation after seizures. Hippocampus 15:281–284

Vogt KE, Nicoll RA (1999) Glutamate and gamma-aminobutyric acid mediate a heterosynaptic depression at mossy fiber synapses in the hippocampus. Proc Natl Acad Sci USA 96:1118–1122

Walker MC, Ruiz A, Kullmann DM (2001) Monosynaptic GABAergic signaling from dentate to CA3 with a pharmacological and physiological profile typical of mossy fiber synapses. Neuron 29:703–715

Yamamoto C (1972) Activation of hippocampal neurons by mossy fiber stimulation in thin brain sections in vitro. Exp Brain Res 14:423–435

Chapter 11
Synaptic Co-Release of ATP and GABA

Functional Characteristics, Modulation and Physiological Implications

S. Hugel, Y.H. Jo, and R. Schlichter

Abstract Over the last 30 years, adenosine 5'-triphosphate (ATP) has been clearly established as a cotransmitter with noradrenaline and acetylcholine in the peripheral nervous system. More recently, ATP was also identified as a cotransmitter in the central nervous system. In neuronal cultures from postnatal rat spinal cord dorsal horn or embryonic chick and postnatal mouse lateral hypothalamus, ATP was surprisingly found to be synaptically coreleased with the inhibitory neurotransmitter GABA, but not with glutamate. In these preparations, ATP activates excitatory cation-permeable receptors (P2X receptors), whereas GABA stimulates anion-permeable inhibitory $GABA_A$ receptors. The corelease of ATP and GABA therefore results in a fast mixed excitatory/inhibitory ATP/GABA cotransmission. Here we review the current knowledge on ATP/GABA cotransmission, the possible interactions of synaptically released ATP and GABA at pre- and postsynaptic sites and discuss the issues related to the role of this mixed cotransmission in the context of physiological and pathological situations.

11.1 Introduction

The possibility that nerve fibres are releasing more than one transmitter has been proposed by Burnstock and colleagues in 1970s (Burnstock 1976), challenging the single transmitter concept known as "Dale's Principle", notwithstanding that Dale never defined it as such (Dale 1935). Since then, numerous studies have demonstrated the existence of synaptic cotransmissions in both the peripheral and the central nervous system (Burnstock 2004). Generally, when two transmitters are released by the same neurone, they have

R. Schlichter, (✉)
Université Louis Pasteur, Institut des Neurosciences Cellulaires et Intégratives (INCI), Centre National de la Recherche Scientifique (CNRS), UMR7168, F-67084 Strasbourg, France
e-mail: schlichter@neurochem.u-strasbg.fr

R. Gutierrez (ed.), *Co-Existence and Co-Release of Classical Neurotransmitters*,
DOI 10.1007/978-0-387-09622-3_11,

either cooperating effects when both are activating ionotropic receptors (e.g. GABA/glycine) or opposite effects with different time-courses inducing a biphasic synaptic current (e.g. glutamate/serotonin) (Johnson 1994). Cotransmissions involving both excitatory and inhibitory fast synaptic components were unknown before the description of a cotransmission involving adenosine 5'-triphosphate (ATP) and γ-aminobutyric acid (GABA) (Jo and Schlichter 1999; Hugel and Schlichter 2000; Jo and Role 2002b), but were subsequently described in the retina for GABA and acetylcholine (ACh) (Duarte et al. 1999) and in the hippocampus for glutamate and GABA (Gutierrez et al. 2003; Gutierrez 2005).

The term "cotransmission" suggests that the two transmitters are released by a single presynaptic neurone and that they are detected by specific receptors located on the postsynaptic neurone. However, both neurotransmitter species are not always stored in the same presynaptic vesicles, and receptors for the two neurotransmitters are not necessarily localized at the same postsynaptic loci. These pre- and postsynaptic heterogeneities offer the possibility of a differential modulation of both components of the cotransmission. Initially, ATP was described as a cotransmitter with noradrenaline (NA) in sympathetic nerves, and acetylcholine (ACh) in parasympathetic nerves (Burnstock 2004, 2006). ATP released by synaptic terminals can activate cation-permeable ionotropic P2X receptors. These receptors are trimers of different subunits ($P2X_1$-$P2X_7$) possessing each two transmembrane segments and displaying no primary sequence homology with other known ionotropic receptors (Khakh and North 2006). The activation of postsynaptic P2X receptors by synaptically released ATP underlies consequently a fast excitatory transmission. Moreover, ATP is rapidly hydrolysed to adenosine 5'-diphosphate (ADP), adenosine 5'-monophosphate (AMP) and adenosine by extracellular ectonucleotidases (Zimmermann 1996; Dunwiddie et al. 1997; Zimmermann 2000). Therefore, in addition to the activation of P2X receptors, the synaptic release of ATP can also induce the subsequent activation of P2Y (by ATP and/or ADP) and adenosine metabotropic receptors.

In this review we will deal with general questions relative to excitatory or inhibitory cotransmissions, such as those related to the storage and the co-release of the cotransmitters. In the context of the issue of codetection, we will focus on the ATP/GABA cotransmission. This point is important because unlike other neurotransmitters, ATP can be metabolized in the synaptic cleft to other purines such as adenosine which also act as neuromodulators. In this respect we will address the issues of codetection by ionotropic and metabotropic receptors located post- and/or presynaptically. We will also discuss how the localization of the receptors within the synapse and the metabolism of ATP might influence the detection of miniature ATPergic postsynaptic currents. Finally, we will envisage the potential role of the ATP/GABA cotransmission in physiological and/or pathological situations. The major points that will be addressed are summarized in Fig. 11.1.

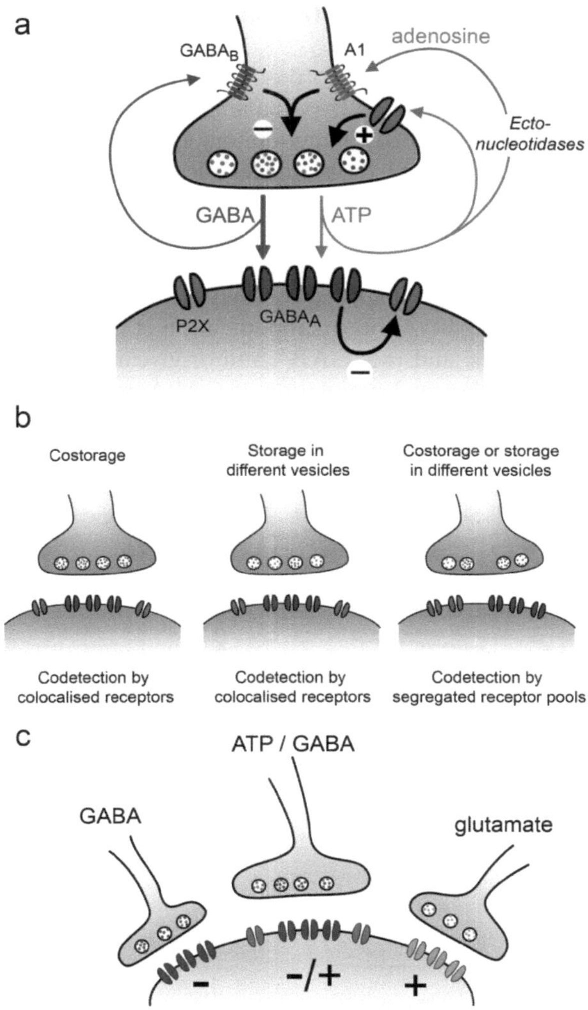

Fig 11.1 The ATP/GABA cotransmission. ATP and GABA are coreleased from the presynaptic terminal and act at postsynaptic ionotropic excitatory ATP receptors (P2X receptors) and inhibitory anionic $GABA_A$ receptors. **(a)** In the dorsal horn of the spinal cord, the synaptic corelease of ATP and GABA is modulated by presynaptic autoreceptors. ATP can facilitate GABA release by acting at presynaptic P2X receptors. Inhibition of corelease involves metabotropic $GABA_B$ receptors and A1 adenosine receptors which act by a partially convergent presynaptic mechanism. Adenosine is generated by the extracellular hydrolysis of ATP by ectonucleotidases. **(b)** Possible anatomical substrates and scenarios of ATP/GABA cotransmission. ATP and GABA might be costored in the same vesicle or stored in different vesicles. Postsynaptic P2X and $GABA_A$ receptors might be colocalized in front of the same release sites or segregated in different synapses. **(c)** The mixed excitatory/inhibitory ATP/GABA cotransmission can finely tune the balance of excitation and inhibition converging on the same neuron by shifting the net equilibrium toward inhibition or excitation, depending on the relative weight of the ATPergic and GABAergic components of the cotransmission. (*See* Color Plate 14)

11.2 ATP/GABA Synaptic Cotransmission

Neurones from postnatal rat spinal cord dorsal horn, and from both embryonic chick and postnatal mouse lateral hypothalamus maintained in dissociated cell culture form functional excitatory and inhibitory synaptic connections. The electrophysiological properties and the characteristics of synaptic transmissions in these cultures are similar to those observed in neurones from native networks (Jo et al. 1998b; Jo et al. 1998a). A subpopulation of cultured neurones is excitatory in nature, involving glutamate as a neurotransmitter. Excitatory glutamatergic transmission is always entirely blocked by AMPA/kainate and NMDA receptor antagonists (Jo et al. 1998b). Another subpopulation of neurones is constituted of inhibitory interneurones, and uses GABA and/or glycine as transmitters (Hugel and Schlichter 2000). Electrical stimulation of GABAergic neurones induces fast evoked postsynaptic currents (ePSCs) that are not always completely blocked by $GABA_A$ receptor antagonists (Fig. 11.2a). The residual ePSC component is however at least partially blocked by P2X receptors antagonists (Jo and Schlichter 1999; Hugel and Schlichter 2000). The GABAergic and the purinergic

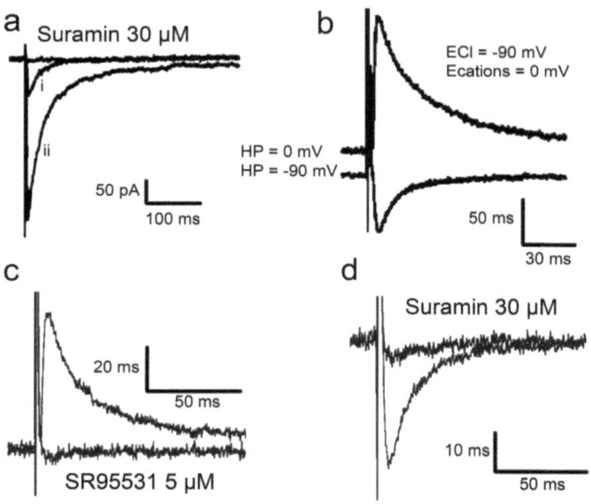

Fig 11.2 Synaptic corelease of ATP and GABA. **(a)** In cultured neonatal neurones from the superficial layers of the dorsal horn of the spinal cord, electrical stimulation of a single presynaptic neurone in the presence of bicuculline (10 μM), strychnine (1 μM), CNQX (10 μM) and D-APV (50 μM) induced an inward postsynaptic current in half of the postsynaptic neurones recorded (trace i). This inward postsynaptic current was blocked by suramin (30 μM). **(b)** Setting E_{Cl} to –90 mV and $E_{cations}$ to 0 mV allowed the isolation of the cationic ATPergic component at a holding potential (HP) of –90 mV and the GABAergic component at a HP of 0 mV. **(c)** The GABAergic component recorded at HP = 0 mV was completely blocked by SR95531 (gabazine; 5 μM), a competitive antagonist of $GABA_A$ receptors. **(d)** The ATPergic component recorded at HP = –90 mV was blocked by suramin (30 μM). *Adapted by permission from Macmillan Publishers Ltd: Nature Neuroscience, Jo and Schlichter, 1999, copyright, 1999*

(ATPergic) components can be easily separated by setting the equilibrium potential for chloride ions (E_{Cl}) at –90 mV and that for cations ($E_{cations}$) at 0 mV (Fig. 11.2b–d). Under these conditions, the P2X antagonist-sensitive ePSC component can be recorded in isolation at a holding potential of –90 mV, revealing the cationic nature of the conductance. A detailed study on the properties of ePSCs showed that both the GABAergic and the P2X receptor-mediated components had the same latency and the same stimulation threshold, demonstrating that a single presynaptic neurone was releasing GABA and ATP. Altogether, these data demonstrate that ATP is coreleased with GABA but never with glutamate in the dorsal horn of the spinal cord and the lateral hypothalamus. This type of corelease is observed only in a subpopulation of GABAergic neurones (Jo and Schlichter 1999; Hugel and Schlichter 2000).

11.2.1 General Considerations on Costorage and Corelease

ATP is present in the cytoplasm of all cell types at concentrations in the millimolar range, and is accumulated in synaptic vesicles (Sperlagh and Vizi 1996). However, the mechanisms involved in vesicular uptake of ATP are still unclear. Recent results obtained on brain synaptosomes suggest that ATP import into synaptic vesicles might involve an ADP/ATP translocase related to the mitochondrial nucleotide translocase (Gualix et al. 1999). Since ATP is present in all synaptic vesicles, it follows that ATP must be costored with classical neurotransmitters and probably with neuropeptides as well. Therefore, ATP and its cotransmitters are likely to be coreleased during fusion of single vesicles with the presynaptic membrane. However, costorage in the same vesicle does not necessarily imply that both cotransmitters are released simultaneously and at the same amounts. Indeed, a recent study on pancreatic β cells indicates that, although different cotransmitters (ATP, serotonin and GABA) are coreleased, a differential release can occur, depending on the properties of the fusion pore (Braun et al. 2007). For example, it appears that GABA and serotonin can easily exit the vesicle during partial opening of the fusion pore (e.g. “kiss and run” exocytosis) whereas ATP does not easily leave the vesicle under these conditions. A similar situation might also apply to synaptic vesicle exocytosis at central synapses (Gandhi and Stevens 2003).

11.2.2 Are ATP and GABA Coreleased From Common or Distinct Synaptic Vesicles?

Although the existence of a pure purinergic transmission has been suggested (Sperlagh and Vizi 1996), it was observed only in a few instances in the CNS (Robertson and Edwards 1998; Pankratov et al. 2007). Instead, in the peripheral nervous system as well as in the central nervous system, ATP appears to be primarily a cotransmitter rather than a principal transmitter (Burnstock 2004). Indeed, in all experiments in which stimulation of a single presynaptic neurone could be performed and electrically-evoked postsynaptic P2X receptor-mediated EPSCs were recorded, the wash out of ionotropic GABA or glutamate receptor antagonists systematically revealed the presence of synaptic GABA or glutamate corelease (Fig. 11.2a) (Jo and Schlichter 1999; Hugel and Schlichter 2000; Mori et al. 2001). Nevertheless, this does not exclude that the release of ATP and of its cotransmitter involves separate vesicular pools and/or separate presynaptic terminals (Fig. 11.1b). The definitive answer to this question relies on the analysis of miniature postsynaptic currents. Separate vesicular storage should generate separate miniature currents, whereas corelease from the same vesicle should generate mixed miniature currents with a component due to ATP and another due to the cotransmitter. Unfortuantely, this issue is extremely difficult to address since it is linked to the problem of detection of ATP by P2X receptors at the postsynaptic level (see below section 11.2.3).

It has been emphasized that a differential modulation of the release of the cotransmitters indicates that the transmitters are stored in different vesicles. In the retina, the involvement of separate vesicle pools releasing ACh and GABA is strengthened by the fact that the Ca^{2+}-dependent release of the two cotransmitters depend on distinct voltage-dependent Ca^{2+} channel types (Duarte et al. 1999). In this preparation, the release of both cotransmitters is also differentially affected by presynaptic adenosine receptors (Duarte et al. 1999). However, as mentioned above, differential release of cotransmitters packaged in the same vesicle might occur, depending on the type of exocytosis, and more precisely of the size of the fusion pore (Gandhi and Stevens 2003; Braun et al. 2007). Therefore, differential modulation of release does not necessarily imply that the transmitters are contained in distinct vesicles. Moreover, in cultured embryonic chick lateral hypothalamic neurones, nicotinic receptor stimulation selectively facilitates synaptic GABA release whereas stimulation of muscarinic receptors increases ATP release (Fig. 11.3) (Jo and Role 2002a). This differential modulation occurs despite the detection of mixed ATP/GABA miniature PSCs in this preparation (Fig. 11.4), which indicates the copackaging and corelease of ATP and GABA from the same vesicles at least at a subset of synapses. Alternatively, the differential modulatory effect of nicotinic and muscarinic agonists might concern a subset of synapses releasing only GABA and ATP, or synapses coreleasing both transmitters but at which only a single detection element (i.e. $GABA_A$ or P2X receptors) is present at the postsynaptic membrane facing the release site.

11.2.3 Codetection of ATP and GABA Corelease by Ionotropic Receptors: The Puzzling Issue of Miniature P2X Receptor-Mediated Postsynaptic Currents

Electrical stimulation of individual presynaptic neurones has revealed the existence of a functional ATPergic synaptic transmission in many structures of the

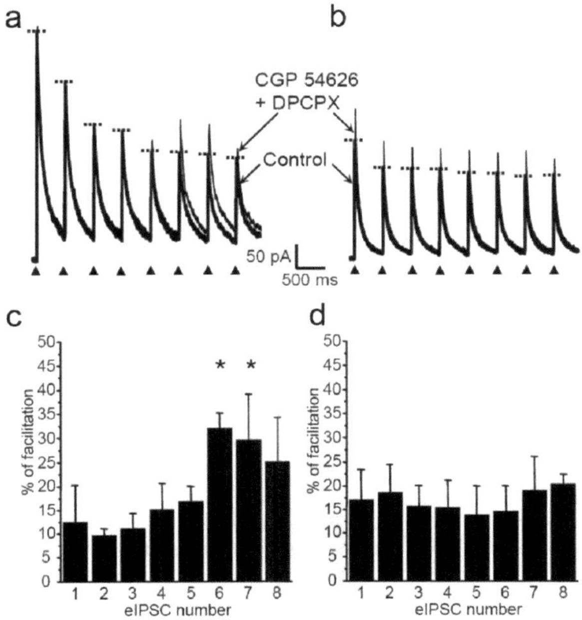

Fig 11.3 Activity-dependent (phasic) activation of presynaptic A_1 and $GABA_B$ autoreceptors in neurones coreleasing GABA and ATP but not in neurones releasing only GABA. Effects of coapplication of the A_1 antagonist (8-Cyclopentyl-1,3-dipropylxanthine, DPCPX, 1 μM) and the $GABA_B$ antagonist (CGP 54626, 1 μM) on the amplitude of eIPSCs evoked by a train of eight pulses (2.5 Hz, arrow heads) applied to the cell body of a presynaptic neurone coreleasing GABA and ATP (**a**) or of a neurone releasing only GABA (**b**). **c-d**. histograms of the percentage of GABAergic eIPSC facilitation by application of DPCPX and CGP 54626 for each of the eight stimulation pulses in neurones coreleasing ATP and GABA (**c**) and in neurones releasing GABA only (**d**). The amplitudes of the effect on the sixth and seventh eIPSCs in (**c**)were found to be significantly different from that on the other eIPSCs (one-way ANOVA with repeated measures, $P < 0.05$ and *post hoc* comparison with Duncan test, $P < 0.05$). There was no statistically significant difference in the case of neurones releasing only GABA (**d**). The HP was 0 mV. Dotted lines indicate the amplitude of the eIPSCs before applications of DPCPX and CGP 54626. *Modified with permission from Hugel and Schlichter, 2003, The Journal of Physiology, Copyright Blackwell Publishing*

peripheral and central nervous system (Burnstock 2004). Surprisingly, spontaneous ATPergic EPSC and particularly ATPergic miniature synaptic currents (mEPSCs) have only rarely been described in the central nervous system. In certain cases, increasing neuronal excitability and the release probability of neurotransmitters with a bath solution containing high concentrations of K^+ and Ca^{2+} allowed the observation of ATPergic EPSCs in the presence of tetrodotoxin (Edwards et al. 1992). Observation of ATPergic mEPSCs may also depend on the preparation and/or on the species, because ATPergic mEPSCs have been recorded in hypothalamic cultured chick embryonic neurones but were undetectable in hypothalamic neurones cultured from postnatal mice

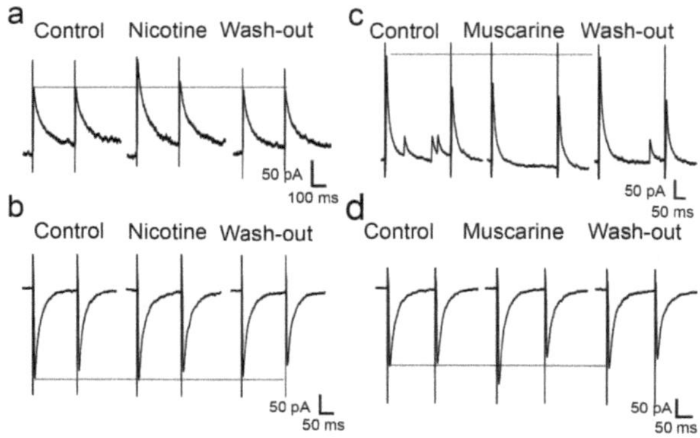

Fig 11.4 Differential cholinergic modulation of ATP/GABA cotransmission in lateral hypothalamus. **a-b** Activation of nicotinic ACh receptors facilitates GABAergic but not ATPergic transmission. **a.** Evoked GABAergic eIPSCs isolated at HP = 0 mV before, during and after application of nicotine (0.5 μM). Nicotine increased the amplitude of eIPSCs and changed the value of the ratio between the amplitude of the two consecutive eIPSCs, indicating a presynaptic action **b.** Evoked ATPergic eEPSCs isolated at HP = −70 mV were not modified by application of nicotine (0.5 μM). **c-d.** Activation of muscarinic ACh receptors depressed GABAergic transmission and facilitated purinergic transmission. **c.** Evoked GABAergic eIPSCs isolated at HP = 0 mV were depressed by application of mucarine (10 μM). **b.** Evoked ATPergic eEPSCs isolated at HP = −70 mV were potentiated by application of mucarine (10 μM). In both cases, a change in the ratio of amplitudes of eIPSCs or eEPSCs was observed indicating a presynaptic action of muscarine. *Modified with permission from Jo and Role, 2002a, The Journal of Neurophysiology, American Physiological Society*

(Jo and Role 2002b). This difficulty to record ATPergic mEPSCs might be explained by peculiarities of ATP fate in the synaptic cleft and/or of the organization of the postsynapse.

11.2.3.1 Ectonucleotidases and ATP Transients in the Synaptic Cleft

Whereas the dwell time of most transmitters within the synaptic cleft is controlled by specific transporters expressed by neurones and/or glial cells, the concentration profile of ATP within the cleft depends on the activity of specific hydrolytic enzymes, as it is the case for ACh. The enzymes metabolising ATP in the extracellular space are grouped under the generic term of ectonucleotidases (Zimmermann 1996, 2000). Interestingly, while degrading ATP in the synaptic cleft, ectonucleotidases are producing agonists for purinergic metabotropic receptors, i.e., ADP, AMP and adenosine. Most ectonucleotidases are anchored in the plasma membrane, but unidentified soluble ectonucleotidases were shown to be coreleased with ATP in the peripheral nervous system (Todorov et al. 1997; Westfall et al. 2002). In acute hippocampal slices, 50% and 96% of ATP is converted to adenosine within 200 ms and 1 s, respectively (Dunwiddie et al. 1997). Therefore, a synchronous release of several vesicles might be required to transiently saturate the ectonucleotidases in order to reach within the synaptic cleft an ATP concentration sufficient to activate postsynaptic P2X receptors. This might also explain why ATPergic mEPSCs are difficult to detect.

Results from studies on the peripheral autonomic nervous system in which ATP is coreleased with ACh or NA, or from neuroendocrine chromaffin cells, have shown that the intravesicular concentration of ATP is generally lower than that of its cotransmitter. Estimated ratios range from 2:1 to 50:1 which represent intravesicular concentrations of ATP between 1 and 200 mM (Sperlagh and Vizi 1996). This suggests that even in the case of complete exocytosis of a vesicle, the concentration transient of ATP in the synaptic cleft will be significantly smaller than that of the cotransmitter. This might explain in part why miniature postsynaptic currents due to the activation of P2X receptors are rarely observed. In addition, the concentration of ATP in the synaptic cleft is likely to be further decreased by the action of extracellular ectonucleotidases which rapidly metabolize ATP (Zimmermann 1996; Dunwiddie et al. 1997; Ghildyal et al. 2006). It can be argued that the enzymatic action of the ectonucleotidases is slow compared to the fast diffusion of ATP in the synaptic cleft. Nevertheless, the binding of ATP to ectonucleotidases (which is a fast process) might act as a buffer and contribute to reduce the concentration of free available ATP in the synaptic cleft. In line with these speculations, it has been recently shown that the ectonucleotidase inhibitor ARL 67156 strongly potentiates purinergic synaptic potentials in the guinea-pig vas deferens (Ghildyal et al. 2006). In this context, it should also be emphasized that since ATP and its nonpeptidic cotransmitter are transported by separate vesicular transporters, a differential regulation of the expression or of the activity of these transporters would result in a change in the ratio of ATP to cotransmitter content of the synaptic vesicles.

11.2.3.2 Distribution of Postsynaptic P2X and $GABA_A$ Receptors

The difficulty to record ATPergic mEPSCs might also be linked to the organization of the postsynapse (Fig. 11.1b). Both morphological and electrophysiological studies indicate an uneven distribution of P2X receptors within the neuronal membrane. P2X receptors are rarely found in excised-patches from ATP-responsive neurones of medial habenula slices (Edwards et al. 1992), but when they are found in outside-out patches from dentate gyrus granule cells, they are grouped, suggesting that P2X receptors are organized in clusters (Wong et al. 2000). In the hippocampus and the cerebellum, $P2X_2$, $P2X_4$ and $P2X_6$ subunits clusters have been localized to the periphery of glutamatergic postsynaptic densities (Rubio and Soto 2001). Such perisynaptic localizations increase the time of diffusion of ATP to its receptors and therefore the probability of ATP hydrolysis by ectonucleotidases.

11.2.3.3 Functional interactions between $GABA_A$ and P2X receptors

Co-activation of $P2X_2$ and $GABA_A$ receptors co-expressed in Xenopus oocytes leads to a functional cross-inhibition (Boue-Grabot et al. 2004) (see section below). Similar observations have been made in sympathetic (Karanjia et al. 2006) and sensory (Sokolova et al. 2001) neurones which naturally express P2X and $GABA_A$ receptors. This phenomenon involves a direct protein-to-protein interaction of the two receptor types (Boue-Grabot et al. 2004). Interestingly, when $GABA_A$ receptors include a γ subunit, which is the case of all synaptic receptors in the dorsal horn and the lateral hypothalamus (Keller et al. 2001; Jo and Role 2002b; Keller et al. 2004), the inhibition becomes unidirectional (Boue-Grabot et al. 2004). Indeed, under these conditions, $GABA_A$ receptors still inhibit the activity of P2X receptors but the reverse is no longer observed. Such an inhibition might occur during synaptic corelease of ATP and $GABA_A$ when spatially close P2X and $GABA_A$ receptors are concomitantly activated.

In neuronal cultures from the chick hypothalamus, a single postsynaptic neurone receiving synaptic inputs from a presynaptic neurone coreleasing ATP and GABA, displays "pure" GABAergic mIPSCs and mixed ATP/GABA mPSCs but "pure" P2X-receptor-mediated ATPergic mEPSCs are never recorded (Jo and Role 2002b). However, after pharmacological blockade of $GABA_A$ receptors with β-hydrastine, P2X-receptor-mediated PSCs are detected. This situation could illustrate the functional inhibition of P2X receptors by the activation of $GABA_A$ receptors during synaptic corelease of ATP and GABA (Boue-Grabot et al. 2004). Preventing the activation of $GABA_A$ receptors would therefore allow the synaptic activation of P2X receptors. Although this seems to be the case in chick hypothalamic neurones (Jo and Role 2002b), it does not apply to other preparations were ATP and GABA corelease has been described. In cultures of the dorsal horn of the spinal cord, blockade of $GABA_A$ receptors by bicuculline did not reveal miniature P2X receptor-mediated EPSCs, but β-hydrastine was not tested in the same situation (Jo and Schlichter 1999). Moreover, since this phenomenon is due to a direct protein-protein interaction, it implies that $GABA_A$ and P2X receptors must be in close spatial proximity.

11.2.4 Codetection of ATP and GABA Corelease by Presynaptic Ionotropic and Metabotropic Receptors

Up to this stage, we have only considered the detection/codetection of synaptically released transmitters/cotransmitters by ionotropic receptors expressed by the postsynaptic neurone. Alternatively, the cotransmitters could be detected by presynaptic ionotropic receptors or/and metabotropic G-protein coupled receptors, i.e $GABA_B$ receptors for GABA and P2Y receptors for ATP/ADP. Moreover, adenosine resulting from rapid degradation of ATP by ectonucleotidases can bind to adenosine receptors (Fig. 11.1a) (Zimmermann

1996; Dunwiddie et al. 1997). The activation of postsynaptic metabotropic receptors by GABA and ATP/adenosine has not been investigated so far in the context of cotransmission. However, we have recently shown that corelease of ATP and GABA might finely regulate their own release by acting synergistically at presynaptic $GABA_B$ and A_1 adenosine receptors (Hugel and Schlichter 2003) (see below).

11.2.4.1 Presynaptic P2X Receptors

P2X receptors are present on the presynaptic terminals of GABAergic neurones but not of glutamatergic neurones in cultures of neonatal spinal cord dorsal horn neurones (Fig. 11.1a) (Hugel and Schlichter 2000). These receptors might be the target of synaptically released ATP but this issue is difficult to address due to the absence of selective antagonists of the P2X receptors expressed in the central nervous system. Nevertheless, it appears that presynaptic P2X receptors are preferentially expressed on GABAergic neurones coreleasing GABA and ATP (Hugel and Schlichter 2000).

11.2.4.2 Presynaptic $GABA_B$ and A1 Autoreceptors

In cultured neurones from the dorsal horn of the spinal cord, the synaptic corelease of ATP and GABA is controlled by both released transmitters and by ATP hyrdrolysis products such as adenosine (Jo and Schlichter 1999; Hugel and Schlichter 2000).

In control conditions, presynaptic $GABA_B$ and A1 receptors are tonically inhibiting both the ATPergic and the GABAergic component of the cotransmission (Hugel and Schlichter 2003). During repetitive stimulation of the presynaptic neurone coreleasing GABA and ATP, $GABA_B$ and A1 autoreceptors are recruited to inhibit evoked transmitter release. This phasic control exerted by the released GABA and adenosine (*via* ATP hydrolysis) is not observed in the case of neurones releasing only GABA (Hugel and Schlichter 2003) (Fig. 11.5). This suggests that $GABA_B$ and A1 receptors are acting in synergy to control presynaptically the ATP/GABA cotransmission, possibly by converging on the same transduction pathway. In addition to presynaptic A1 receptors, we have shown that postsynaptic A1 receptors are inhibiting selectively $GABA_A$ receptors mediated currents in dorsal horn neurones (Jo and Schlichter 1999).

Presynaptic adenosine, P2X and $GABA_B$ receptors seem therefore able to sense and to regulate the corelease of ATP and GABA (Hugel and Schlichter 2000, 2003). In the hippocampus, synaptically released GABA is also detected by perisynaptic astrocytes which express $GABA_B$ receptors and respond to the synaptic release of GABA by an elevation in the intracellular calcium concentration (Kang et al. 1998). ATP has recently been established as an important gliotransmitter, and astrocytes express functional P2X and P2Y receptors (Haydon 2001; Nedergaard et al. 2003; Haydon and Carmignoto 2006). These

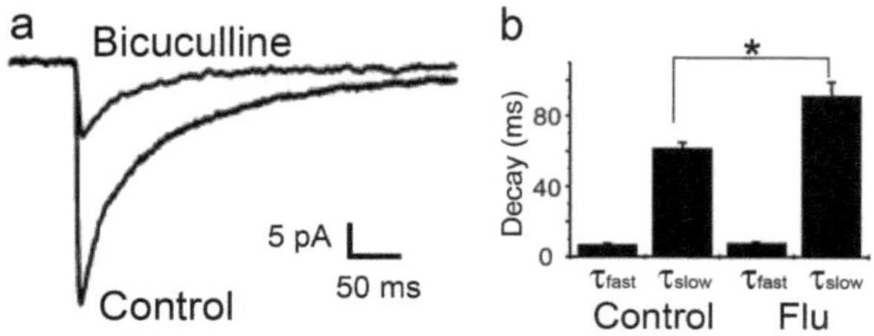

Fig 11.5 Mixed ATPergic/GABAergic spontaneous miniature PSCs recorded in cultured chick lateral hypothalamus neurones. **(a)** averaged traces of bicuculline-sensitive and bicuculline-resistant mPSCs. **(b)** The decay time of the mixed events was best-fitted by a biexponential function. Flunitrazepam (Flu, 100 nM), a positive allosteric modulator of $GABA_A$ receptors had no effect on the time constant of the fast component (τ_{fast}) whereas it increases the time constant of the slow exponential function (τ_{slow}), confirming the GABAergic nature of the slow component. Modified from Jo and Role, 2002b. *The Journal of Neuroscience, Copyright 2002 by the Society for Neuroscience*

receptors participate in the propagation of calcium waves in the astrocytic network (Haydon 2001; Nedergaard et al. 2003; Haydon and Carmignoto 2006). However, it is not known whether ATP released from neurones can activate P2X or P2Y receptors expressed by perisynaptic astrocytes.

11.2.4.3 Conclusion on Corelease Costorage and Codetection

As discussed above, synaptic corelease of ATP and GABA from the same presynaptic neurone has been clearly established in cultures of rat dorsal horn spinal cord neurones and lateral hypothalamic neurones of chick and mouse. However, the issue of ATP/GABA corelease from the same vesicles is difficult to assess in the absence of recordable miniature P2X-receptor-mediated PSCs in most of preparations. It is hoped that conditions revealing such miniature currents can be found in order to fully answer this question which is of fundamental importance for understanding the role of cotransmissions involving ATP, as well as the general organization of pre- and postsynaptic elements in this context.

11.3 Physiological Role of ATP/GABA Cotransmission: From Facts to Speculations

An important question concerning synaptic ATP and GABA corelease is related to its role under physiological and/or pathological situations. In particular, it is important to know whether this phenomenon is only transiently observed during development in immature systems or if it is still present in adult organisms.

11.3.1 ATP/GABA Cotransmission During Development and in the Adult

Purines act as signalling molecules in the CNS (Fields and Stevens 2000; Fields and Burnstock 2006; Haydon and Carmignoto 2006) but they play also an important role as trophic factors for neurones and glial cells (Rathbone et al. 1999). Therefore it is legitimate to ask whether synthesis and release of purines is prominent only in developing organisms. So far, ATP and GABA corelease has been described in cultures or slices from immature animals. However, in the dorsal spinal cord or the lateral hypothalamus of adult rodents P2X receptors are still expressed (Collo et al. 1996) along with $GABA_A$ receptors, suggesting that release and codetection of ATP and GABA release might still be possible in these regions in adults. Moreover, expression of functional P2X receptors increases during postnatal development in lamina V of the spinal cord (Shiokawa et al. 2006). We have preliminary evidence that a functional synaptic transmission involving ATP and postsynaptic P2X receptors still occurs in 3 to 4 weeks old rats (unpublished observations), but the issue of corelease of ATP with GABA could not be addressed for the moment. Indeed, identification of ATP/GABA cotransmission in the absence of detectable miniature P2X receptor-mediated PSCs requires to identify and to selectively stimulate electrically a single presynaptic neurone. This is difficult in slices because: (1) even minimal local extracellular stimulation will activate several presynaptic fibres due to the dense fibre network in the superficial layers of the spinal cord and (2) unitary connections within lamina II of the spinal cord are difficult to identify rendering paired recordings problematic (Lu and Perl 2003).

11.3.2 Role in Synaptogenesis and in the Modulation of $GABA_A$ Receptor Function

An important process taking place during development is the formation of functional synaptic contacts. This phenomenon seems to depend on local increases in intracellular calcium at postsynaptic sites where synapses are to be established. During this phase of synaptogenesis, $GABA_A$ receptor activation triggers a membrane depolarization which facilitates the activation of voltage dependent Ca^{2+} channels (Fitzgerald 2005). It is only later in development that $GABA_A$ receptor-mediated responses become hyperpolarising, a phenomenon due to the progressive expression of the K^+/Cl^- cotransporter KCC2 (Owens et al. 1996; Rivera et al. 1999; Ben-Ari 2002; Baccei and Fitzgerald 2004). Therefore, at early stages at which the $GABA_A$ responses are still depolarising, activation of P2X receptors by ATP coreleased with GABA at future synaptic sites could amplify the rise in intracellular calcium concentration at the post-synaptic membrane, thereby facilitating the establishment and the stabilization of synaptic contacts. Once $GABA_A$ responses are hyperpolarising, activation of

P2X receptors might finely tune the activity of $GABA_A$ receptors via calcium-dependent mechanisms. Indeed, $GABA_A$ receptors are directly modulated by increases in intracellular free calcium concentration and the sign of the modulation depends on the amplitude of the calcium signal (Inoue et al. 1986; Llano et al. 1991; Mouginot et al. 1991). In addition, intracellular calcium can activate calcium-dependent protein kinases and phosphatases which are known to modulate $GABA_A$ receptors (Moss and Smart 1996, 2001). Therefore, P2X receptors might provide a means for activity- and calcium-dependent modulation of synaptic $GABA_A$ receptors. In a slightly different context, ATP might also act in concert with GABA in facilitating the integration and differentiation of new-born neurones (Owens and Kriegstein 2002).

11.3.3 Cross-inhibition of P2X and $GABA_A$ Receptors

It has been shown recently that P2X and $GABA_A$ receptors display a reciprocal functional inhibition due to direct protein-protein interactions between the two receptors (Sokolova et al. 2001; Boue-Grabot et al. 2004; Karanjia et al. 2006). Interestingly, the presence of a γ subunit, which confers sensitivity to benzodiazepines, in the composition of the $GABA_A$ receptor allows inhibition of P2X receptors by $GABA_A$ receptors but precludes inhibition of $GABA_A$ receptors by P2X receptor activation. In the dorsal horn of the spinal cord and in the lateral hypothalamus, synaptic $GABA_A$ receptors are positively modulated by benzodiazepines indicating that they include a γ subunit (Chery and de Koninck 1999; Jo and Role 2002b; Keller et al. 2004). This situation will certainly limit the action of P2X receptors on synaptic $GABA_A$ receptors which are in close association with them. However, ultrastructural studies have shown that P2X receptors have rather a perisynaptic localization (Rubio and Soto 2001) and it is not known whether these perisynaptic P2X receptors are in close association with γ subunit-containing $GABA_A$ receptors. By contrast, extrasynaptic $GABA_A$ receptors do generally not include γ subunits (Mohler et al. 2004) rendering a reciprocal inhibition between extrasynaptic $GABA_A$ and P2X receptors possible. Such an interaction could occur during spillover of GABA and ATP and might finely regulate, either positively or negatively, the excitability of neurones in which such an interaction occurs (Semyanov et al. 2004). Under these circumstances, the sign of the modulatory effect on excitability will essentially depend on the balance between the two transmitters, i.e. the relative activation of both types of receptors.

11.3.4 ATP/GABA Cotransmission and Pathological Situations

One could speculate that ATP and GABA corelease is prominent during early development, that this corelease might "disappear" at later stages, and could subsequently "reappear" under particular physiological and/or pathological conditions. There are examples for such a situation in the case of other synaptic coreleases. For instance, synaptic corelease of GABA and glutamate is observed at synapses between mossy fibre and CA3 pyramidal neurones in the developing hippocampus (Gutierrez et al. 2003; Gutierrez 2005). These synapses subsequently become purely glutamatergic, but following epileptic seizures the mixed phenotype reappears due to the reexpression of glutamic acid decarboylase, the biosynthetic enzyme for GABA, in the mossy fibre terminals (Gutierrez 2000, 2002, 2005). In this case, the cotransmission plasticity is of presynaptic origin. In the same line, GABA and glycine corelease is present at a subset of inhibitory synapses in laminae I and II of the dorsal horn of the spinal cord during the early postnatal period (Chery and de Koninck 1999; Keller et al. 2004). Subsequently, these synapses become exclusively glycinergic in lamina I and GABAergic and glycinergic in lamina II, i.e. in lamina II one detects only "pure" GABAergic or "pure" glycinergic events in the same neurone but never mixed GABA/glycine synaptic currents (Chery and de Koninck 1999; Keller et al. 2004). The "disappearance" of mixed GABA/glycine mIPSCs is not due to a "disappearance" of corelease but most probably to the redistribution of synaptic $GABA_A$ receptors which become extrasynaptic at the synapses that corelease GABA and glycine. Indeed, application of positive allosteric modulators of $GABA_A$ receptors such as benzodiazepines or $3\alpha5\alpha$-reduced neurosteroids leads to the reappearance of mixed GABA/glycine events (Chery and de Koninck 1999; Keller et al. 2004). In this case, the apparent downregulation of cotransmission involves postsynaptic modifications. A similar "reappearance" of mixed GABA/glycine events is observed in lamina II following the induction of an inflammatory pain and this phenomenon is related to the stimulation the local production of $3\alpha5\alpha$-reduced neurosteroids (Poisbeau et al. 2005). One could therefore speculate that ATP/ GABA cotransmission might be subjected to similar regulations.

The phenotype of a mixed ATP/GABA synapse will depend on: (1) the relative amount of GABA and ATP released, (2) the density and the properties of the postsynaptic receptors activated and (3) the functional interactions (cross-inhibition) between postsynaptic P2X and $GABA_A$ receptors. Our observations suggest that under resting conditions, the $GABA_A$ component is always dominant indicating that the phenotype of these synapses is inhibitory. However, under conditions of differential release of the two transmitters or of differential regulation of postsynaptic P2X and $GABA_A$ receptors, the inhibitory weight of the synapses is likely to change. In an extreme situation, one could even envisage that the synapse becomes excitatory when the P2X component becomes dominant with respect to the $GABA_A$ component. This could occur during plastic states associated with inflammatory or neuropathic pain (see below) and could in part explain the changes that take place in the processing of nociceptive and non-nociceptive messages in the dorsal horn of the spinal cord in the absence of major changes in anatomical connections. It is important to underline that changes in the processing of somatosensory information by the dorsal horn neuronal networks do no necessarily imply that the sign of the mixed ATP/GABA synapse has to switch to an excitatory phenotype. Indeed, the postsynaptic neurone receiving these mixed synapses also receives excitatory and inhibitory synaptic inputs mediated by glutamate and GABA or glycine (Fig. 11.1c). The postsynaptic neurone integrates these inputs and the mixed (ATP/GABA) synapses contribute to the balance between excitation and inhibition. In the case of a reduction in the inhibitory power of the mixed synapse (relative increase of the P2X component), this would result in a net excitation of the postsynaptic neurone due to a deficit in inhibition. In cultures of neurones from the lateral hypothalamus, Jo and Role have described a differential modulation of the GABAergic and the purinergic components of the ATP/GABA cotransmission by stimulation of nicotinic and muscarinic ACh receptors, respectively (Fig. 11.3) (Jo and Role 2002a). This modulation takes place at the presynaptic level. A differential control of this cotransmission would also occur in the case of a differential modulation of postsynaptic P2X and $GABA_A$ receptors. We have recently shown that a peripheral inflammation stimulates the production of $3\alpha5\alpha$-reduced neurosteroids in lamina II of the dorsal horn of the spinal cord (Poisbeau et al. 2005). These $3\alpha5\alpha$-reduced neurosteroids are selective positive allosteric modulators of $GABA_A$ receptors (Belelli and Lambert 2005; Schlichter et al. 2006) that do not affect the function P2X receptors (unpublished observation). Therefore, the production of $3\alpha5\alpha$-reduced neurosteroids would selectively potentiate the GABAergic component of the ATP/GABA cotransmission, leaving the purinergic component unaffected. However, one cannot exclude the production of other neurosteroids such as dehydroepiandrosterone (DHEA) and its derivatives in lamina II (Kibaly et al. 2005). Interestingly DHEA potentiates the activity of P2X receptors including the $P2X_2$ subunit and has either no effect on or inhibits $GABA_A$ receptor-mediated currents (De Roo et al. 2003). The $P2X_2$ subunit is expressed in lamina II of the dorsal horn of the spinal cord and the lateral hypothalamus (Collo et al. 1996) and we have described the presence of post- and presynaptic P2X receptors having the properties of $P2X_2$-containing receptors in cultures of superficial dorsal horn neurones (Hugel and Schlichter 2000). Therefore, in case of a local production of DHEA, a selective potentiation of the P2X receptor-mediated component can be expected whereas the GABAergic component will be unaffected or inhibited. These considerations suggest that a differential modulation of the GABAergic and the purinergic components of the ATP/GABA cotransmission is theoretically possible. An important issue to follow is to know under which physiological or pathological situations such differential modulations are likely to occur. As discussed above, changes in the local production of neurosteroids might achieve such conditions. $GABA_A$ receptors play an important role in the processing of pain messages in the dorsal horn of the spinal cord because intrathecal administration of the $GABA_A$ receptor antagonist bicuculline induces states of hyperalgesia and allodynia (Millan 1999; Yaksh 1999). Similarly, P2X receptors are implicated in the detection and transmission of peripheral pain messages. However, up to now, studies on the role of P2X receptors in nociception have essentially concentrated on receptors expressed by primary afferent neurones and little is known about the plasticity in the expression and the function of P2X receptors in the dorsal horn of the spinal cord. There are indications that these receptors might be implicated in inflammatory pain processing in the dorsal horn (Stanfa et al. 2000). In the case of neuropathic pain, neuronal P2X receptors seem to be less involved (Stanfa et al. 2000) but $P2X_4$ receptors expressed by microglial cells play an important role in the development of the pathological situation (Tsuda et al. 2003).

Interestingly, P2X and $GABA_A$ receptors are also colocalized on the central terminals of primary nociceptors within the spinal cord (Labrakakis et al. 2003). Therefore, corelease of ATP and GABA might contribute to the fine tuning of synaptic glutamate release at the first relay of nociceptive messages in the dorsal horn of the spinal cord.

11.4 Conclusion

The simultaneous corelease and postsynaptic codetection of ATP with GABA has been demonstrated in cultures from two different structures of the central nervous system: the dorsal horn of the spinal cord and the lateral hypothalamus. The lack of miniature ATPergic PSC together with the difficulty to identify pairs of synaptically connected neurones has, for the moment, precluded the identification of corelease in slice preparations from spinal cord and hypothalamus. It is however probable that this corelease persists in the adult dorsal horn of the spinal cord where it might participate in the modulation of neuronal excitability observed in inflammatory pain situations. In such a context, the modulation of ATP/GABA cotransmission might contribute to the modulation or alteration of the processing of nervous messages without reorganization of the anatomical connections. In particular, it will be interesting to determine whether extracellular mediators (e.g. inflammatory mediators) released in these structures under specific physiological or pathological situations will reveal, potentiate or differentially modulate this intriguing mixed excitatory/inhibitory cotransmission.

Acknowledgments The authors would like to acknowledge financial support from Centre National de la Recherche Scientifique (CNRS), French Ministry of Education and Research, University Louis Pasteur, Institut UPSA de la Douleur, DRTC and American Diabetes Association.

List of abbreviations

ACh	Acetylcholine
ADP	adenosine 5'-diphosphate
AMP	adenosine 5'-monophosphate
ATP	adenosine 5'-triphosphate

CNQX	6-cyano-7-nitroquinoxaline-2,3-dione
D-APV	D-amino-phosphonovaleric acid
E_{Cl}	equilibrium potential for Cl^- ions
$E_{cations}$	equilibrium potential for cations
ePSC	electrically-evoked postsynaptic current
eEPSC	electrically-evoked excitatory postsynaptic current
eIPSC	electrically-evoked inhibitory postsynaptic current
DHEA	dehydroepiandrosterone
DPCPX	8-Cyclopentyl-1,3-dipropylxanthine
GABA	γ-aminobutyric acid
HP	holding potential
mPSC	miniature postsynaptic current
mEPSC	miniature excitatory postsynaptic current
mIPSC	miniature inhibitory postsynaptic current
NA	noradrenaline
NMDA	N-methyl-D-Aspartic acid

References

Baccei ML, Fitzgerald M (2004) Development of GABAergic and glycinergic transmission in the neonatal rat dorsal horn. J Neurosci 24:4749–4757

Belelli D, Lambert JJ (2005) Neurosteroids: endogenous regulators of the GABA(A) receptor. Nat Rev Neurosci 6:565–575

Ben-Ari Y (2002) Excitatory actions of gaba during development: the nature of the nurture. Nat Rev Neurosci 3:728–739

Boue-Grabot E, Toulme E, Emerit MB et al. (2004) Subunit-specific coupling between gamma-aminobutyric acid type A and P2X2 receptor channels. J Biol Chem 279:52517–52525

Braun M, Wendt A, Karanauskaite J et al. (2007) Corelease and differential exit via the fusion pore of GABA, serotonin, and ATP from LDCV in rat pancreatic beta cells. J Gen Physiol 129:221–231

Burnstock G (1976) Do some nerve cells release more than one transmitter? Neuroscience 1:239–248

Burnstock G (2004) Cotransmission. Curr Opin Pharmacol 4:47–52

Burnstock G (2006) Historical review: ATP as a neurotransmitter. Trends Pharmacol Sci 27:166–176

Chery N, de Koninck Y (1999) Junctional versus extrajunctional glycine and GABA(A) receptor-mediated IPSCs in identified lamina I neurons of the adult rat spinal cord. J Neurosci 19:7342–7355

Collo G, North RA, Kawashima E et al. (1996) Cloning OF P2X5 and P2X6 receptors and the distribution and properties of an extended family of ATP-gated ion channels. J Neurosci 16:2495–2507

Dale HH (1935) Pharmacology and nerve endings. Proc Roy Soc Med 18:319–322

De Roo M, Rodeau JL, Schlichter R (2003) Dehydroepiandrosterone potentiates native ionotropic ATP receptors containing the P2X2 subunit in rat sensory neurones. J Physiol 552:59–71

Duarte CB, Santos PF, Carvalho AP (1999) Corelease of two functionally opposite neurotransmitters by retinal amacrine cells: experimental evidence and functional significance. J Neurosci Res 58:475–479

Dunwiddie TV, Diao L, Proctor WR (1997) Adenine nucleotides undergo rapid, quantitative conversion to adenosine in the extracellular space in rat hippocampus. J Neurosci 17:7673–7682

Edwards FA, Gibb AJ, Colquhoun D (1992) ATP receptor-mediated synaptic currents in the central nervous system. Nature 359:144–147

Fields RD, Stevens B (2000) ATP: an extracellular signaling molecule between neurons and glia. Trends Neurosci 23:625–633

Fields RD, Burnstock G (2006) Purinergic signalling in neuron-glia interactions. Nat Rev Neurosci 7:423–436

Fitzgerald M (2005) The development of nociceptive circuits. Nat Rev Neurosci 6:507–520

Gandhi SP, Stevens CF (2003) Three modes of synaptic vesicular recycling revealed by single-vesicle imaging. Nature 423:607–613

Ghildyal P, Palani D, Manchanda R (2006) Post- and prejunctional consequences of ecto-ATPase inhibition: electrical and contractile studies in guinea-pig vas deferens. J Physiol 575:469–480

Gualix J, Pintor J, Miras-Portugal MT (1999) Characterization of nucleotide transport into rat brain synaptic vesicles. J Neurochem 73:1098–1104

Gutierrez R (2000) Seizures induce simultaneous GABAergic and glutamatergic transmission in the dentate gyrus-CA3 system. J Neurophysiol 84:3088–3090

Gutierrez R (2002) Activity-dependent expression of simultaneous glutamatergic and GABAergic neurotransmission from the mossy fibers in vitro. J Neurophysiol 87:2562–2570

Gutierrez R (2005) The dual glutamatergic-GABAergic phenotype of hippocampal granule cells. Trends Neurosci 28:297–303

Gutierrez R, Romo-Parra H, Maqueda J et al. (2003) Plasticity of the GABAergic phenotype of the "glutamatergic" granule cells of the rat dentate gyrus. J Neurosci 23:5594–5598

Haydon PG (2001) GLIA: listening and talking to the synapse. Nat Rev Neurosci 2:185–193

Haydon PG, Carmignoto G (2006) Astrocyte control of synaptic transmission and neurovascular coupling. Physiol Rev 86:1009–1031

Hugel S, Schlichter R (2000) Presynaptic P2X receptors facilitate inhibitory GABAergic transmission between cultured rat spinal cord dorsal horn neurons. J Neurosci 20:2121–2130

Hugel S, Schlichter R (2003) Convergent control of synaptic GABA release from rat dorsal horn neurones by adenosine and GABA autoreceptors. J Physiol 551:479–489

Inoue M, Oomura Y, Yakushiji T et al. (1986) Intracellular calcium ions decrease the affinity of the GABA receptor. Nature 324:156–158

Jo YH, Schlichter R (1999) Synaptic corelease of ATP and GABA in cultured spinal neurons. Nat Neurosci 2:241–245

Jo YH, Role LW (2002a) Cholinergic modulation of purinergic and GABAergic cotransmission at in vitro hypothalamic synapses. J Neurophysiol 88:2501–2508

Jo YH, Role LW (2002b) Coordinate release of ATP and GABA at in vitro synapses of lateral hypothalamic neurons. J Neurosci 22:4794–4804

Jo YH, Stoeckel ME, Schlichter R (1998a) Electrophysiological properties of cultured neonatal rat dorsal horn neurons containing GABA and met-enkephalin-like immunoreactivity. J Neurophysiol 79:1583–1586

Jo YH, Stoeckel ME, Freund-Mercier MJ et al. (1998b) Oxytocin modulates glutamatergic synaptic transmission between cultured neonatal spinal cord dorsal horn neurons. J Neurosci 18:2377–2386

Johnson MD (1994) Synaptic glutamate release by postnatal rat serotonergic neurons in microculture. Neuron 12:433–442

Kang J, Jiang L, Goldman SA et al. (1998) Astrocyte-mediated potentiation of inhibitory synaptic transmission. Nat Neurosci 1:683–692

Karanjia R, Garcia-Hernandez LM, Miranda-Morales M et al. (2006) Cross-inhibitory interactions between GABAA and P2X channels in myenteric neurones. Eur J Neurosci 23:3259–3268

Keller AF, Breton JD, Schlichter R et al. (2004) Production of 5alpha-reduced neurosteroids is developmentally regulated and shapes GABA(A) miniature IPSCs in lamina II of the spinal cord. J Neurosci 24:907–915
Keller AF, Coull JA, Chery N et al. (2001) Region-specific developmental specialization of GABA-glycine cosynapses in laminas I-II of the rat spinal dorsal horn. J Neurosci 21:7871–7880
Khakh BS, North RA (2006) P2X receptors as cell-surface ATP sensors in health and disease. Nature 442:527–532
Kibaly C, Patte-Mensah C, Mensah-Nyagan AG (2005) Molecular and neurochemical evidence for the biosynthesis of dehydroepiandrosterone in the adult rat spinal cord. J Neurochem 93:1220–1230
Labrakakis C, Tong CK, Weissman T et al. (2003) Localization and function of ATP and GABAA receptors expressed by nociceptors and other postnatal sensory neurons in rat. J Physiol 549:131–142
Llano I, Leresche N, Marty A (1991) Calcium entry increases the sensitivity of cerebellar Purkinje cells to applied GABA and decreases inhibitory synaptic currents. Neuron 6:565–574
Lu Y, Perl ER (2003) A specific inhibitory pathway between substantia gelatinosa neurons receiving direct C-fiber input. J Neurosci 23:8752–8758
Millan MJ (1999) The induction of pain: an integrative review. Prog Neurobiol 57:1–164
Mohler H, Fritschy JM, Crestani F et al. (2004) Specific GABA(A) circuits in brain development and therapy. Biochem Pharmacol 68:1685–1690
Mori M, Heuss C, Gahwiler BH et al. (2001) Fast synaptic transmission mediated by P2X receptors in CA3 pyramidal cells of rat hippocampal slice cultures. J Physiol 535:115–123
Moss SJ, Smart TG (1996) Modulation of amino acid-gated ion channels by protein phosphorylation. Int Rev Neurobiol 39:1–52
Moss SJ, Smart TG (2001) Constructing inhibitory synapses. Nat Rev Neurosci 2:240–250
Mouginot D, Feltz P, Schlichter R (1991) Modulation of GABA-gated chloride currents by intracellular $Ca2+$ in cultured porcine melanotrophs. J Physiol 437:109–132
Nedergaard M, Ransom B, Goldman SA (2003) New roles for astrocytes: redefining the functional architecture of the brain. Trends Neurosci 26:523–530
Owens DF, Kriegstein AR (2002) Is there more to GABA than synaptic inhibition? Nat Rev Neurosci 3:715–727
Owens DF, Boyce LH, Davis MB et al. (1996) Excitatory GABA responses in embryonic and neonatal cortical slices demonstrated by gramicidin perforated-patch recordings and calcium imaging. J Neurosci 16:6414–6423
Pankratov Y, Lalo U, Verkhratsky A et al. (2007) Quantal release of ATP in mouse cortex. J Gen Physiol 129:257–265
Poisbeau P, Patte-Mensah C, Keller AF et al. (2005) Inflammatory pain upregulates spinal inhibition via endogenous neurosteroid production. J Neurosci 25:11768–11776
Rathbone MP, Middlemiss PJ, Gysbers JW et al. (1999) Trophic effects of purines in neurons and glial cells. Prog Neurobiol 59:663–690
Rivera C, Voipio J, Payne JA et al. (1999) The $K+/Cl-$ co-transporter KCC2 renders GABA hyperpolarizing during neuronal maturation. Nature 397:251–255
Robertson SJ, Edwards FA (1998) ATP and glutamate are released from separate neurones in the rat medial habenula nucleus: frequency dependence and adenosine-mediated inhibition of release. J Physiol 508 (Pt 3):691–701
Rubio ME, Soto F (2001) Distinct Localization of P2X receptors at excitatory postsynaptic specializations. J Neurosci 21:641–653
Schlichter R, Keller AF, De Roo M et al. (2006) Fast nongenomic effects of steroids on synaptic transmission and role of endogenous neurosteroids in spinal pain pathways. J Mol Neurosci 28:33–51
Semyanov A, Walker MC, Kullmann DM et al. (2004) Tonically active GABA A receptors: modulating gain and maintaining the tone. Trends Neurosci 27:262–269

Shiokawa H, Nakatsuka T, Furue H et al. (2006) Direct excitation of deep dorsal horn neurones in the rat spinal cord by the activation of postsynaptic P2X receptors. J Physiol 573:753–763

Sokolova E, Nistri A, Giniatullin R (2001) Negative cross talk between anionic GABAA and cationic P2X ionotropic receptors of rat dorsal root ganglion neurons. J Neurosci 21:4958–4968

Sperlagh B, Vizi ES (1996) Neuronal synthesis, storage and release of ATP. Sem Neurosci 8:175–186

Stanfa LC, Kontinen VK, Dickenson AH (2000) Effects of spinally administered P2X receptor agonists and antagonists on the responses of dorsal horn neurones recorded in normal, carrageenan-inflamed and neuropathic rats. Br J Pharmacol 129:351–359

Todorov LD, Mihaylova-Todorova S, Westfall TD et al. (1997) Neuronal release of soluble nucleotidases and their role in neurotransmitter inactivation. Nature 387:76–79

Tsuda M, Shigemoto-Mogami Y, Koizumi S et al. (2003) P2X4 receptors induced in spinal microglia gate tactile allodynia after nerve injury. Nature 424:778–783

Westfall DP, Todorov LD, Mihaylova-Todorova ST (2002) ATP as a cotransmitter in sympathetic nerves and its inactivation by releasable enzymes. J Pharmacol Exp Ther 303:439–444

Wong AY, Burnstock G, Gibb AJ (2000) Single channel properties of P2X ATP receptors in outside-out patches from rat hippocampal granule cells. J Physiol 527 Pt 3:529–547

Yaksh TL (1999) Spinal systems and pain processing: development of novel analgesic drugs with mechanistically defined models. Trends Pharmacol Sci 20:329–337

Zimmermann H (1996) Biochemistry, localization and functional roles of ecto-nucleotidases in the nervous system. Prog Neurobiol 49:589–618

Zimmermann H (2000) Extracellular metabolism of ATP and other nucleotides. Naunyn Schmiedebergs Arch Pharmacol 362:299–309

Chapter 12
The Co-Release of Glutamate and Acetylcholine in the Vertebrate Nervous System

Wen-Chang Li

Abstract Co-release of acetylcholine (ACh) and glutamate has been found in a number of cases including *Xenopus* tadpole spinal cord and hindbrain interneurons, neonatal mouse motoneuron central synapses and rat basal forebrain cholinergic neurons. It is not clear at present whether the co-release is restricted to certain developmental stages. The significance of co-release of both excitatory transmitters may include complex interactions between cholinergic and glutamatergic transmissions during normal functions or development. In Xenopus tadpole spinal cord, co-released ACh from glutamatergic excitatory interneurons (dINs) activate the nicotinic receptors which may help maintain tonic NMDA receptor mediated membrane potential depolarization that is critical for persistent swimming. Nicotinic excitation in early development may also help facilitate the maturation of glutamatergic transmission at dIN synapses.

12.1 Short Background to Neurotransmitter Co-release

An increasing number of cases of co-release of small molecule transmitters have been identified recently. Following the first case where physiological evidence revealed co-release of GABA and glycine in neonatal rat spinal interneurons (Jonas et al. 1998), GABA and ACh (Zheng et al. 2004), GABA and ATP (Jo and Role 2002), glutamate and ACh (Li et al. 2004a; Mentis et al. 2005; Nishimaru et al. 2005), glutamate and GABA (Gutierrez 2002) (but see Uchigashima et al. 2007) and even GABA, glycine and glutamate (Gillespie et al. 2005) have been reported. More extensive co-release of neurotransmitters is suggested by immunocytochemical evidence of co-localization (see reviews: Chery and De Koninck 1999; Trudeau 2004) but has yet to be supported by physiological evidence.

The brainstem and spinal cord are the regions where co-release of classical neurotransmitters including glutamate and ACh has been frequently found

W.-C. Li (✉)
School of Biological, Bute Building, University of St Andrews, St Andrews, Fife, KY16, 9TS, UK
e-mail: wl21@st-andrews.ac.uk

R. Gutierrez (ed.), *Co-Existence and Co-Release of Classical Neurotransmitters*, 225
DOI 10.1007/978-0-387-09622-3_12, © Springer Science+Business Media, LLC 2009

(Jonas et al. 1998; Li et al. 2004a; Nabekura et al. 2004; Gillespie et al. 2005; Mentis et al. 2005; Nishimaru et al. 2005). However, spinal cord functions in advanced vertebrates are poorly understood, making it a challenging task to reveal the significance of transmitter co-release. This chapter will briefly review the cases where co-release of glutamate and ACh has been found or suggested and will discuss the functional significance of this co-release.

12.2 Co-release of Glutamate and ACh – History and Evidence

12.2.1 Vertebrate Neuromuscular Junction

One of the first reports of co-release of glutamate and ACh came from the study of the electric organ of a Mediterranean species of ray called *Torpedo*. *Torpedo* electric organs can generate high currents to disorient their fish prey. The electric organ cells are modified muscle cells innervated by electromotor nerve cells in the spinal cord. Release of glutamate was observed from synaptosomes prepared from the cholinergic electric terminals in the *Torpedo* electric organ (Vyas and Bradford 1987; Israel et al. 1993). Using synaptosomes and proteoliposomes reconstituted with mediatophore lipids and loaded with equivalent amounts of glutamate and ACh, it was reported that when the external Ca^{2+} level was low, release of ACh was preferred. When Ca^{2+} level was above 3 mM, glutamate release became predominant. However, electrical stimulation of the *Torpedo* electric nerve in electric organ slices failed to reveal significant release of glutamate (Israel et al. 1993). Whether glutamate participates in the transmission in vivo therefore has not been conclusive.

At the frog tadpole neuromuscular junction, coexistence of glutamate and ACh was first found by Fu et al. in 1998. They reported glutamate-like immunoreactivity in cultured *Xenopus* tadpole motoneurons (Fu et al. 1998). Double-staining also revealed immunoreactivity for both glutamate and choline acetyltransferase (ChAT) in the same cultured nerve-myocyte following recordings, suggesting both transmitters coexist in motoneuron terminals. In addition, glutamate can potentiate spontaneous ACh release by activating presynaptic glutamate kainate/quisqualate and N-methyl-D-aspartate (NMDA) receptors in the same preparation (Fu et al. 1995), assigning a potential role for co-released glutamate from the conventional cholinergic terminals. However, the evidence for endogenous release of glutamate was limited and no iontropic glutamatergic receptor currents were recorded on the postsynaptic muscle cells.

Before synapses are formed, *Xenopus* embryo spinal neurons have spontaneous sharp rises in intracellular Ca^{2+} levels (Ca^{2+} spikes) following action potentials (Spitzer and Ribera, 1998). By artificially altering the Ca^{2+} spike activities in early development, Borodinsky et al. (2004) revealed surprisingly wide-spread neurotransmitter phenotype plasticity. They further examined the transmitters that motoneurons express and the transmitter receptors on the muscle cells (Borodinsky and Spitzer 2007) following similar manipulations. Indeed, glutamatergic apart from GABAergic and glycinergic transmission were seen to coexist with cholinergic neurotransmission at the *Xenopus* neuromuscular junction, supporting a novel activity-dependent transmitter-receptor matching process proposed by the authors (also see relevant chapter in this book).

Enriched glutamate-like immunoreactivity was also found at young rat motor nerve terminals (Waerhaug and Ottersen 1993) but no fast glutamatergic components were found in the neuromuscular transmission (Nishimaru et al. 2005).

12.2.2 Xenopus Tadpole Swimming Interneurons

The first direct evidence of co-release of glutamate and ACh came from investigation of excitatory interneurons with ipsilateral descending axons (descending interneurons, dINs) in the young *Xenopus* tadpole spinal cord and hindbrain (Li et al. 2004a). Hatchling *Xenopus* tadpoles at around 2 days (stage 37/38) from fertilization (Fig. 12.1a) have long been used by Roberts and his colleagues as a simple model vertebrate for studying mechanisms controlling swimming (Roberts 2000). The neuropharmacology of dINs was first tested in the mid-1980s and shown to be glutamatergic (Dale and Roberts 1985). Unwittingly, the tubocurarine, which was used to immobilize tadpoles during experiments by blocking neuromuscular junction nicatinic receptors (nAChRs), also blocked central nAChRs (as well as GABA receptors). This unfortunately postponed the discovery of classic transmitter co-release at dIN synapses for nearly two decades.

Fig. 12.1 *Xenopus* tadpole and the co-release of glutamate and ACh from dINs. **(a)** Drawing of a 2-day-old tadpole when most physiology studies have been carried out. **(b)** A simultaneous recording showing an impulse in a dIN and 3 superimposed unitary EPSCs in a CPG neuron (cN) . dIN was recorded in current clamp mode and was made to fire impulses by positive current injections. cN was recorded in voltage-clamp mode at –60 mV. **(c)** dIN spike evoked EPSCs could only be blocked by a mixture of glutamate and ACh antagonists. Different combinations of antagonists isolated nAChR, AMPAR and NMDAR currents, respectively. NBQX blocks AMPARs, D-AP5 blocks NMDARs and DHβe blocks nAChRs *Modified from Li et al. (2004a) with permission*

The pharmacology of dIN synapses was re-examined recently following two changes in experimental conditions. Firstly, α-bungarotoxin, which has better specificity for neuromuscular nAChRs, has been used as the immobilizing agent in this preparation since the early 1990s. The second change is the successful application of the whole-cell patch-clamp recording method (Li et al. 2002). *Xenopus* tadpole neurons at early developmental stages are self-sufficient, being filled with nutrients in the form of yolk platelets. The circulatory system is therefore insignificant and this makes the young *Xenopus* tadpole nervous system highly accessible. Careful dissections can be made to expose neuronal cell bodies and allow simultaneous recordings from neuron pairs while system functions are largely undisturbed. Surprisingly, dIN transmission was shown to have both fast nicotinic and glutamatergic components using simultaneous whole-cell recordings.

Simultaneous recordings were made between dINs and central pattern generator (CPG) neurons with dINs in current-clamp and CPG neurons in voltage-clamp modes. dINs only fire a single action potential to current injection (Li et al. 2006). This facilitated the analysis of the pharmacological nature of the unitary excitatory postsynaptic currents (EPSCs) produced by single impulses. When Mg^{2+} was removed from the recording saline, long-lasting NMDAR currents could be seen at membrane potentials clamped close to normal resting levels (Fig. 12.1b). The use of glutamatergic antagonists alone revealed some residual currents that could only be blocked by further adding nAChR antagonists, indicating that single dINs co-released both glutamate and ACh at their dIN-CPG neuron synapses. The glutamatergic transmission has both fast AMPAR and slow NMDAR components. In terms of time courses, nAChR currents are slightly longer than the AMPAR but shorter than the NMDAR ones (Fig. 12.1c).

Analyses were then carried out on spontaneous miniature EPSCs (mEPSCs) recorded in postsynaptic neurons by using TTX to block action potentials. The miniature synaptic currents are believed to be currents activated by quantal transmitter release from single presynaptic vesicles. In simultaneous recordings, nAChR and AMPAR currents in dINs produced unitary EPSCs having similar fast time courses, but NMDAR currents have slow dynamics (Fig. 12.1c). The AMPAR currents were blocked with NBQX to simplify the analysis and mEPSCs with mixed nAChR and NMDAR components were identified (Fig. 12.2). They had a fast rise similar to nAChR currents and a slow fall similar to NMDAR currents. Because all mEPSCs could be blocked by combined application of glutamate and ACh antagonists, the mixed mEPSCs were mediated by glutamate and ACh receptors. The frequency of spontaneous mEPSCs is very low in tadpole spinal neurons at stage 37/38, making it unlikely that mixed mEPSCs are coincidentally occurring separate vesicle events. The most plausible explanation for these mixed currents was the co-activation of postsynaptic NMDARs and nAChRs by glutamate and ACh co-released from single presynaptic vesicles.

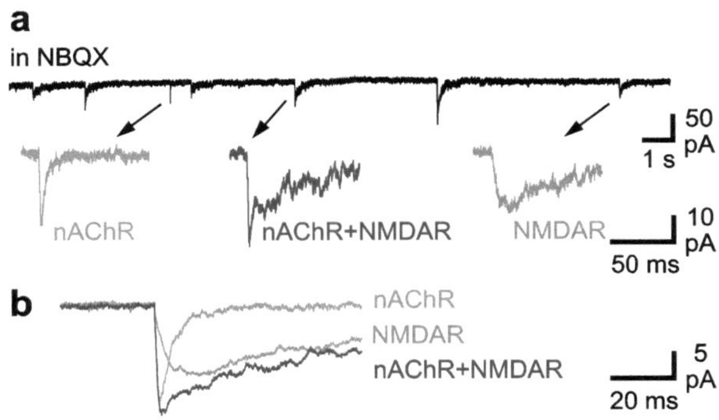

Fig. 12.2 Analysis of mEPSCs shows glutamate and ACh are co-released from single vesicles. **(a)** When glutamate AMPA receptor blocker NBQX was used, three types of mEPSCs were identified (arrows point to individual examples). **(b)** Averaged shape of the three types of mEPSCs. The time course of green and orange traces matched that of nAChR and NMDAR mEPSCs separated pharmacologically. The brown trace has a fast rise like nAChR and a fall like NMDAR currents, hence are most likely mixed nAChR + NMDAR responses. *Figure modified from Li et al. (2004a) with permission.* (*See* Color Plate 15)

12.2.3 Mammalian Motoneuron Central Synapses

Although there had been evidence suggesting glutamate might be co-released with ACh from the neuromuscular junction in *Xenopus* or the neuromuscular-equivalent electric organ in electric fish, better evidence of co-release of both transmitters recently was recently obtained at neonatal mouse motoneuron central synapses on Renshaw cells. Renshaw cells are glycinergic spinal inhibitory interneurons which are activated by motoneuron central axon collaterals and send feedback inhibition onto motoneurons (Fig. 12.3a). Although in adult cat, it was shown a long time ago that mecamylamine, an nAChR blocker, could only partially bock Renshaw cells' response to backfiring motoneurons (Noga et al. 1987), the pharmacological nature of the residual response had been surprisingly neglected ever since. On the other hand, glutamate transporter mRNA and glutamate-like immunoreactivity were reported in presumed mammalian spinal motoneurons (Meister et al. 1993). It is only recently that two separate laboratories investigated this issue using similar approaches in neonatal mouse spinal cord (Mentis et al. 2005; Nishimaru et al. 2005).

Rodent ventral roots can be dissected out and stimulated to backfire motoneurons. Meanwhile, the ability to sustain a completely isolated neonatal mouse spinal cord in vitro facilitated recording from small interneurons which previously was difficult. By backfiring motoneuron axons, intracellular recordings can be made from Renshaw cells to analyze the pharmacology of motoneuron central synapses. Both studies found that the motoneuron inputs onto Renshaw cells were only blocked by a combination of cholinergic and glutamatergic blockers (Fig. 12.3), indicating the release of both glutamate and

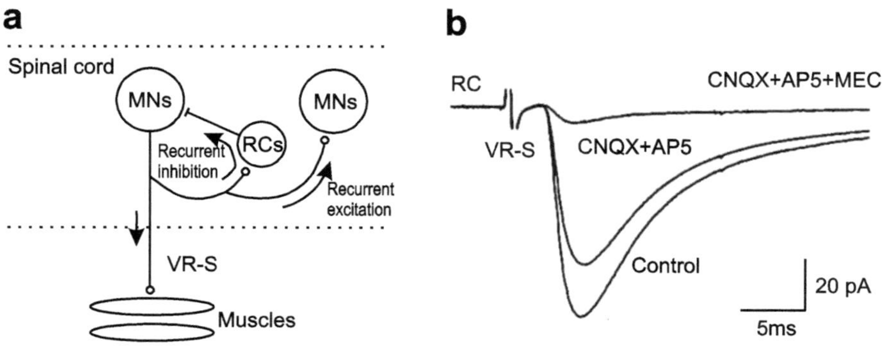

Fig. 12.3 co-release of glutamate and ACh from neonatal mouse spinal motoneurons (MN). **(a)** Mammal motoneurons have peripheral axons that form the ventral roots and central axons that excite other motoneurons and Renshaw cells (RC). In an isolated spinal cord preparation, ventral roots can be electrically stimulated (VR-S) to backfire motoneurons activating the central synapses. **(b)** Superimposed responses in a Renshaw cell following ventral root stimulation was only blocked by joint application of glutamate and ACh antagonists. CNQX blocks non-NMDA glutamate receptors and AP5 blocks NMDA receptors. Mecamylamine (MEC) is a nAChR blocker. *Figure modified from Nishimaru et al. (2005) with permission*

ACh at motoneuron central synapses. Anatomically, the presence of glutamate vesicular transporters (VGLUTs) in neuronal axon terminals has been regarded as the most reliable method in identifying glutamatergic transmission (Takamori 2006). Interestingly, neither study identified levels of known glutamate transporter mRNA or immunoreactivity in motoneurons suitable to confirm their glutamatergic phenotype. This may be explained if there is an unidentified glutamate transporter or the glutamatergic currents are activated by L-aspartate or L-homocysteic acid (Gundersen et al. 1998; Mentis et al. 2005).

12.2.4 Basal Forebrain Cholinergic Neurons

In higher brain areas, there is evidence to support the co-release of glutamate and ACh from some basal forebrain neurons which project to cerebral cortex and hippocampus. Co-release of glutamate and ACh was reported in a ChAT affinity-purified preparation of cortical cholinergic synaptosomes by Docherty et al. (1987). Like most neural pathways in higher vertebrates, it is difficult to carry out in-depth studies in neurotransmission in situ. To overcome access difficulty, Allen et al. (2006) used micro-island cultures of single rat basal forebrain neurons and allowed them to make autaptic connections onto themselves.

This simplified micro-island culture preparation allowed direct examination of the transmitters released from basal forebrain neurons. Simultaneous recordings were made from basal forebrain neurons and skeletal myoballs positioned close to their neurites acting as an artificial postsynaptic cell. Following brief stimulation of the basal forebrain neurons, fast glutamate receptor currents were recorded in the basal forebrain neuron and fast nAChR currents in the detector myoball, presenting clear evidence that basal forebrain neurons can release both transmitters. Activation of presynaptic muscarinic ACh receptors can suppress the release of both glutamate and ACh; producing a negative feedback control and indicating complex interactions between the co-released transmitters (Allen et al. 2006). The co-released ACh may activate both presynaptic and postsynaptic muscarinic responses *in situ* (Dudar and Szerb 1969; Allen et al. 2006).

12.2.5 Other Possible Locations

The sparse distribution of cholinergic neurons in the central nervous system, complicated modes of action of both cholinergic and glutamatergic transmissions, and the difficulty in obtaining direct physiological evidence using simultaneous recordings may be the reason that co-release of glutamate and ACh has not yet been established in more brain areas. Coexpression of VGLUT1, VGLUT2 or both with ChAT mRNAs was found in juvenile rat septum and striatum and probably cortex and hippocampus (Danik et al. 2005). VGLUT3 was co-expressed with vesicular ACh transporter (VACHT) in the striatum (Gras et al. 2002; Schafer et al. 2002; Boulland et al. 2004). Physiological recordings are needed to check if co-release of glutamate and ACh takes place in any of these areas.

Many responses of a mixed transmitter nature have been interpreted as the result of the activation of different types of neurons. In some cases (Gillespie et al. 2005; Allen et al. 2006), later evidence has suggested they are mediated by transmitters co-released from the same axon terminals. Stimulating the pedunculopontine tegmental area can evoke fast nAChR and iontropic glutamatergic responses in the substantia nigra pars compacta which could not be distinguished temporally in rat (Forster and Blaha 2003). Interestingly, glutamate and ACh were shown to co-localize in the same afferents from the same nucleus in squirrel monkey (Lavoie and Parent 1994), suggesting that co-release of glutamate and ACh may mediate the mixed responses.

Lastly, presynaptic modulation of transmitter release by nAChRs is widespread in the CNS (see review Dani and Bertrand, 2007). ACh might be co-released with other transmitters, including glutamate in some of these cases.

12.3 Functional Significance of Glutamate and ACh Co-release

In contrast to the exciting findings of new cases of co-release, the functional meaning of such co-release has not been clearly established. Some studies suggest co-release might be a transitional, developmental phase of neurotransmitter phenotype switching in individual neurons (Walker et al. 2001; Nabekura et al. 2004; Zheng et al. 2004; Gillespie et al. 2005). Other studies show that the expression of certain neurotransmitters could be linked to circuit activity levels (Belousov et al. 2001; Gutierrez 2002; Borodinsky et al. 2004; Borodinsky and Spitzer 2007), pointing to an activity-dependent neurotransmitter phenotype plasticity. In addition, co-released neurotransmitters can activate specific pre- and postsynaptic receptors to co-ordinate synapse functions, especially when metabotropic receptors are involved (Aguayo et al. 2004).

The difficulty in disclosing the importance of co-release largely results from our limited understanding of system functions and the development of the underlying neuronal circuits in higher vertebrates. Also, studying co-release often requires technically challenging simultaneous recordings from the pre- and postsynaptic neurons. This review will start with a brief account of why co-release of glutamate and ACh takes place. Then emphasis will be on the discussion of possible functions of co-release from *Xenopus* tadpole swimming interneurons where system functions and the contributions of each type of neuron are well documented.

12.3.1 Is Co-release a Snapshot of Developing Neurotransmission?

The transient use of neurotransmitters seems to be a wide-spread phenomenon during development. Coexistence of different transmission systems is often observed and undergoes developmental changes. During development, GABA in glutamatergic dentate gyrus granule cells (Walker et al. 2001), ACh in GABAergic amacrine cells (Zheng et al. 2004), glutamate and GABA in glycinergic synapses from the medial nucleus of the trapezoid body (MNTB) to the lateral superior olive (LSO) (Nabekura et al. 2004; Gillespie et al. 2005) disappear or dramatically weaken. Co-release of GABA and glycine was seen in both unitary and mIPSCs in neonatal rat spinal cord (Jonas et al. 1998). Later, in the 30- to 60-day-old rat spinal cord, co-release could be detected in evoked IPSCs but only glycinergic miniature IPSCs were recorded. This was shown to be due to GABA receptors being located extrasynaptically (Chery and De Koninck 1999; Keller et al. 2001). Most co-release cases that have been identified so far are from developing nervous systems, implying that a switch in transmitter phenotype in individual neurons is a common phenomenon during normal neural development.

Co-release of glutamate and ACh has not been thoroughly studied in the context of development for any case. In cultured developing *Xenopus* motoneuron terminals where glutamate may be co-released with ACh, glutamate can strengthen the spontaneous ACh release (Fu et al. 1995). This effect declines or disappears in older *Xenopus* tadpoles (Liou et al. 1996), which may result from glutamatergic transmission being switched off. Interestingly, co-release of glutamate and ACh at mouse motoneuron central synapses was suggested to continue into adulthood (Nishimaru et al. 2005) because an anatomical study by Herzog et al. (2004) found VGLUTs in adult rodent motoneuron central synapses onto Renshaw cells. Also, in adult cat, mecamylamine could only partially block Renshaw cell responses to backfiring motoneurons (Noga et al. 1987).

There are also many cases where co-release is yet to be confirmed physiologically but a possible transient presence of either glutamatergic or cholinergic transmission has been reported. In rat brain, VGLUT2 and VGLUT3 are only expressed at high levels in embryonic or postnatal stages. VGLU3 can co-localize extensively with other neurotransmitter transporters including VAChT in developing rat striatum (Boulland et al. 2004). ACh might be co-released with other transmitters in some of the cases. In human, ChAT expression was high in prefrontal cortex and cerebellum in the foetus or infant but decreased to low levels in the adult (Court et al. 1993). The ChAT decline occurs concurrently with rising glutamate receptor levels (Kwong et al. 2000), implying some transient use of ACh as a transmitter during development. Overall, it seems likely that either ACh or glutamate transmission can be developmentally transient, leading to the observation of their co-release only at certain developmental stages.

Why is this transient expression of transmitters needed in development? There is evidence that early embryonic circuits have different neural transmission pathways from mature ones which can underlie spontaneous embryonic neuronal activities. When the circuits are rewired, some neurons may simply switch their transmitter phenotype. Co-release of transmitters may be a snapshot of such a switch of phenotype. Early vertebrate embryos from teleosts to mammals have similar spontaneous motor behavior (Hamburger 1963). In both embryonic chick and rodent spinal cord, early spontaneous activities are dependent on GABAergic and cholinergic transmission (O'Donovan et al. 1998; Milner and Landmesser 1999; Hanson and Landmesser 2003). When the spinal circuit matures, spinal locomotion outputs become dependent on mainly glycinergic and glutamatergic transmission (Milner and Landmesser 1999). Whether this transmitter change is due to differentiation and death of alternative phenotypes or the individual neurons switching their phenotypes, or both, is not clear. Spontaneous activities have also been found in neonatal rat hippocampus CA1 region (Garaschuk et al. 1998) and retina (Feller et al. 1996; Zhou and Zhao 2000). It has been shown that rabbit retinal waves (E24 to P4) are transiently dependent on nicotinic transmission from starburst amacrine cells which co-release ACh and GABA (Zheng et al. 2004) at this stage, suggesting at least that switching of transmitter phenotypes in individual neurons is present. These early spontaneous neuronal activities mediated by some transiently expressed neurotransmitters may be important in activity-dependent circuit refinement (Kim and Kandler 2003; Malenka and Bear 2004) and axon path finding (Hanson and Landmesser 2004).

Although vertebrate neuromuscular junctions have long been proposed to release glutamate as well as ACh (Israel et al. 1993; Meister et al. 1993; Fu et al. 1998), no co-release has been found in physiological conditions. At neuromuscular synapses of crustacean and insects (Atwood 1982), glutamate is the transmitter. It is possible that glutamate transmission at vertebrate neuromuscular junctions is an evolutionary relic which will become functional when normal physiological conditions are disturbed, e.g., in cell cultures (Fu et al. 1998) or system excitability is artificially enhanced or depressed through Ca^{2+} spikes (Borodinsky and Spitzer 2007).

The abundant number of different neurotransmitters and their receptors has offered many ways for various neurotransmitters to interact with each other in the nervous system. One way for transmitters to interact is through feedback control of transmitter release via presynaptic receptors. Both glutamate and ACh can enhance ACh release from cultured *Xenopus* embryo motoneuron terminals via their presynaptic iontropic receptors (Fu et al. 1995; Fu and Liu 1997). In contrast, in micro-island cultured basal forebrain neurons, the ACh modulation of glutamate and ACh release is through presynaptic muscarinic receptors (Allen et al. 2006). In these cases, co-released transmitters from the same terminals may allow better temporal or spatial control of effects than those from different synaptic inputs.

12.3.2 The Swimming Circuit and Role of dINs

In comparison to other cases where co-release of glutamate and ACh has been found, a lot more is known about *Xenopus* tadpole dINs and their role in the spinal cord and hindbrain circuits that generate swimming rhythms (Roberts 2000; Li et al. 2004a; Li et al. 2006). Tadpole swimming involves alternating contractions of swimming muscles on both sides of the body. The central pattern generator that controls it comprises four types of neurons (Fig. 12.4a). Motoneurons (mns) have central synapses onto other motoneurons (Perrins and Roberts, 1995) and peripheral axons that control swimming muscles. The peripheral axons form loose bundles (ventral roots) that run between muscle segments. Suction electrodes placed on the muscles can pick up motoneuron axon impulses in the ventral roots after the tadpole is immobilized and fictive swimming is initiated by skin stimulation. During swimming commissural interneurons (cINs) produce glycinergic reciprocal inhibition of the CPG neurons on the opposite side of the spinal cord. This crossing inhibition is critical in coordinating CPG activity on the left and right sides (Dale 1985). Ascending interneurons (aINs) produce recurrent inhibition of both sensory pathway neurons and CPG neurons on the same side. In the sensory pathway, such inhibition can produce phase-locked gating-out of sensory inputs that arrive in phase with the ipsilateral ventral root discharge (Sillar and Roberts 1988; Li et al. 2002). The inhibition of CPG neurons by aINs can limit CPG neuron firing on each swimming cycle and

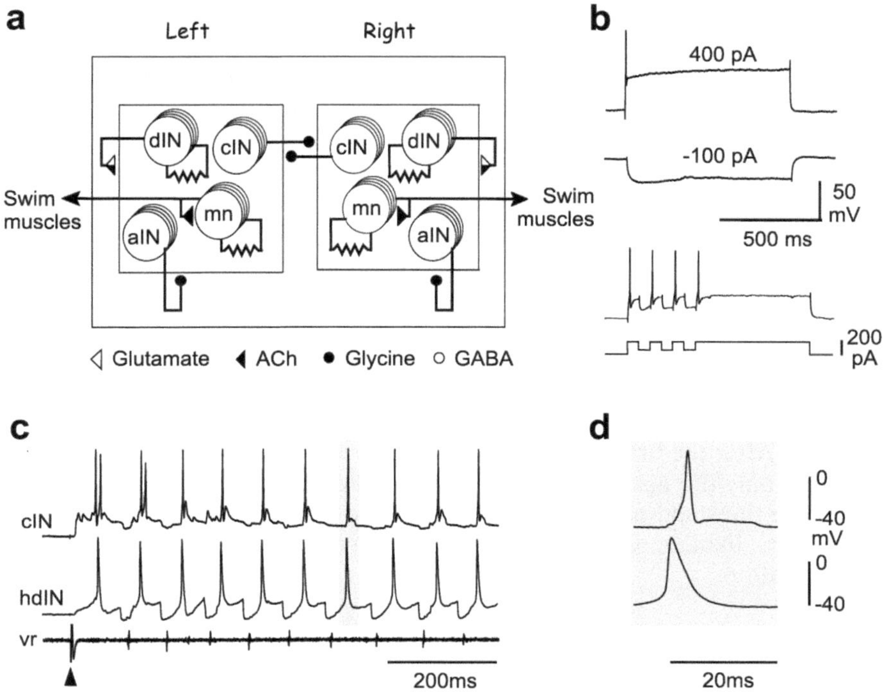

Fig. 12.4 Tadpole swimming CPG, dIN rebound firing and its activity in swimming. **(a)** The tadpole swimming CPG consists of two half centers each containing 4 types of neurons. Motoneurons (mn) innervate the swimming muscles. cINs inhibit neurons in the other half center. aINs feedback inhibition onto the ipsilateral CPG neurons. dINs excite ipsilateral CPG neurons. Resistor symbols represent electrical coupling among mns and dINs. **(b)** dIN rebound firing requires depolarization. Upper (400 pA): dIN only fires a single spike at the beginning of a long current pulse (threshold: 140 pA). Middle (–100 pA): dIN does not fire following negative current injection at rest. Lower: dIN can fire repetitively on rebound when depolarized (for more details see text). **(c)** ventral root (vr) and simultaneous whole-cell recordings illustrate the beginning of a fictive swimming episode started by skin stimulation (arrow head). Note the plateau-like depolarization in the dIN which is close to –40 mV even in the middle of swimming cycles. **(d)** Stretched cIN and dIN spikes on the same swimming cycle from the shaded area in **c.** to show the difference in their spike widths. *Figure modified from Li et al. (2006) with permission*

lead to more synchronous motoneuron firing and swift contractions in swimming muscles (Li et al. 2004b).

dINs form a continuous column of neurons that extends from the hindbrain throughout the spinal cord (we call those in the hindbrain hdINs). As we have seen, they co-release glutamate and ACh to directly excite other CPG neurons on the same side including other dINs in more caudal positions (Dale and Roberts 1985; Li et al. 2004a; Li et al. 2006). During tadpole fictive swimming, dINs only fire a single spike on each swimming cycle (the time from one ventral root discharge to the next, Fig. 12.4c). Our recent extensive studies of dINs have shown that they are the actual neurons that drive firing in other CPG neurons during swimming (unpublished results). In the caudal hindbrain area, some dINs also have ascending axons which allow them to make reciprocal excitatory synaptic connections. This excitation, partly mediated by the long duration NMDAR component, is critical in maintaining prolonged swimming activity after brief sensory stimuli (Li et al. 2006).

Interestingly, dINs also have unique electrical properties in comparison to other CPG neurons. These include relatively positive resting membrane potentials, wide action potentials, single firing to long positive current injection steps, extensive electrical coupling within the population (unpublished results) and rebound firing (Li et al. 2006). Unlike typical rebound firing following release of inhibition seen in many preparations (e.g. see Wilson et al. 2005), dINs do not show depolarization after negative current injection at rest which can lead to impulses. After the first spike at the onset of a long positive current injection, dINs can only fire again if their membrane potential is briefly brought below dIN firing threshold presumably to remove the inactivation of Na^+ channels (Fig. 12.4b). In other words, dIN rebound firing requires sustained background depolarization.

12.3.3 Significance of Glutamate and ACh Co-release in Tadpole Swimming

Having identified the role of dINs in the tadpole swimming circuit and their electrical properties, how can we fit the co-release of glutamate and ACh from dIN synapses into dIN functions? Like other cases of co-release, the functions of these two fast-acting neurotransmitters are yet to be revealed. Co-expression of inhibitory and excitatory transmitters can be induced in tadpole spinal cord neurons by artificially changing the rate of spontaneous Ca^{2+} spike activities (Borodinsky et al. 2004; Borodinsky and Spitzer, 2007). Whether the co-release of glutamate and ACh from dINs in normally developing animals is also controlled by Ca^{2+} spike activity is not yet known. Based on the available knowledge of dIN properties, dIN functions in the swimming circuit, and other preliminary data we have gained so far, I will discuss the potential significance of co-release of glutamate and ACh in tadpole swimming and its development.

In normal tadpole swimming, nAChR depolarization may help dINs maintain the plateau-like depolarization that is critical for maintaining persistent swimming by contributing to the voltage-dependent unblocking of NMDARs. As shown above, dIN rebound firing requires both brief inhibition and sustained depolarization. During tadpole swimming, phasic IPSPs from reciprocal inhibitory cINs on the other side are the main source of such inhibition. One important source of depolarization, once swimming has been initiated, is NMDAR activation. As is widely reported, NMDAR responses show a voltage-dependent Mg^{2+} block. Although nAChR and AMPAR currents both have shorter time courses than NMDARs, the nAChR ones have slower dynamics than the AMPARs (Li et al. 2004a). The cellular capacitance will further prolong depolarization in the membrane potentials produced by dIN EPSCs. This longer depolarization may help remove the Mg^{2+} block of NMDARs and be responsible for maintaining the tonic depolarization during swimming that is needed for dIN rebound firing. In addition, dINs have longer duration spikes than other CPG neurons and more positive resting potentials, both further facilitating the removal of Mg^{2+} block of NMDARs. The combined effect of these factors will lead to dIN membrane potentials being more depolarized during swimming, increasing the likelihood that dINs will fire rebound spikes following brief contralateral inhibition and carry the swimming activity into another cycle.

Any developmental changes in dIN transmitter release remain to be determined. Although there is good physiological evidence of co-release of glutamate and ACh from dINs at stage 37/38 tadpoles, information about dIN to CPG synapses at earlier and later stages is lacking at the moment. Co-release of glutamate and ACh can be readily revealed in unitary EPSCs in dIN-CPG paired recordings in the hindbrain or rostral spinal cord (Li et al. 2004a). Recording mEPSCs in the same region, however, did not reveal currents with mixed NMDAR and nAChR components (my unpublished results). mEPSCs recorded from more caudal CPG neurons, in contrast, can have both prominent NMDAR and nAChR components. Because rostral spinal cord and hindbrain neurons and synapses develop earlier than the more caudal ones (Hayes and Roberts 1974), this longitudinal difference in unitary and mEPSCs indicates an intriguing developmental change. It is also tempting to speculate that the failure to see co-release in mEPSCs in rostral neurons is due to nAChRs being located extrasynaptically. However, a gradual decrease in mEPSC size was found in the tadpole RB to sensory pathway transmission (Rohrbough and Spitzer 1999) and many other systems in development, so the failure may also be explained simply by the small size of quanta. Further studies are needed to test these different developmental possibilities.

The maturation of glutamatergic synapses is proposed to require Ca^{2+} influx through NMDARs following coordinated pre- and postsynaptic neuronal activities (Dan and Poo 2004). Although there is controversy about the presence of AMPARs in nascent glutamatergic synapses, it is agreed that correlated pre- and postsynaptic activities can translate the nascent synapses into mature synapses with functional AMPARs (see the latest review Groc et al. 2006). This process involves Ca^{2+} influx through NMDARs that leads to long-term potentiation or depression of synaptic strength (LTP/LTD, see the latest review Dan and Poo 2004). Though this model of glutamatergic synapse development has been found in an increasing number of cases, the natural source of the depolarization that can relieve Mg^{2+} block of NMDARs has rarely been identified. In theory, concurrently activated or early depolarizing GABA and glycinergic transmission could be such a source, although the presence of

NMDARs is often suggested to contribute to the development of inhibitory transmissions instead (Ben-Ari 2002; Gillespie et al. 2005). However, in the tadpole spinal cord, GABAergic and glycinergic responses in CPG neurons were hyperpolarizing (Bixby and Spitzer 1984) early in development.

It is possible that nAChR activation in early development may produce the first depolarization that is needed in the development of glutamatergic transmission at dIN synapses. In the absence of GABA/glycine depolarization in the early developing tadpole swimming circuit, nicotinic transmission may provide an initial depolarization that opens NMDARs to allow Ca^{2+} influx and facilitate the development of glutamatergic transmission at the dIN-CPG synapses. When the circuit activity is properly established and there is sufficient AMPAR current in the synaptic inputs, the nAChRs may phase out with development. Interestingly, the effects of nicotinic transmission on short- or long-term synaptic plasticity at glutamatergic synapses can depend on its relative timing (Ji et al. 2001; Ge and Dani 2005), but it is not known if there is any developmental change in this type of interaction.

There are still a lot of unanswered questions about dIN transmission even in this simple tadpole spinal cord. For example, metabotropic glutamate receptors (mGLuR) have recently been suggested to regulate *Xenopus* tadpole swimming (Chapman and Sillar 2007) but their synaptic location is not known. mGLuRs are involved in regulating synapse formation in the nigrostriatal dopamine pathway (Plenz and Kitai 1998) and the expression of GABAergic transmission at dentate gyrus granules cells in kindling experiments (Gutierrez 2002). In both cases glutamate was suggested to be the co-transmitter. Experiments need to be done to see if glutamate release by dINs activates mGLuRs. In addition, the action of ACh in tadpole swimming via muscarinic receptors was previously identified (Panchin et al. 1991), but the source of this action has not been clarified. The direct evidence of recordings from pairs of motoneurons indicates that they only make central cholinergic synapses with other motoneurons (Perrins and Roberts 1995). In contrast, dINs are known to excite all types of CPG neurons (Li et al. 2006). They may be the source of muscarinic control of tadpole swimming. Furthermore, we still do not know if there are presynaptic nAChRs on the dIN synapses. The presence of nAChRs in the presysnaptic terminals is widespread in mammal brain (Dani 2001). A fuller investigation of dIN synaptic transmission, together with good understanding of tadpole swimming rhythm generation mechanisms, is needed at different developmental stages.

12.4 Concluding Remarks

As a result of difficulties in producing direct physiological evidence, clear co-release of fast neurotransmitters has only been demonstrated at synapses with easier access in the vertebrate CNS. Immunocytochemical staining of transmitter-specific proteins and the identification of their molecular transcripts in higher vertebrates have ear-marked more brain areas where co-release of glutamate and ACh may well take place. However, in some cases, even the most intensive use of these double labeling methods failed to detect colocalization where co-release had been shown physiologically (Mentis et al. 2005; Nishimaru et al. 2005). The actual number of cases of co-release of glutamate and ACh and other transmitters may exceed most researchers' expectations. A lot more detailed anatomical and physiological investigation is needed to understand the function of these neurons which use both transmitters and to determine the role of co-release in development and normal system functions.

Ultimately, we want to understand how neurotransmitter phenotypes are specified and controlled in development and pathological conditions. Pioneering work by Borodinsky and Spitzer has identified one activity factor, spontaneous Ca^{2+} spikes, that are involved in balancing the proportion of excitatory and inhibitory transmitter phenotypes in the young *Xenopus* tadpole spinal cord (Borodinsky et al. 2004; Borodinsky and Spitzer, 2007). Apart from the normal co-release of glutamate and ACh from excitatory interneurons, my preliminary evidence suggests that GABA and glycine are also co-released in this preparation. High accessibility for physiological recordings and the potential for molecular manipulations will make the young *Xenopus* tadpole spinal cord an attractive place to address this important issue.

Acknowledgment I thank Drs Steve Soffe and Alan Roberts for their helpful comments, the Wellcome Trust and the Royal Society for their support, and Tim Colborn and Jenny Maxwell for technical assistance.

Reference

Aguayo LG, van Zundert B, Tapia JC, Carrasco MA, Alvarez FJ (2004) Changes on the properties of glycine receptors during neuronal development. Brain Res Brain Res Rev 47:33–45

Allen TGJ, Abogadie FC, Brown DA (2006) Simultaneous Release of Glutamate and Acetylcho-line from Single Magnocellular "Cholinergic" Basal Forebrain Neurons. J Neurosci 26:1588–1595

Atwood HL (1982) Synapses and neurotransmitters. New York: Academic Press

Belousov AB, O'Hara BF, Denisova JV (2001) Acetylcholine becomes the major excitatory neurotransmitter in the hypothalamus in vitro in the absence of glutamate excitation. J Neurosci 21:2015–2027

Ben-Ari Y (2002) Excitatory actions of gaba during development: the nature of the nurture. Nat Rev Neurosci 3:728–739

Bixby JL, Spitzer NC (1984) The appearance and development of neurotransmitter sensitivity in Xenopus embryonic spinal neurones in vitro. J Physiol 353:143–155

Borodinsky LN, Spitzer NC (2007) Activity-dependent neurotransmitter-receptor matching at the neuromuscular junction. Proc Natl Acad Sci U S A 104:335–340

Borodinsky LN, Root CM, Cronin JA, Sann SB, Gu X, Spitzer NC (2004) Activity-dependent homeostatic specification of transmitter expression in embryonic neurons. Nature 429:523–530

Boulland JL, Qureshi T, Seal RP, Rafiki A, Gundersen V, Bergersen LH, Fremeau RT, Jr., Ed-wards RH, Storm-Mathisen J, Chaudhry FA (2004) Expression of the vesicular glutamate transporters during development indicates the widespread corelease of multiple neurotransmitters. J Comp Neurol 480:264–280

Chapman RJ, Sillar KT (2007) Modulation of a spinal locomotor network by metabotropic glutamate receptors. Eur J Neurosci 26:2257–2268

Chery N, De Koninck Y (1999) Junctional versus Extrajunctional Glycine and GABAA Receptor-Mediated IPSCs in Identified Lamina I Neurons of the Adult Rat Spinal Cord. J Neurosci 19:7342–7355

Court JA, Perry EK, Johnson M, Piggott MA, Kerwin JA, Perry RH, Ince PG (1993) Regional patterns of cholinergic and glutamate activity in the developing and aging human brain. Brain Res Dev Brain Res 74:73–82

Dale N (1985) Reciprocal inhibitory interneurones in the Xenopus embryo spinal cord. J Physiol (London) 363:61–70

Dale N, Roberts A (1985) Dual component amino - acid - mediated synaptic potentials: excitatory drive for swimming in Xenopus embryos. J Physiol (London) 363:35–59

Dan Y, Poo MM (2004) Spike timing-dependent plasticity of neural circuits. Neuron 44:23–30

Dani JA (2001) Overview of nicotinic receptors and their roles in the central nervous system. Biol Psychiatry 49:166–174

Dani JA, Bertrand D (2007) Nicotinic acetylcholine receptors and nicotinic cholinergic mechanisms of the central nervous system. Annu Rev Pharmacol Toxicol 47:699–729

Danik M, Cassoly E, Manseau F, Sotty F, Mouginot D, Williams S (2005) Frequent coexpression of the vesicular glutamate transporter 1 and 2 genes, as well as coexpression with genes for choline acetyltransferase or glutamic acid decarboxylase in neurons of rat brain. J Neurosci Res 81:506–521

Docherty M, Bradford HF, Wu JY (1987) Co-release of glutamate and aspartate from cholinergic and GABAergic synaptosomes. Nature 330:64

Dudar JD, Szerb JC (1969) The effect of topically applied atropine on resting and evoked cortical acetylcholine release. J Physiol 203:741–762

Feller MB, Wellis DP, Stellwagen D, Werblin FS, Shatz CJ (1996) Requirement for cholinergic synaptic transmission in the propagation of spontaneous retinal waves. Science 272:1182–1187

Forster GL, Blaha CD (2003) Pedunculopontine tegmental stimulation evokes striatal dopamine efflux by activation of acetylcholine and glutamate receptors in the midbrain and pons of the rat. Eur J Neurosci 17:751–762

Fu WM, Liu JJ (1997) Regulation of acetylcholine release by presynaptic nicotinic receptors at developing neuromuscular synapses. Mol Pharmacol 51:390–398

Fu WM, Liou JC, Lee YH, Liou HC (1995) Potentiation of neurotransmitter release by activation of presynaptic glutamate receptors at developing neuromuscular synapses of Xenopus. J Physiol 489 (Pt 3):813–823

Fu WM, Liou HC, Chen YH, Wang SM (1998) Coexistence of glutamate and acetylcholine in the developing motoneurons. Chin J Physiol 41:127–132

Garaschuk O, Hanse E, Konnerth A (1998) Developmental profile and synaptic origin of early network oscillations in the CA1 region of rat neonatal hippocampus. J Physiol 507 (Pt 1):219–236

Ge S, Dani JA (2005) Nicotinic Acetylcholine Receptors at Glutamate Synapses Facilitate Long-Term Depression or Potentiation. J Neurosci 25:6084–6091

Gillespie DC, Kim G, Kandler K (2005) Inhibitory synapses in the developing auditory system are glutamatergic. Nat Neurosci 8:332–338

Gras C, Herzog E, Bellenchi GC, Bernard V, Ravassard P, Pohl M, Gasnier B, Giros B, El Mestikawy S (2002) A third vesicular glutamate transporter expressed by cholinergic and sero-toninergic neurons. J Neurosci 22:5442–5451

Groc L, Gustafsson B, Hanse E (2006) AMPA signalling in nascent glutamatergic synapses: there and not there! Trends Neurosci 29:132–139

Gundersen V, Chaudhry FA, Bjaalie JG, Fonnum F, Ottersen OP, Storm-Mathisen J (1998) Synaptic vesicular localization and exocytosis of L-aspartate in excitatory nerve terminals: a quantitative immunogold analysis in rat hippocampus. J Neurosci 18:6059–6070

Gutierrez R (2002) Activity-dependent expression of simultaneous glutamatergic and GABAergic neurotransmission from the mossy fibers in vitro. J Neurophysiol 87:2562–2570

Hamburger V (1963) Some Aspects Of The Embryology Of Behavior. Q Rev Biol 38:342–365

Hanson MG, Landmesser LT (2003) Characterization of the circuits that generate spontaneous episodes of activity in the early embryonic mouse spinal cord. J Neurosci 23:587–600

Hanson MG, Landmesser LT (2004) Normal patterns of spontaneous activity are required for correct motor axon guidance and the expression of specific guidance molecules. Neuron 43:687–701

Hayes BP, Roberts A (1974) The distribution of synapses along the spinal cord of an amphibian embryo: an electron microscope study of junction development. Cell & Tissue Res 153:227–244

Herzog E, Landry M, Buhler E, Bouali-Benazzouz R, Legay C, Henderson CE, Nagy F, Dreyfus P, Giros B, El Mestikawy S (2004) Expression of vesicular glutamate transporters, VGLUT1 and VGLUT2, in cholinergic spinal motoneurons. Eur J Neurosci 20:1752–1760

Israel M, Lesbats B, Bruner J (1993) Glutamate and acetylcholine release from cholinergic nerve terminals, a calcium control of the specificity of the release mechanism. Neurochem Int 22:53

Ji D, Lape R, Dani JA (2001) Timing and location of nicotinic activity enhances or depresses hippocampal synaptic plasticity. Neuron 31:131–141

Jo Y-H, Role LW (2002) Coordinate Release of ATP and GABA at In Vitro Synapses of Lateral Hypothalamic Neurons. J Neurosci 22:4794–4804

Jonas P, Bischofberger J, Sandkuhler J (1998) Corelease of two fast neurotransmitters at a central synapse. Science 281:419–424

Keller AF, Coull JA, Chery N, Poisbeau P, De Koninck Y (2001) Region-specific developmental specialization of GABA-glycine cosynapses in laminas I-II of the rat spinal dorsal horn. J Neurosci 21:7871–7880

Kim G, Kandler K (2003) Elimination and strengthening of glycinergic/GABAergic connections during tonotopic map formation. Nat Neurosci 6:282–290

Kwong WH, Chan WY, Lee KK, Fan M, Yew DT (2000) Neurotransmitters, neuropeptides and calcium binding proteins in developing human cerebellum: a review. Histochem J 32:521–534

Lavoie B, Parent A (1994) Pedunculopontine nucleus in the squirrel monkey: distribution of cholinergic and monoaminergic neurons in the mesopontine tegmentum with evidence for the presence of glutamate in cholinergic neurons. J Comp Neurol 344:190–209

Li WC, Soffe SR, Roberts A (2002) Spinal inhibitory neurons that modulate cutaneous sensory pathways during locomotion in a simple vertebrate. J Neurosci 22:10924–10934

Li WC, Soffe SR, Roberts A (2004a) Glutamate and acetylcholine corelease at developing synapses. Proc Natl Acad Sci U S A 101:15488–15493

Li WC, Higashijima S, Parry DM, Roberts A, Soffe SR (2004b) Primitive roles for inhibitory interneurons in developing frog spinal cord. J Neurosci 24:5840–5848

Li WC, Soffe SR, Wolf E, Roberts A (2006) Persistent Responses to Brief Stimuli: Feedback Excitation among Brainstem Neurons. J Neurosci 26:4026–4035

Liou HC, Yang RS, Fu WM (1996) Potentiation of spontaneous acetylcholine release from motor nerve terminals by glutamate in Xenopus tadpoles. Neuroscience 75:325–331

Malenka RC, Bear MF (2004) LTP and LTD: an embarrassment of riches. Neuron 44:5–21

Meister B, Arvidsson U, Zhang X, Jacobsson G, Villar MJ, Hokfelt T (1993) Glutamate transporter mRNA and glutamate-like immunoreactivity in spinal motoneurones. Neuroreport 5:337–340

Mentis GZ, Alvarez FJ, Bonnot A, Richards DS, Gonzalez-Forero D, Zerda R, O'Donovan MJ (2005) Noncholinergic excitatory actions of motoneurons in the neonatal mammalian spinal cord. Proc Natl Acad Sci U S A 102:7344–7349

Milner LD, Landmesser LT (1999) Cholinergic and GABAergic inputs drive patterned spontaneous motoneuron activity before target contact. J Neurosci 19:3007–3022
Nabekura J, Katsurabayashi S, Kakazu Y, Shibata S, Matsubara A, Jinno S, Mizoguchi Y, Sasaki A, Ishibashi H (2004) Developmental switch from GABA to glycine release in single central synaptic terminals. Nat Neurosci 7:17–23
Nishimaru H, Restrepo CE, Ryge J, Yanagawa Y, Kiehn O (2005) Mammalian motor neurons corelease glutamate and acetylcholine at central synapses. Proc Natl Acad Sci U S A 102:5245–5249
Noga BR, Shefchyk SJ, Jamal J, Jordan LM (1987) The role of Renshaw cells in locomotion: antagonism of their excitation from motor axon collaterals with intravenous mecamylamine. Exp Brain Res 66:99–105
O'Donovan MJ, Chub N, Wenner P (1998) Mechanisms of spontaneous activity in developing spinal networks. J Neurobiol 37:131–145
Panchin Yu Y, Perrins RJ, Roberts A (1991) The action of acetylcholine on the locomotor central pattern generator for swimming in Xenopus embryos. J Exp Biol 161:527–531
Perrins R, Roberts A (1995) Cholinergic and electrical motoneuron-to-motoneuron synapses contribute to on-cycle excitation during swimming in Xenopus embryos. J Neurophysiol 73:1005–1012
Plenz D, Kitai ST (1998) Regulation of the nigrostriatal pathway by metabotropic glutamate receptors during development. J Neurosci 18:4133–4144
Roberts A (2000) Early functional organization of spinal neurons in developing lower vertebrates. Brain Res Bull 53:585–593
Rohrbough J, Spitzer NC (1999) Ca(2+)-permeable AMPA receptors and spontaneous presynaptic transmitter release at developing excitatory spinal synapses. J Neurosci 19:8528–8541
Schafer MK, Varoqui H, Defamie N, Weihe E, Erickson JD (2002) Molecular cloning and functional identification of mouse vesicular glutamate transporter 3 and its expression in subsets of novel excitatory neurons. J Biol Chem 277:50734–50748
Sillar KT, Roberts A (1988) A neuronal mechanism for sensory gating during locomotion in a vertebrate. Nature 331:262–265
Spitzer NC, Ribera AB (1998) Development of electrical excitability in embryonic neurons: mechanisms and roles. J Neurobiol 37:190–197
Takamori S (2006) VGLUTs: 'Exciting' times for glutamatergic research? Neurosci Res 55:343
Trudeau LE (2004) Glutamate co-transmission as an emerging concept in monoamine neuron function. J Psychiatry Neurosci 29:296–310
Uchigashima M, Fukaya M, Watanabe M, Kamiya H (2007) Evidence against GABA release from glutamatergic mossy fiber terminals in the developing hippocampus. J Neurosci 27:8088–8100
Vyas S, Bradford HF (1987) Co-release of acetylcholine, glutamate and taurine from synaptosomes of Torpedo electric organ. Neurosci Lett 82:58
Waerhaug O, Ottersen OP (1993) Demonstration of glutamate-like immunoreactivity at rat neuromuscular junctions by quantitative electron microscopic immunocytochemistry. Anat Embryol (Berl) 188:501–513
Walker MC, Ruiz A, Kullmann DM (2001) Monosynaptic GABAergic Signaling from Dentate to CA3 with a Pharmacological and Physiological Profile Typical of Mossy Fiber Synapses. Neuron 29:703
Wilson JM, Hartley R, Maxwell DJ, Todd AJ, Lieberam I, Kaltschmidt JA, Yoshida Y, Jessell TM, Brownstone RM (2005) Conditional Rhythmicity of Ventral Spinal Interneurons Defined by Expression of the Hb9 Homeodomain Protein. J Neurosci 25:5710–5719
Zheng J-j, Lee S, Zhou ZJ (2004) A Developmental Switch in the Excitability and Function of the Starburst Network in the Mammalian Retina. Neuron 44:851
Zhou ZJ, Zhao D (2000) Coordinated transitions in neurotransmitter systems for the initiation and propagation of spontaneous retinal waves. J Neurosci 20:6570–6577

Chapter 13
Colocalization and Cotransmission of Classical Neurotransmitters: An Invertebrate Perspective

Mark W. Miller

Abstract Once considered a curiosity, the notion that individual neurons can contain more than one classical neurotransmitter has gained increasing credibility in recent years. Several contributions to the growing recognition of classical neurotransmitter colocalization and cotransmission originate from studies using invertebrate nervous systems. Some of these model systems contain large identified neurons that contribute to well-understood circuits and networks. They therefore enable investigators to pose questions that are presently beyond the technical limitations of experimental approaches to mammalian brain function. This chapter reviews our current understanding of classical neurotransmitter colocalization and cotransmission in invertebrates. It focuses on identified neurons that could enable assessment of cotransmitter contributions to synaptic signals and neural network function. Major gaps in our present conception of classical neurotransmitter colocalization and cotransmission are emphasized, with an aim toward stimulating further study of their physiological and functional consequences.

13.1 Introduction

> *"It looks as though Mother Nature just threw a handful of neurotransmitters at the nervous system and worked with them wherever they landed."*
>
> I. Kupfermann, personal communication

In his comprehensive review of cotransmission, Kupfermann (1991) attempted to discern general principles governing combinations of classic, or conventional neurotransmitters, with neuropeptides and other substances that he classified as "unusual neurotransmitters". He concluded that "it has

M.W. Miller (✉)
Institute of Neurobiology and Department of Anatomy & Neurobiology, University of Puerto Rico, 201 Blvd del Valle, San Juan, Puerto Rico 00901

R. Gutierrez (ed.), *Co-Existence and Co-Release of Classical Neurotransmitters*, DOI 10.1007/978-0-387-09622-3_13, © Springer Science+Business Media, LLC 2009 not proven possible to derive any simple rules that describe the observed combinations of cotransmitters". In the years since that article appeared, instantiations of neurotransmitter colocalization and cotransmission have increased considerably. As reviewed in this volume, these observations include instances in which classical or conventional neurotransmitters have been reported to be colocalized. Despite the increased support for its occurrence, however, many fundamental questions concerning neurotransmitter colocalization and cotransmission remain unanswered. This contribution reviews advances from studies using invertebrate nervous systems that hold promise for addressing some of the outstanding questions concerning colocalization and cotransmission of classical neurotransmitters. Evidence of colocalization in neurons that participate in well-characterized circuits is emphasized, as these models can provide opportunities to determine cotransmitter contributions to synaptic signaling, circuit function, and ultimately to behavior.

Historically, our understanding of neurotransmitters and synaptic mechanisms has benefited greatly from investigations using the large neurons found in many invertebrates (reviewed in Gerschenfeld 1973; Kandel 1975; Kuffler et al. 1984). Moreover, numerous insights into cotransmission have emerged from studies on invertebrates (Adams and O'Shea 1983; Marder et al. 1995; Nusbaum et al. 2001). Readers who are not steeped in this literature will recognize all of the classical neurotransmitters considered below as major mediators of synaptic signals in the mammalian CNS. Moreover, the biosynthetic pathways, mechanisms of release, and signal transduction pathways are, for the most part, not esoteric or idiosyncratic to the invertebrates. In view of these precedents and commonalities, it may be anticipated that findings obtained with the invertebrates can inform and facilitate efforts to decipher the functions of classical neurotransmitter colocalization and cotransmission in the mammalian brain.

Some invertebrate neurons can be recognized in all members of a particular species based upon their position, size, branching pattern, synaptic connections, firing patterns, intrinsic membrane properties, and neurotransmitter phenotype. In some instances, corresponding neurons may be found in other related species (Kandel 1979; Croll 1987). Studies of neuronal circuits composed of such "identified neurons" have advanced our understanding of numerous nervous system operations, including sensorimotor integration, central pattern generation, and plasticity (Getting 1989; Pearson 1993; Marder and Calabrese 1996). This chapter will thus emphasize the evidence for classical neurotransmitter colocalization and cotransmission in identified invertebrate neurons that contribute to neural circuit function. This focus reflects the conviction that the principles governing the operation of such "simpler" neural networks are pertinent to more complex systems.

13.2 "Giant" Serotonergic Cells

Some of the earliest studies suggesting colocalization of classical neurotransmitters in any nervous system were conducted using exceptionally large serotonergic neurons that are found in mollusks (see Burnstock 1976; Kupfermann 1991). Although distinct nomenclatures signify these neurons in different species [*Lymnaea stagnalis*: Cerebral Giant Cell (CGC); *Aplysia californica*: Metacerebral Cell (MCC); *Planorbis corneus*: Giant Serotonergic Cell (GSC), *Helix aspersa*: Giant Cerebral Neuron (GCN)] they all exhibit common structural and functional properties (see Weiss and Kupfermann 1976; Kandel 1979; Croll 1987). In all instances, their cell bodies are among the largest in the CNS. They are located in the cerebral ganglion and they project to the circuits that generate feeding behaviors. Although their serotonergic phenotype is a defining feature in all species, some observations suggesting classical cotransmitter colocalization have been reported in specific cases.

Using radioenzymatic micromethods on individual neurons dissected from *Aplysia*, Brownstein et al. (1974) measured synthesis of both serotonin and histamine in neuron C-1 (later designated the MCC). These investigators noted that their values for histamine concentrations were two orders of magnitude lower than their serotonin measurements. Subsequent studies showed that such measurements could be contaminated by the presence of presynaptic fibers and terminals that invaginate the somata of *Aplysia* neurons (Ono and McCaman 1984; see also Osborne 1979; 1984). Moreover, immunohistochemical studies did not detect histamine in the MCC of *Aplysia* (Elste et al. 1990). The MCC was found to receive innervation from a cerebral neuron that was known to be histaminergic (C2; Ono and McCaman 1980), increasing the likelihood that the Brownstein et al. (1974) originated from presynaptic sources.

Using immunohistochemistry on adjacent sections of the CGC of *Lymnaea*, Boer et al. (1984) reported colocalization of serotonin and dopamine-like immunoreactivities. In a subsequent study that mapped dopamine immunoreactivity and glyoxylic acid fluorescence in the *Lymnaea* CNS, Elekes et al. (1991) did not find evidence for the presence of DA in the CGCs. Notably, different antibodies against dopamine were used in these two investigations. Additional studies that examined the distribution of catecholamines in *Lymnaea* and additional mollusks likewise failed to detect them in the CGC or its homologs (Salimova et al. 1987; Hernádi et al. 1993; Hernádi and Elekes 1995; Sakharov et al. 1996; Croll 2001).

Thus, some of the earliest assertions of classical neurotransmitter colocalization in the giant serotonergic neurons were subsequently refuted. They are included in this survey primarily due to their impact on later studies. These investigations demonstrated how classical neurotransmitter colocalization could be erroneously inferred when based upon data obtained with a single method of detection. Consequently, they served to increase the stringency of criteria applicable to demonstrations of colocalization and cotransmission in invertebrate neurons.

13.3 Cholinergic/Serotonergic Mechanosensory Neurons

A comprehensive series of studies conducted by Katz, Harris-Warrick and coworkers provided evidence for colocalization and cotransmission of acetylcholine and serotonin in specific sensory neurons of crabs (Katz et al. 1989; Katz and Harris-Warrick 1989, 1990a, 1991; Kiehn and Harris-Warrick 1992). The gastropyloric receptor (GPR) cells, a set of four peripheral mechanosensory neurons (bilaterally paired GPR1 and GPR2 cells) in the stomatogastric nervous system of *Cancer borealis* (Jonah crab) and *Cancer irroratus* (rock crab), are activated by tension at the gastropyloric border of the foregut. They project to the stomatogastric ganglion (STG), an intensively studied central pattern generator (CPG) neuronal circuit that controls foregut movements. Measurements of choline acetyltransferase in the nerve containing the axon of GPR2 (the gastropyloric nerve, *gpn*) *en route* to its peripheral innervation were significantly above background measurements (obtained from another nerve that is not thought to contain cholinergic fibers; Katz et al. 1989). Moreover, pharmacological experiments showed that the rapid excitatory postsynaptic potentials (EPSPs) evoked by the GPRs in specific STG neurons were blocked by several nicotinic antagonists, including *d*-tubocurarine, decamethonium, hexamethonium, and mecamylamine (Katz and Harris-Warrick 1989, 1990). These data led to the proposal that ACh serves as a GPR neurotransmitter (Katz et al. 1989; Katz and Harris-Warrick 1989). In this respect, the GPRs resemble prototypical crustacean mechanosensory neurons, which use ACh as their neurotransmitter to evoke rapid EPSPs via nicotinic-like receptors in target CNS neurons (Barker et al. 1972; Hildebrand et al. 1974; Miller et al. 1992).

Immunohistochemical observations indicated that the GPRs of *Cancer* also contained serotonin and that they provided the sole source of serotonergic innervation to the STG (Katz et al. 1989; see also Beltz et al. 1984). Moreover, in addition to their rapid cholinergic signaling, the GPRs were found to exert slow modulatory actions within the STG. (Katz and Harris-Warrick 1989, 1990a; Kiehn and Harris-Warrick 1992). These modulatory GPR effects varied among the different STG neurons, with some targets responding with tonic inhibition and others responding with tonic excitation, rhythmic bursting, or plateau potentials. Each of these modulatory effects was also evoked by exogenous 5-HT introduced via bath application (Katz and Harris-Warrick 1989, 1990a) or by puffing directly to the cells (Zhang and Harris-Warrick 1994). Finally, serotonergic agonists were shown to evoke each effect and specific serotonergic antagonists were shown to block them. Each effect exhibited distinct pharmacological profiles and in all cases the ability of antagonists to block the actions of exogenous serotonin was in agreement with their ability to block the modulatory responses produced by stimulating the GPRs. These pharmacological data provided strong support for a modulatory role of serotonin in signaling by the GPRs.

In sum, these studies provided biochemical, anatomical, and pharmacological evidence for colocalization and cotransmission of acetylcholine and serotonin in the GPR neurons. The known mechanosensory function of these neurons, and their characterized projections to specific identified neurons within the STG, enabled these investigators to dissect their signaling into a rapid cholinergic component, and a slow serotonergic component, with variable effects that differed according to the receptor/transduction mechanism activated in each target.

13.4 Dopaminergic/Serotonergic Neurosecretory Cells

A complex neuron, termed the "L-cell" (Selverston et al. 1976), has been described in several crustacean species, including the crabs *Carcinus maenas* (Cooke and Goldstone 1970) and *Callinectes sapidus* (Wood and Derby 1996; Fort et al. 2004), the lobsters *Panulirus interruptus* (Kushner and Maynard 1977), *Homarus gammarus* (Cournil et al. 1984, 1994), and *Homarus americanus* (Siwicki et al. 1987; Pulver et al. 2003) and the crayfish *Oronectes rusticus* (Tierney et al. 2003). One L-cell is located in each commissural ganglion, a small aggregation of neurons located on the connective that joins the brain to the remainder of the CNS. In those species in which its anatomy has been described, the L-cell projects to the brain, where small collaterals innervate the tritocerebral neuropil. The main axon of the L-cell then reverses its course and projects in the posterior direction past its ganglion of origin, to the thoracic nervous system. Upon reaching the thoracic ganglia, the L-cell axon (termed the A fiber by Maynard, 1961) turns sharply to exit the CNS via the first segmental nerve (SN1; Cooke and Goldstone 1970; Fort et al. 2004). It projects via SN1 to the pericardial organs (POs), major neurosecretory structures that flank the heart, where it ramifies into many smaller fibers with varicose terminals positioned to release its products into the pericardial sinus. In *Callinectes sapidus*, it was proposed that a branch of the L-cell leaves the POs and projects to the heart (Fort et al. 2004), where it terminates within the cardiac ganglion (CG), a small (9 neurons) aggregate of cells that produces the neurogenic crustacean heartbeat. The extensive projections of the L-cell thus enable it to influence (1) circuits within the brain, (2) the heartbeat via its projection to the CG, and (3) systems and tissues throughout the organism that are responsive to circulating neurohormones.

The physiological properties of the L-cell were examined in the lobster *Homarus gammarus* (Robertson and Moulins 1981). It was found that the L-cell firing pattern reflected the influence of four distinct foregut rhythms and it was postulated that it provided a corollary discharge reflecting this motor output. The L-cell was also proposed to act in a feedback capacity, modifying the stomatogastric nervous system that controls the foregut via neurohormonal release from its terminals in the PO (Robertson and Moulins

1981). Synaptic actions of the L-cell have not been studied. In contrast to its neurosecretory role which has been known for some time, its projections to sites where it may exert more direct synaptic actions, e.g., the tritocerebrum (Tierney et al. 2003) and the cardiac ganglion (Fort et al. 2004) have only been recently disclosed.

A defining feature of the L-cell in all species examined to date is its catecholaminergic phenotype. Originally demonstrated using histofluorescent methods in *Carcinus maenas* (Cooke and Goldstone 1970) and *Panulirus interruptus* (Kushner and Maynard 1977), the presence of catecholamines in the L-cell was subsequently shown using antibodies to dopamine in *Homarus gammarus* (Cournil et al. 1984) and *Callinectes sapidus* (Wood and Derby 1996). Antibodies to tyrosine hydroxylase (TH), the rate-limiting enzyme in the catecholamine biosynthetic pathway have been shown to label the L-cell in *Homarus gammarus* (Cournil et al. 1984, 1994), *Callinectes sapidus* (Wood and Derby 1996; Fort et al. 2004) and *Homarus americanus* (Pulver et al. 2003).

Biochemical approaches also support the presence of catecholamines in the L-cell. Initially, radioenzymatic assays showed that the L-cell of *Panulirus interruptus* contained and accumulated dopamine (Kushner and Barker 1983). High performance liquid chromatography (HPLC) with electrochemical detection also showed high levels of DA in extracts of L-cell somata isolated from *Homarus gammarus* (Cournil et al. 1984). Importantly, the biochemical methods did not detect significant quantities of norepinephrine, suggesting that dopamine is the primary, and possibly the only, catecholamine neurotransmitter in crustaceans (Sullivan et al. 1977; Barker et al. 1979; Cooke and Sullivan 1982). Together, the accumulated evidence supports the conclusion that dopamine is present in the L-cells of all crustacean species that have thus far been examined.

In addition to dopamine, the L-cells of various decapod species also contain cotransmitters. In contrast to the apparent ubiquity of DA, however, the L-cell cotransmitter complement exhibits substantial species variability. In the lobster *Homarus americanus* (Siwicki et al. 1987) and the crabs *Cancer irroratus* and *Cancer borealis* (Marder et al. 1986) it contains the pentapeptide proctolin. In *Homarus gammarus* (Cournil et al. 1984), several crayfish species (Tierney et al. 1999), and the prawn *Macrobrachium rosenbergii* (Sosa et al. 2002), the following observations indicate that serotonin serves as an L-cell cotransmitter: (1) Immunohistochemical experiments by Cournil et al. (1984) on serial sections of the *Homarus gammarus* commissural ganglion showed colocalization of dopamine and serotonin in the L-cell. These investigators also found that levels (4×10^{-4} M) of serotonin in isolated L-cell somata measured by radioimmunoassay were comparable to dopamine concentrations (2×10^{-4} M) measured using HPLC. Finally, it was reported that L-cells identified using electrophysiological criteria exhibited serotonin immunoreactivity (Cournil et al. 1984). (2) Tierney et al. (1999) identified a large serotonin-immunoreactive neuron in the commissural ganglia of seven crayfish species. They proposed that this neuron corresponds to the L-cell and, in at least one case (*Pacifasticus*

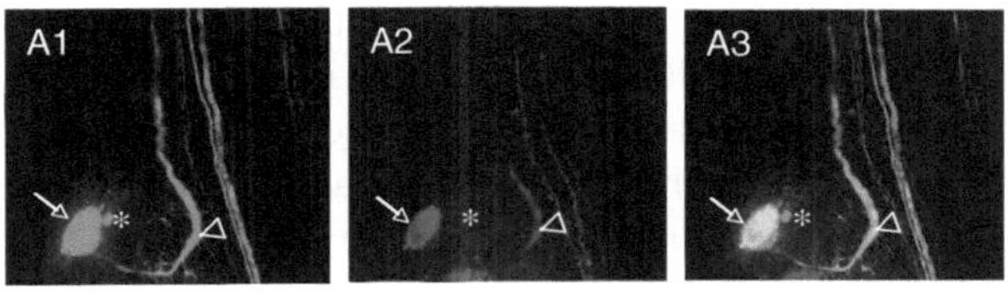

Fig. 13.1 Colocalization of tyrosine hydroxylase-like immunoreactivity and serotonin-like immunoreactivity in the L-cell of the blue crab, *Callinectes sapidus*. (A1) THli in the L-cell (*arrow*) and a second small cell body (*asterisk*; detected with a mouse monoclonal primary antibody and Alexa 488 goat anti-mouse secondary antibody). Note that the axon of the L-cell is constricted as it passes through the commissural ganglion, but that it then widens (*arrowhead*) after entering the circumesophageal connective to ascend to the brain (see text and Fort et al. 2004 for overall L-cell structure). (A2) 5HTli in the same preparation shown in *a1*. 5HTli was observed in the L-cell soma (*arrow*) and in the axon ascending in the connective (*arrowhead*; visualized with rabbit polyclonal antibody and Alexa 546 goat anti-rabbit secondary antibody). It was not detected in the initial segment of the L-cell. 5HTli was not seen in the small THli neuron (*asterisk*) or in other THli fibers, supporting the deduction that icolocalization of markers does not reflect an artifact of marker 'bleedthrough'. *Calibration bar* = 100 μm, applies to *A1–A3*. (*See* Color Plate 16)

leniusculus) colocalization with TH-like immunoreactivity was demonstrated in whole mount ganglia. (3) In double labeling whole mount experiments conducted on the prawn, *Macrobrachium rosenbergii*, Sosa et al. (2002) found serotonin and TH-like immunoreactivities in a large CG neuron that exhibited the anatomical features of the L-cell.

We used double-labeling immunohistochemical methods to assess dopamine/serotonin colocalization in the L-cells of the blue crab *Callinectes sapidus* (García et al. 2007). Previous studies using antibodies to dopamine and TH demonstrated the presence of DA in the *Callinectes* L-cell (Wood and Derby 1996; Fort et al. 2004). When double-labeling (TH and serotonin) experiments were performed, serotonin-like immunoreactivity was observed in the L-cell soma (Fig. 13.1). As observed in other species (Cournil et al. 1984; Tierney et al. 1999; Sosa et al. 2002) the serotonin-like immunoreactivity was less intense than that observed for TH.

13.5 Cholinergic/GABAergic Interneurons in *Aplysia*

The neuronal network that controls consummatory feeding behaviors in the marine mollusk *Aplysia* has been the subject of intensive study aimed toward disclosing principles of motor system organization and plasticity (Kupfermann 1974a,b; Elliott and Susswein 2002; Cropper et al. 2004). In their original identification of interneurons that could contribute to initiating patterned activity in the feeding motor system of *Aplysia*, Susswein and Byrne (1988) designated one such cell B34[1] (see also Hurwitz et al. 1994). Subsequent investigations demonstrated that B34 possesses multi-action synaptic capabilities, exerting excitatory synaptic connections on certain follower neurons and inhibitory actions on others (Hurwitz et al. 1997). The observation that its excitatory synaptic actions were blocked by hexamethonium (10^{-4} M) led to the proposal that acetylcholine acts as the neurotransmitter of B34 (Fig. 13.2a; from Hurwitz et al. 2003). Subsequently, B34 was found to contain GABA-like immunoreactivity and its rapid inhibitory IPSPs to other targets were shown to be blocked by picrotoxin (Fig. 13.2b; from Jing et al. 2003). It was therefore proposed that B34 could be

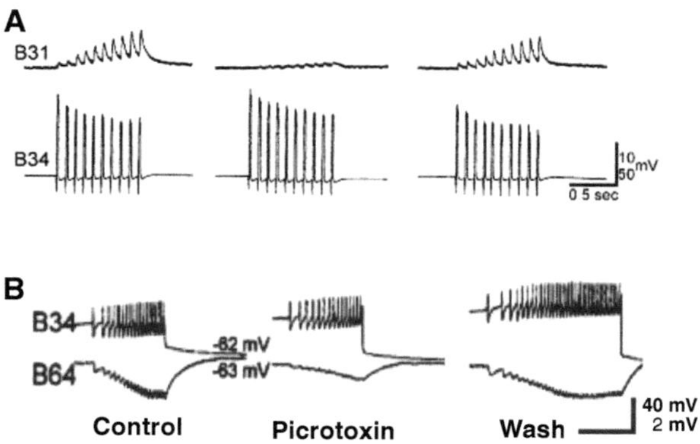

Fig. 13.2 Pharmacological observations indicating cholinergic and GABAergic signaling by B34, an interneuron in the feeding motor circuitry of *Aplysia.* (A) (*reprinted from Hurwitz et al.* 2003) *left*: firing B34 (*lower record*) for 1 s at a frequency of 10 Hz evoked a train of facilitating EPSPs in identified postsynaptic neuron B31 (*upper record*). *A, middle*: the EPSPs were blocked by hexamethonium (5×10^{-4} M). *A, right*: the effect of hexamethonium was reversed following washout of the drug. (B) (*reprinted from Jing et al.* 2003) *left*: firing B34 (*upper record*) evoked a train of IPSPs in identified postsynaptic neuron B64 (*lower record*). *B, middle*: the IPSPs were diminished by picrotoxin (1 mM). *B, right*: the effect of picrotoxin was reversed following washout of the drug. *A* and *B* were both performed in a raised divalent saline that attenuates polysynaptic signaling. In *B*, the postsynaptic membrane potential was pre-set to a level 15 mV more depolarized than rest (*V*m indicated for each neuron in *left* panel) to enhance IPSP amplitudes

[1] The neuronal network that generates *Aplysia* consummatory behaviors is located primarily in the buccal and cerebral ganglia. These ganglia have a bilaterally symmetrical organization and all neurons discussed in this article occur as pairs, one in each hemiganglion, unless otherwise noted. Cell nomenclature denotes the ganglion in which the cell body is located (Buccal in the case of B34). Numerals convey nominal information only, and do not specify neuron structure, function, or phenotype.

using the two distinct neurotransmitters, ACh and GABA respectively, to exert its rapid excitatory and inhibitory synaptic actions (Jing et al. 2003).

The divergent synaptic actions of B34 can be interpreted in the context of its proposed participation in feeding motor programs (Hurwitz et al. 1997; Jing and Weiss 2001; Jing et al. 2003; Cropper et al. 2004). Such motor programs are always initiated by a protraction of the tongue-like radula that is followed by a phase of radula retraction. B34 fires during the phase of radula protraction and its synaptic actions include excitation of protraction interneurons, and inhibition of interneurons that generate the antagonistic movement of retraction. Its cholinergic signaling is therefore thought to enhance and prolong radula protraction, while its GABAergic inhibitory signaling delays the onset of radula retraction. The efficiency of controlling two sequential phases of a motor program with a single neuron can be readily appreciated, as it will ensure that motor signals specifying the two antagonistic movements do not overlap. Any advantage that may be conferred by implementing such control with two distinct neurotransmitters, however, remains to be determined (see discussion below, under Overview and Future Directions).

13.6 Dopaminergic/GABAergic Interneurons in *Aplysia*

The neuronal network that controls feeding in *Aplysia* can be configured to perform multiple consummatory behaviors. Such multifunctionality is achieved via recruitment of particular interneurons that specify distinct motor patterns (Kupfermann and Weiss 2001). Two such interneurons, B20 and B65, were initially identified on the basis of their ability to elicit coordinated rhythmic motor programs from the feeding network (Teyke et al. 1993; Kabotyanski et al. 1998). Aldehyde fluorescence histology showed B20 and B65 to be catecholaminergic and pharmacological data supported their dopaminergic signaling (Teyke et al. 1993; Kabotyanski et al. 1998).

Subsequently, a survey of GABA-immunoreactive neurons in the central nervous system of *Aplysia* revealed GABAli cells with morphological similarities (size, shape, position, projections) to B20 and B65 (Díaz-Ríos et al. 1999). These observations prompted a systematic investigation in which GABA-catecholamine colocalization was tested using four independent protocols: (1) nerve backfill combined with GABAli, (2) FaGlu histochemistry combined with GABAli, (3) THli combined with GABAli, and (4) electrophysiological identification combined with GABAli (Díaz-Ríos et al. 2002). This study demonstrated that colocalization of GABA and DA markers was limited to five neurons in the entire CNS of *Aplysia*; the paired B20 cells, the paired B65 cells, and one unpaired neuron that has not as yet been identified

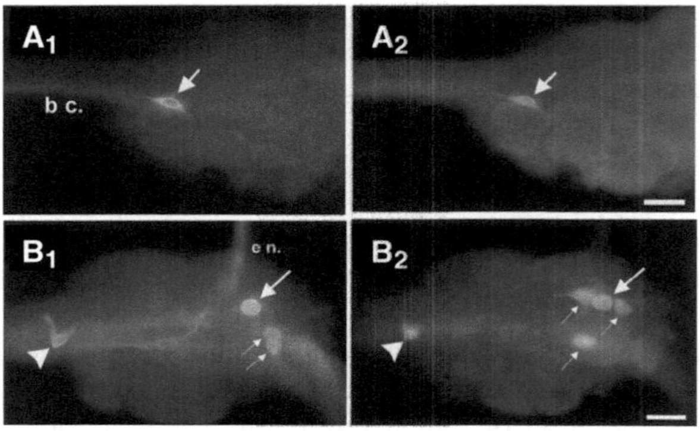

Fig. 13.3 Colocalization of THli and GABAli in neurons B20 and B65 of *Aplysia californica* (reprinted with permission from Díaz-Ríos et al. 2002). (A1): THli was observed in a single neuron, B20 (*arrow*), on the rostral surface of each buccal hemiganglion (only the left hemiganglion is shown). (A2): GABAli in the same preparation as *a1*. GABAli was also localized to the B20 neuron (*cf.* arrows in *a1* and *a2*). Scale bar: 40 μm applies to *a1* and *a2*. (B1): THli was observed in four neurons on the caudal surface of each buccal hemiganglion (the right hemiganglion is shown); one unpaired cell (*arrowhead*) near the buccal commissure and three cells (*arrows*) in the lateral region of the ganglion. (B2): GABAli in the same preparation as *B1*. GABAli was localized to the unpaired neuron (*cf. arrowheads* in *b1* and *b2*). It was also present in four lateral neurons, one of which corresponded to B65, the THli labeled in panel *b1* (*cf.* large arrows in *b1* and *b2*). Scale bar: 40 μm applies to *b1* and *b2*

(Fig. 13.3; from Díaz-Ríos et al. 2002). The presence of GABAli in B65 was confirmed independently by Jing et al. (2003).

Following the demonstration of DA-GABA colocalization in B20 and B65, experiments were performed aimed toward identifying the neurotransmitters mediating their synaptic signaling (Due et al. 2004; Díaz-Rios and Miller 2005, 2006). The rapid EPSPs from B65 and B20 to specific followers were occluded by dopamine, but not GABA, and blocked by the dopamine antagonist sulpiride (Due et al. 2004; Díaz-Ríos and Miller 2005). It was therefore proposed that these rapid EPSPs were mediated by dopamine. GABA, acting through $GABA_B$-like receptors, was shown to modulate the rapid dopaminergic EPSPs in a target specific manner (Díaz-Ríos and Miller 2005, 2006). To date, there is no evidence for inhibitory signaling or rapid GABAergic PSPs originating from B20 or B65 (*cf.* B34 above).

The GABAergic modulation of signaling by B20 was further examined in studies aimed toward disclosing the contributions of GABA to various forms of synaptic plasticity (Svensson et al. 2004; Díaz-Ríos and Miller 2006). GABA was found to potentiate inward currents produced by dopamine on specific postsynaptic targets (Svensson et al. 2004) and GABA was proposed to potentiate three forms of synaptic plasticity; short-term potentiation (Svensson et al.

2004), facilitation, and summation (Díaz-Ríos and Miller 2006). In all cases, data supported a postsynaptic action of GABA that was mediated via $GABA_B$-like receptors.

In sum, the available data indicate that GABA and DA are colocalized in a limited number of neurons that are highly influential in promoting and shaping the feeding motor programs of *Aplysia*. Convergent and divergent rapid excitatory synaptic signaling from these neurons is mediated by dopamine. In the synapses that have been studied, GABA could modify the rapid dopaminergic signals via postsynaptic $GABA_B$-like receptors or presynaptic receptors. To gain additional support for GABA-DA colocalization and cotransmission, we have begun to explore whether similar patterns occur in related mollusks (Fig. 13.4). On the rostral surface of the buccal ganglion of *Dolabrifera dolabrifera*, another member of the Aplysiidae family, GABA-DA colocalization was observed in a bilateral pair of neurons (one shown in Fig. 13.4a1–3) near the buccal commissure. The size, position, and branching pattern of these cells suggest that they correspond to the B20 interneurons of *Aplysia*. A second pair of GABA-DA neurons was observed more laterally and closer to the caudal surface of the *Dolabrifera* buccal ganglion (one shown in Fig. 13.4b1–3), in a position corresponding to the B65 interneuron of *Aplysia*.

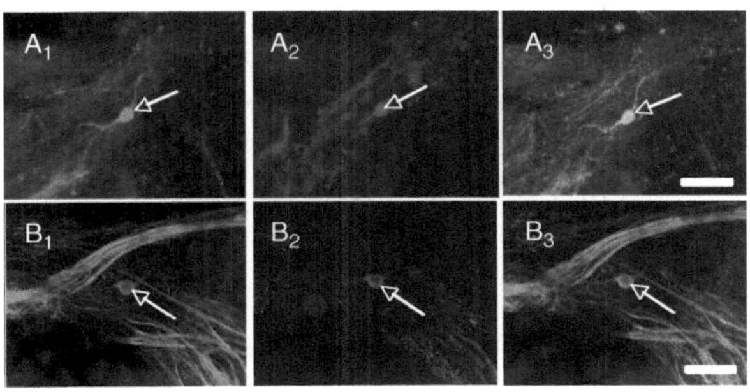

Fig. 13.4 Colocalization of TH-like immunoreactivity and GABA-like immunoreactivity in the buccal ganglion of *Dolabrifera dolabrifera*. (*A1*) A single neuron (*arrow*) on the rostral surface was marked with an antibody against tyrosine hydroxylase (mouse monoclonal; Alexa 488 goat anti-mouse secondary). (*A2*) When the same preparation was processed for GABA-like immunoreactivity (rabbit polyclonal; Alexa 546 goat anti-rabbit secondary), the same neuron (*arrow*) was marked. (*A3*) The labeled neuron (*arrow*) appears yellow in an overlay of panels *A1* and *A2*. *Calibration bar*: 100 μm applies to all *A* panels. (*B1*) A neuron (*arrow*) in the lateral region of the caudal surface of each buccal hemiganglion (only the right is shown) was marked with an antibody against TH. (*B2*) When the same preparation was processed for GABA-like immunoreactivity, the same neuron (*arrow*) was marked. (*B3*) The labeled neuron (*arrow*) appears yellow in an overlay of panels *B1* and *B2*. *Calibration bar*: 100 μm applies to all *B* panels. (*See* Color Plate 16)

13.7 Overview

It is clear from this synopsis that large gaps exist in our present understanding of colocalization and cotransmission of classical neurotransmitters in invertebrates. Although the data reviewed allow few definitive conclusions to be drawn, they do permit some inferences and speculation that can guide further study.

1. The number of neurons in which two classical neurotransmitters are colocalized tends to be small in comparison to the number in which they are not. Far from being obligatory, the colocalization of two particular neurotransmitters appears to occur rarely in the invertebrate nervous systems that have been examined in greatest detail. In the case of GABA and dopamine in *Aplysia*, each neurotransmitter is present in more than fifty central neurons (Croll 2001; Díaz-Ríos et al. 1999), and yet their colocalization has only been observed in five cells (Díaz-Ríos et al. 2002). Similarly, while the central nervous systems of decapod crustaceans contain up to one hundred serotonergic and dopaminergic neurons (Beltz and Kravitz 1983; Tierney et al. 2003), the overlap of these two systems has only been reported to occur in the L-cell.

2. It is not possible to extract rules concerning which classical neurotransmitters are more or less likely to be paired as cotransmitters. In the limited number of examples described, GABA was paired with DA and ACh, ACh was paired with GABA and serotonin, serotonin was paired with ACh and dopamine, and dopamine was paired with serotonin and GABA. It may be anticipated that other pairings will emerge as additional markers for classical neurotransmitters become available (see following text).

3. Classical neurotransmitter colocalization can occur in a variety of neuron types. This article has described the coexistence of classical neurotransmitters in neurosecretory cells, proprioceptors, and interneurons within a central pattern generator circuit. Interestingly, although neuropeptide cotransmitters are utilized extensively by the motor neurons of invertebrates, the presence of multiple classical transmitters in motor neurons has not yet been described.

13.8 Future Directions

Each of the neurons considered in this chapter participates in an intensively investigated motor circuit. In most instances, direct postsynaptic targets are known, and the contributions of these neurons to motor pattern generation or regulation can be evaluated. These systems should therefore present exceptional opportunities for studying the functional contributions of classical neurotransmitter colocalization and cotransmission to synaptic integration and circuit operation.

1. How can classical cotransmitters broaden the signaling capacity of individual neurons? Cotransmitters can expand signaling in the temporal or spatial domains. Temporally, classical cotransmitters may enable a neuron to influence a particular postsynaptic target on multiple timescales. Spatially, divergent synaptic cotransmission can increase the number of postsynaptic cells that a single neuron can influence. Each of these forms of enhanced signaling depends upon the nature and distribution of the receptors that are present on the neurons that receive signals from a cotransmitter-containing neuron (Marder et al. 1995; Marder 1999). In principle, given a sufficient variety of receptors, they could be achieved using a single multi-action neurotransmitter (see Gardner and Kandel 1972; Katz and Frost 1995).

A complementary model of cotransmitter function proposes that multiple released substances may activate common or convergent signaling pathways with distinct efficacies (Brezina and Weiss 1997a,b). As a result of their varying capabilities to stimulate these pathways, combinations of coactive transmitters can achieve a range or precision of signaling that could not be accomplished by any of the transmitters acting alone. Notably, this hypothesis specifies a computational benefit imparted by cotransmitters that may not be readily achieved by additional classes of postsynaptic receptors.

Finally, cotransmission by classical neurotransmitters may enable the signaling of a neuron to be precisely regulated by its own previous activity. In contrast to peptide cotransmitters, which are typically thought to be packaged in large dense core vesicles that possess distinct release properties, impulse-mediated cotransmission of classical neurotransmitters could be envisioned to produce specific stoichiometries of signaling molecules within the synaptic cleft. One consequence of such cotransmission could be the modification of short-term synaptic plasticity during repetitive firing (see Svensson et al. 2004; Díaz-Ríos and Miller 2006).

2. Do classical cotransmitters enhance the ability of individual neurons to regulate motor circuits? Several of the neurons considered in this article are found at the boundaries between sensory and motor systems. They are thus effectively positioned to control motor patterns via signals that exert multiple actions on central pattern generator circuits. It has been proposed that such motor system regulation can be achieved via two general architectures. When motor system regulation is imposed by neurons that are not *sensu stricto* participants in the CPG, it is termed extrinsic (Kupfermann 1979; Morgan et al. 2000). When it derives from neurons that are themselves elements of the CPG (Katz and Frost 1996), or from motor neurons (Cropper et al. 1987), it is designated intrinsic. The invertebrate neurons that are proposed to utilize multiple classical cotransmitters embody both of these major regulatory motifs. While the L-cell and the GPR neurons can exert strong effects on the motor circuits that they regulate, their activity is not required for the motor patterns to be expressed. The interneurons of the *Aplysia* feeding network (B34, B20, and B65), on the other hand, are more deeply embedded within the buccal CPG that they regulate.

Consistent with its potential contributions to synaptic signaling considered above, cotransmission by colocalized classical neurotransmitters could contribute to motor system regulation in both temporal and spatial domains. Temporally, the consequence of signaling on multiple timescales to CPG operation was emphasized by Getting (1989), who recognized how such signals could enable individual interneurons to influence multiple phases of motor activity. Spatially, cotransmission may enable an individual interneuron to efficiently achieve coactivation of combined populations of motor neurons, each of which, under other conditions, may be individually activated by interneurons that only utilize one neurotransmitter. Finally, direct synchronous excitation and inhibition of motor neurons whose coactivation is inconsistent with adaptive motor patterns could produce a level of precision in phase timing and phase transitions that could not be achieved with interposed interneurons.

13.9 Conclusions

Clearly, much investigation will be required to validate and clarify the instances of classical neurotransmitter colocalization and cotransmission that have been described in this chapter. Recent genomic and neuromic initiatives with invertebrate model organisms promise to provide additional markers that will facilitate demonstrations of colocalized classical neurotransmitters in these systems (Moroz et al. 2006; Schulz et al. 2007 see also www.NeuronBank.org). Emerging technologies and approaches should also provide tools to explore important cell biological questions, including transmitter biosynthesis, sorting, compartmentalization, release, and uptake in invertebrate neurons that contain more than one classical neurotransmitter (Fuller et al. 1998; Anderson and Ewing 1999; Martí et al. 2006, 2007).

Even in the simple systems considered here, it may be appreciated how the adaptive operation of neural networks can be enhanced if multiple signals are implemented at precisely the same time (synchrony) or in precisely the same place (convergence). In our present understanding of information transfer in nervous systems, the synapse represents the highest level of both temporal and spatial precision, and synaptic cotransmission can clearly achieve levels of synchrony and convergence that could not be accomplished by two independent neurons utilizing a single neurotransmitter. Thus, although we can speculate about benefits that colocalization and cotransmission of classical neurotransmitters could confer upon a neuron's contribution to circuit function, the pressures and constraints that would lead to this resolution in specific neurons, and not in others, remain unknown. In view of nature's inherently conservative approach to problem-solving in neural systems, however, it may be expected that analogous computational challenges were met with similar resolutions in more complex nervous systems, including our own brains.

Acknowledgments Supported by: the National Institutes of Health: NIGMS MBRS: GM-08224 and NCRR RCMI G12 RR03051; the National Science Foundation: DBI-0115825.

References

Adams ME, O'Shea M (1983) Peptide cotransmitter at a neuromuscular junction. Science 221:286–288
Anderson BB, Ewing AG (1999) Chemical profiles and monitoring dynamics at an individual nerve cell in *Planorbis corneus* with electrochemical detection. J Pharm Biomed Anal 19:15–32
Barker DL, Herbert E, Hildebrand JG, Kravitz EA (1972) Acetylcholine and lobster sensory neurones. J Physiol Lond 226:205–229
Barker DL, Kushner PD, Hooper NK (1979) Synthesis of dopamine and octopamine in the crustacean stomatogastric nervous system. Brain Res 161:99–113
Beltz BS, Kravitz EA (1983) Mapping of serotonin-like immunoreactivity in the lobster nervous system. J Neurosci 3:585–602
Beltz B, Eisen JS, Flamm RE, Harris-Warrick RM, Hooper SL, Marder E (1984) Serotonergic innervation and modulation of the stomatogastric ganglion of three decapod crustaceans (*Panulirus interruptus*, *Homarus americanus*, and *Cancer irroratus*). J Exp Biol 109:35–54
Boer HH, Schot LP, Steinbusch HW, Montagne C, Reichelt D (1984) Co-existence of immunoreactivity to anti-dopamine, anti-serotonin and anti-vasotocin in the cerebral giant neuron of the pond snail *Lymnaea stagnalis*. Cell Tissue Res 238:411–412
Brezina V, Weiss KR (1997a) Analyzing the functional consequences of transmitter complexity. Trends Neurosci 20:538–543
Brezina V, Weiss KR (1997b) Functional consequences of divergence and convergence in physiological signaling pathways. Mol Psychiatry 2:9–11
Brownstein MJ, Saavedra JM, Axelrod J, Zeman GH, Carpenter DO (1974) Coexistence of several putative neurotransmitters in single identified neurons of *Aplysia*. Proc Natl Acad Sci 71:4662–4665
Burnstock G (1976) Do some nerve cells release more than one transmitter? Neuroscience 1:239–248
Cooke IM, Goldstone MW (1970) Fluorescence localization of monoamines in crab neurosecretory structures. J Exp Biol 53:651–668
Cooke IM, Sullivan RE (1982) Hormones and neurosecretion. In: Atwood HL, Sandeman DC (eds) The Biology of Crustacea, Vol 3, Neurobiology: Structure and Function, Academic Press, New York
Cournil I, Geffard M, Moulins M, Le Moal M (1984) Coexistence of dopamine and serotonin in an identified neuron of the lobster nervous system. Brain Res 310:397–400
Cournil I, Helluy SM, Beltz BS (1994) Dopamine in the lobster *Homarus gammarus*. I. Comparative analysis of dopamine and tyrosine hydroxylase immunoreactivities in the nervous system of the juvenile. J Comp Neurol 344:455–469
Croll RP (1987) Identified neurons and cellular homologies. In: Ali MA (ed) Nervous Systems in Invertebrates. New York: Plenum
Croll RP (2001) Catecholamine-containing cells in the central nervous system and periphery of *Aplysia californica*. J Comp Neurol 441:91–105
Cropper EC, Lloyd PE, Reed W, Tenenbaum R, Kupfermann I, Weiss KR (1987) Multiple neuropeptides in cholinergic motor neurons of *Aplysia*: evidence for modulation intrinsic to the motor circuit. Proc Natl Acad Sci USA:3486–3490
Cropper EC, Evans CG, Hurwitz I, Jing J, Proekt A, Romero A, Rosen SC (2004) Feeding neural networks in the mollusk *Aplysia*. Neurosignals 13:70–86
Díaz-Ríos M, Miller MW (2005) Rapid dopaminergic signaling by interneurons that contain markers for catecholamines and GABA in the feeding circuitry of *Aplysia*. J Neurophysiol 93:2142–2156
Díaz-Ríos M, Miller MW (2006) Target-specific regulation of synaptic efficacy in the feeding central pattern generator of *Aplysia*: Potential substrates for behavioral plasticity? Biol Bull 210:215–229

Díaz-Ríos M, Suess E, Miller MW (1999) Localization of GABA-like immunoreactivity in the central nervous system of *Aplysia californica*. J Comp Neurol 413:255–270
Díaz-Ríos M, Oyola E, Miller MW (2002) Colocalization of g-aminobutyric acid-like immunoreactivity and catecholamines in the feeding network of *Aplysia californica*. J Comp Neurol 445:29–46
Due MR, Jing J, Weiss KR (2004) Dopaminergic contributions to modulatory functions of a dual-transmitter interneuron in *Aplysia*. Neurosci Letters 358:53–57
Elekes K, Kemenes G, Hiripi L, Geffard M, Benjamin PR (1991) Dopamine-immunoreactive neurons in the central nervous system of the pond snail, *Lymnaea stagnalis*. J Comp Neurol 307:214–224
Elliott CJH, Susswein AJ (2002) Comparative neuroethology of feeding control in mollusks. J Exp Biol 205:877–896
Elste A, Koester J, Shapiro E, Panula P, Schwartz JH (1990) Identification of histaminergic neurons in *Aplysia*. J Neurophysiol 64:736–744
Fort TJ, Brezina V, Miller MW (2004) Modulation of an integrated central pattern generator-effector system: dopaminergic regulation of cardiac activity in the blue crab *Callinectes sapidus*. J Neurophysiol 92:3455–3470
Fuller RR, Moroz LL, Gillette R, Sweedler JV (1998) Single neuron analysis by capillary electrophoresis with fluorescence spectroscopy. Neuron 20:173–181
García-Crescioni KB, Miller MW, Stern E, Brezina V (2007) Serotonergic regulation of heartbeat in the crab *Callinectes sapidus*: actions on the CPG and periphery. Program No. 289.14. 2007 Neuroscience Meeting Planner. San Diego, CA: Society for Neuroscience, Online
Gardner D, Kandel ER (1972) Diphasic postsynaptic potential: a chemical synapse capable of mediating conjoint excitation and inhibition. Science 176:675–678
Gerschenfeld HM (1973) Chemical transmission in invertebrate central nervous systems and neuromuscular junctions. Physiol Rev 53:1–119
Getting PA (1989) Emerging principles governing the operation of neural networks. Annu Rev Neurosci 12:185–204
Hernádi L, Juhos S, Elekes K (1993) Distribution of tyrosine-hydroxylase-immunoreactive and dopamine-immunoreactive neurons in the central nervous system of the snail *Helix pomatia*. Cell Tissue Res 274:503–513
Hernádi L, Elekes K (1995) Neurons with different immunoreactivity form clusters in the CNS of *Helix pomatia*. Acta Biol Hung 46:271–280
Hildebrand JG, Townsel JG, Kravitz EA (1974) Distribution of acetylcholine, choline, choline acetyltransferase and acetylcholinesterase in regions and single identified axons of the lobster nervous system. J Neurochem 23:951–963
Hurwitz I, Goldstein RS, Susswien AJ (1994) Compartmentalization of pattern-initiation and motor function in the B31 and B32 neurons of the buccal ganglia of *Aplysia californica*. 71:1514–1527
Hurwitz I, Kupfermann I, Susswein AJ (1997) Different roles of B63 and B34 that are active during the protaction phase of buccal motor programs in *Aplysia californica*. J Neurophysiol 78:1305–1319
Hurwitz I, Kupfermann I, Weiss KR (2003) Fast synaptic connections from CBIs to pattern-generating neurons in *Aplysia*: Initiation and modification of motor programs. J Neurophysiol 89:2120–2136
Jing J, Vilim FS, Wu J-S, Park J-H, Weiss KR (2003) Concerted GABAergic actions of *Aplysia* feeding interneurons in motor program specification. J Neurosci 23:5283–5294
Jing J, Weiss KR (2001) Neural mechanisms of motor program switching in *Aplysia*. J Neurosci 21:7349–7362
Kabotyanski EA, Baxter DA, Byrne JH (1998) Identification and characterization of catecholaminergic neuron B65, which initiates and modifies patterned activity in the buccal ganglia of *Aplysia*. J Neurophysiol 79:605–621
Kandel ER (1975) Cellular basis of behavior. WH Freeman, San Francisco

Kandel ER (1979) Behavioral biology of *Aplysia*. WH Freeman, San Francisco
Katz PS, Frost WN (1995) Intrinsic neruomodulation in the *Trtonia* swim CPG: serotonin mediates both neuromodulation and neurotransmission by the dorsal swim interneurons. J Neurophysiol 74:2281–2294
Katz PS, Frost WN (1996) Intrinsic neuromodulation: altering neuronal circuits from within. Trends Neurosci 19:54–61
Katz PS, Harris-Warrick RM (1989) Serotonergic/cholinergic muscle receptor cells in the crab stomatogastric nervous system. II. Rapid nicotinic and prolonged modulatory effects on neurons in the stomatogastric ganglion. J Neurophysiol 62:571–581
Katz PS, Harris-Warrick RM (1990a) Neuromodulation of the crab pyloric central pattern generator by serotonergic/cholinergic proprioceptive afferents. J Neurosci 10:1495–1512
Katz PS, Harris-Warrick RM (1990b) Actions of identified neuromodulatory neurons in a simple motor system. Trends in Neurosci 13:367–373
Katz PS, Eigg MH, Harris-Warrick RM (1989) Serotonergic/cholinergic muscle receptor cells in the crab stomatogastric nervous system. I. Identification and characterization of the gastropyloric receptor cells. J Neurophysiol 62:558–570
Kiehn O, Harris-Warrick RM (1992) Serotonergic stretch receptors induce plateau properties in a crustacean motor neuron by a dual-conductance mechanism. J Neurophysiol 68:485–495
Kuffler SW, Nichols JG, Martin AR (1984) From neuron to brain, (2nd ed). Sinauer Press, Sunderland MA
Kupfermann I (1974a) Feeding behavior in *Aplysia*: a simple system for the study of motivation. Behav Biol 10:1–26
Kupfermann I (1974b) Dissociation of the appetitive and consummatory phases of feeding behavior in *Aplysia*: a lesion study. Behav Biol 10:89–97
Kupfermann I (1979) Modulatory actions of neurotransmitters. Annu Rev Neurosci. 2:447–465
Kupfermann I (1991). Functional studies of cotransmission. Physiol Revs 71:683–732
Kupfermann I, Weiss KR (2001) Motor program selection in simple model systems. Curr Opin Neuobiol 11:673–677
Kushner PD, Barker DL (1983) A neurochemical description of the dopaminergic innervation of the stomatogastric ganglion of the spiny lobster. J Neurobiol 14:17–28
Kushner PD, Maynard EA (1977) Localization of monoamine fluorescence in the stomatogastric nervous system of lobsters. Brain Res 129:13–28
Marder E (1999) Neural signalling: Does colocalization imply cotransmission? Curr Biol 9:R809–811
Marder E, Calabrese RL (1996) Principles of rhythmic motor pattern generation. Physiol Rev 76:687–717
Marder E, Hooper SL, Siwicki KK (1986) Modulatory action and distribution of the neuropeptide proctolin in the crustacean stomatogastric nervous system. J Comp Neurol 243:454–467
Marder E. Christie AE, Kilman VL (1995) Functional organization of cotransmission systems: lessons from small nervous systems. Invert Neurosci 1:105–112
Martí AA, Li X, Jockusch S, Li Z, Raveendra B, Kalachikov S, Russo JJ, Morozova I, Sathyanarayanan VP, Ju J, Turro NJ (2006) Pyrene binary probes for unambiguous detection of mRNA using time-resolved fluorescence spectroscopy. Nucleic Acids Res 34:3161–3168
Martí AA, Jockusch S, Stevens N, Ju J, Turro NJ (2007) Fluorescent hybridization probes for sensitive and selective DNA and RNA detection. Acc Chem Res 40:402–409
Maynard DM (1961b) Thoracic neurosecretory structures in Brachyura. II. Secretory neurons. Gen Comp Endocrinol 1:237–263
Miller MW, Vu E, Krasne FB (1992) Cholinergic transmission at the first synapse mediating the crayfish lateral giant escape reaction. J Neurophysiol 68:2174–2184

Morgan PT, Perrins R, Lloyd PE, Weiss KR (2000) Intrinsic and extrinsic modulation of a single central pattern generating circuit. J Neurophysiol 84:1186–1193

Moroz LL, Edwards JR, Puthanveettil SV, Kohn AB, Ha T, Heyland A, Knudsen B, Sahni A, Yu F, Liu L, Jezzini S, Iannucculli W, Chen M, Nguyen T, Sheng H, Shaw R, Kalachikov S, Panchin YV, Farmerie W, Russo JJ, Ju J, Kandel ER (2006) Neuronal transcriptome of *Aplysia*: neuronal compartments and circuitry. Cell 127:1453–1467

Nusbaum MP, Blitz DM, Swensen AM, Wood D, Marder E (2001) The roles of co-transmission in neural network function. Trends Neurosci 24:146–154

Ono JK, McCaman RE (1980) Identification of additional histaminergic neurons in *Aplysia*: Improvement of single cell isolation techniques for in tandem physiological and chemical studies. Neuroscience 5:835–840

Ono JK, McCaman RE (1984) Immunocytochemical localization and direct assays of serotonin-containing neurons in *Aplysia*. Neuroscience 11:549–560

Osborne NN (1979) Is Dale's principle valid? Trends Neurosci 2:73–75

Osborne NN (1984) Putative neurotransmitters and their coexistence in gastropod mollusks. In: Chan-Palay V, Palay SL (eds) Coexistence of Neuroactive Substances in Neurons. Wiley. New York

Pearson KG (1993) Common principles of motor control in vertebrates and invertebrates. Annu Rev Neurosci 16:265–297

Pulver SR, Thirumalai V, Richards KS, Marder E (2003) Dopamine and histamine in the developing stomatogastric system of the lobster *Homarus americanus*. J Comp Neurol 462:400–414

Robertson RM, Moulins M (1981) A corollary discharge of total foregut motor activity is monitored by a single interneurone in the lobster *Homarus gammarus*. J Physiol Paris 77:823–827

Sakharov DA, Voronezhskaya EE, Nezlin L, Baker MW, Elekes K, Croll RP (1996) Tyrosine hydroxylase-negative, dopaminergic neurons are targets for transmitter-depleting action of haloperidol in the snail brain. Cell Mol Neurobiol 16:451–461

Salimova NB, Sakharov DA, Milosevic I, Turpaev TM, Rakic L (1987) Monoamine-containing neurons in the *Aplysia* brain. Brain Res 400:285–299

Schulz DJ, Goaillard JM, Marder EE (2007) Quantitative expression profiling of identified neurons reveals cell-specific constrants of highly variable levels of gene expression. Proc Natl Acad Sci USA 104:13187–13191

Selverston AI, Russell DF, Miller JP, King DG (1976) The stomatogastric nervous system: structure and function of a small neural network. Prog Neurobiol 7:215–290

Siwicki KK, Beltz BS, Kravitz EA (1987) Proctolin in identified serotonergic, dopaminergic, and cholinergic neurons in the lobster, *Homarus americanus*. J Neurosci 7:522–532

Sosa MA, Hernández CM, Rivera N, Rolon S (2002) Tyrosine hydroxylase and FMRFamide immunohistochemistry in the CNS of the freshwater prawn *Macrobrachium rosenbergii*. Program No. 59.2. *Abstract Viewer/Itinerary Planner*. Washington, DC: Society for Neuroscience, online

Sullivan RE, Friend BJ, and Barker DL (1977) Structure and function of spiny lobster ligamental nerve plexuses: Evidence for synthesis, storage and secretion of biogenic amines. J Neurobiol 8:581–605

Susswein AJ, Byrne JH (1988) Identification and characterization of neurons initiating patterned neural activity in the buccal ganglia of *Aplysia*. J Neurosci 8:2049–2061

Svensson E, Proekt A, Weiss KR (2004) Complementary effects of co-localized dopamine and GABA on synaptic transmission in the *Aplysia* feeding network. Program No. 537.4. *Abstract Viewer/Itinerary Planner*. Washington, DC: Society for Neuroscience, 2004. Online

Teyke T, Rosen SC, Weiss KR, Kupfermann I (1993) Dopaminergic neuron B20 generates rhythmic neuronal activity in the feeding motor circuitry of *Aplysia*. Brain Res 630:226–237

Tierney AJ, Godleski MS, Rattananont P (1999) Serotonin-like immunoreactivity in the stomatogastric nervous systems of crayfishes from four genera. Cell Tissue Res 295:537–551

Tierney AJ, Kim T, Abrams R (2003) Dopamine in crayfish and other crustaceans: Distribution in the central nervous system and physiological functions. Microsc Res Tech 60:325–335

Weiss KR, Kupfermann I (1976) Homology of the giant serotonergic neurons (metacerebral cells) in *Aplysia* and pulmonate moluscs. Brain Res 117:33–49

Wood DE, Derby CD (1996) Distribution of dopamine-like immunoreactivity suggests a role for dopamine in the courtship display behavior of the blue crab, *Callinectes sapidus*. Cell Tissue Res 285:321–330

Zhang B, Harris-Warrick RM (1994) Multiple receptos mediate the modulatory effects of serotonergic neurons in a small neural network. J exp Biol 190:55–77

Chapter 14
E pluribus unum: Out of Many, One

Post-synaptic Integration of Co-transmitters

R. Gutiérrez and J. A. Arias-Montaño

Abstract E pluribus unum, from many (signals), one (response) implies the reception of different time-locked and spatially restricted messages that should be integrated in the post-synaptic cell to produce a response. All neurons have receptors for a variety of neurotransmitters, which are distributed along the dendritic, somatic and axonal compartments, and these should match with the neurotransmitter phenotype of the neurons that impinge onto the different compartments. However, the integration of the messages conveyed by two or more substances that are co-released by single terminals will only be possible if the receptors for these substances are in the synaptic or perisynaptic site. The nature of the neurotransmitter and above all of the receptors (i.e., cationic or anionic ionotropic, or metabotropic) determines the direction and time course of the response. If the neurotransmitter phenotype of the presynaptic neuron is subject to developmental and activity-dependent plasticity, so must be the receptors in the postsynaptic cells. The interplay between the presynaptic message and the postsynaptic reception apparatus should be concerted to produce a response, especially if the nature of the presynaptic message is plastic.

E pluribus unum, out of many (signals), one (response). Is this not what we are ultimately looking at, the rationale behind the co-expression and co-release of classical neurotransmitters? As Dr. Hökfelt pointed out in his illustrative preface, the concept of a "classical neurotransmitter" cannot be equated to that of a "main signaling molecule". Maybe the neurotransmitters should no longer be regarded as "classical" or "not classical" and maybe our classic view of neurotransmission needs to be redefined. In any case, more than one message is generally received by the post-synaptic cell and they probably cooperate to form a single message. These messages are conveyed by "fast" or "slow" acting neurotransmitters, their receptors can be ionotropic or metabotropic, and the

R. Gutiérrez (✉)
Department of Physiology, Biophysics and Neurosciences, Center for Research and Advanced Studies of the National Polytechnic Institute, Apartado Postal 14-740, México, D.F. 07000
e-mail: grafacl@fisio.cinvestav.nx

R. Gutierrez (ed.), *Co-Existence and Co-Release of Classical Neurotransmitters*,
DOI 10.1007/978-0-387-09622-3_14, © Springer Science+Business Media, LLC 2009

cell must then make sense of the different components in order to generate a single response. Indeed, although it is clear that neurons receive multiple signals all the time, as they have receptors for many chemical substances, in this chapter we will concentrate on the time-locked messages that are conveyed by single cells or even single synapses.

Neurotransmitters are co-released and act on different sets of receptors that initiate intracellular cascades that may interact with one another (cross-talk). Alternatively, they produce the flow of different ionic species that can be directly translated into a modification of the membrane potential and that are ultimately integrated to produce the cellular response. Hence, the presence or absence, the level of expression, and the functional state of the receptors for the released neurotransmitters in the "listener" cell will determine its final output. As we have already seen (see Borodinsky and Spitzer in this volume), the functional expression of neurotransmitter receptors not only involves the transcription and translation of the appropriate genes but also, the localization of the assembled receptors in the cell membrane. For many different receptor subunits, these processes depend on the electrical activity in diverse postsynaptic cells. Moreover, these phenomena are relevant not only during the formation of the nervous system but also, in some circumstances, in the adult.

Does the co-release of classical neurotransmitters confer an advantage to neuronal communication? Several chapters in this volume shed some light onto this question. Thus, we will try to summarize these advantages and add some considerations that we feel should be taken into account in the future research in this field.

Throughout this book it has become clear that the concerted action of co-released neurotransmitters is fundamental to shape developing circuits (see Gillespie and Kandler, Gutiérrez, Safiulina et al., Luther and Birren in this volume). At this circuit level, co-release of multiple transmitters coupled with the expression of the distinct transmitter receptors in target cells ensures the exclusivity of signaling and removes a potential restriction on the structural complexity of circuits.

We will start by pointing out some of the possible immediate cellular effects that the co-released neurotransmitters may exert. Two (or more) neurotransmitters released at a restricted site will enable rapid synergistic or antagonistic actions in a given target cell. This implies that the concentration of the substances released will be high at a given time at a restricted site. This will not only have an immediate repercussion on the post-synaptic site, where fast modulation can be exerted, but also on the pre-synaptic site. Most neurons possess receptors in their terminals or in the preterminal regions of their axons for the very neurotransmitters that they release. This probably contributes to the highly efficient modulation of fast pre- and post-synaptic responses by directly regulating co-release. Moreover, because the receptors determine the action of the released neurotransmitters, the (combined) properties of the different receptors that co-localize will open the way for new forms of plasticity at both pre- and post-synaptic sites.

For instance, the co-existence of two neurotransmitters at synaptic contacts where there is no receptor for one of the transmitters and conversely, the co-existence of receptors to two neurotransmitters apposed to terminals lacking one of the transmitters has been documented (Bekkers, 2005; Rao et al., 2000). These synapses are therefore silent to one neurotransmitter but eventually, they can become active to both through plastic changes at either side of the synapse. Indeed, the activation of $GABA_A$ receptors apposed to terminals that co-release GABA and glycine has been reported (Lu et al., 2008). Whether other receptors can be activated by two different neurotransmitters is not known.

The existence of different pre-synaptic receptors at the sites where co-release takes place can also account for differential regulation of the release of either neurotransmitter (Bergersen et al., 2003; Gutierrez, 2003, 2005; Gillespie et al. in this volume; Chandler et al., 2003). Again, this process can be regulated by developmental or by activity-dependent mechanisms, as different receptors or receptor conformations can be expressed at different times during life. Presynaptically, synapses using vesicles with varying proportions of two neurotransmitters (Bergersen et al., 2003) could achieve a more fine tuned information transfer than that generally achieved through quantal release of a single-neurotransmitter (Somogyi, 2006).

One of the consequences of co-transmission is the regulation of the time course of the post-synaptic conductances activated once the presynaptic cell has conveyed the mixed message. In mixed glycinergic-GABAergic synapses, each amino acid activates its respective receptor, opening conductances with a different time course and shaping the overall response of the post-synaptic cell (Jonas et al., 1998; Gillespie et al., 2005). Another way in which the interaction of the neurotransmitters produces shifts in either response alone is by activating the same receptor, as seen when each agonist activates the $GABA_A$ receptor in a different way (Lu et al., 2008). Thus, the relative amount of glycine and GABA co-released from the presynaptic interneuron could precisely regulate the time course of the post-synaptic conductance. As discussed earlier, co-transmission would also facilitate feedback control of transmitter release by presynaptic $GABA_B$ receptors, which may not be possible at pure glycinergic synapses. Finally, co-transmission may enhance compensatory mechanisms where genetic defects of the GlyR subunit exist (Jonas et al., 1998; see also Dieudonné and Dana in this volume).

As can be inferred, not only does the different time course of the conductances opened by the different neurotransmitters released determine the behavior of the membrane potential of the post-synaptic membrane but also, the direction of the conductances. For instance, the co-release of an excitatory (ATP) and an inhibitory (GABA) neurotransmitter at the same synapse could have important functional implications. Indeed, it may represent a reversible switch between the inhibition and excitation of a given synapse without any anatomical reorganization of the neuronal circuitry. In the peripheral autonomic nervous system, ATP is co-released with acetylcholine or noradrenaline, which have an excitatory effect (Zimmermann, 1994; Burnstock, 1997).

However, while ATP activates cationic ionotropic receptors and may act as a fast excitatory neurotransmitter within the superficial dorsal horn of the spinal cord, GABA activates chloride channels acting as an inhibitory transmitter (Jo and Schlichter, 1999).

The mossy fiber-CA3 synapse utilizes glutamate and GABA for fast neurotransmission and their actions are synergistic during early development (see Gutiérrez; Safiulina et al. in this volume; Romo-Parra et al., 2008). GABA initially has a depolarizing effect that permits the activation of NMDA receptors by glutamate (Ben-Ari et al., 1997). As development progresses, GABA shifts its effect from depolarizing to hyperpolarizing and when GABA no longer depolarizes, the mossy fibers shut off their GABAergic phenotype (Gutiérrez in this volume; Safiulina in this volume; Romo-Parra et al., 2008). Interestingly, Gillespie et al. (this volume) have shown that the release of glutamate in a mixed gly-GABA-glutamatergic synapse is shut off at the end of development, persisting the gly-GABA transmission. This synergism may also occur in the case of acetylcholine (ACh) and glutamate co-release in the developing spinal cord, where the lack of AMPA receptors is compensated by the depolarization provoked by ACh, enabling glutamate to open NMDA channels (Li et al., 2004).

In the adult rodent GABA is mostly hyperpolarizing and glutamate depolarizing. However, after a period of enhanced excitability, the activation of mossy fibers produces synchronous activation of $GABA_A$ and glutamate receptors in pyramidal cells of the CA3, which in the end produces inhibition of this region. By contrast, interneurons respond differently and are readily excited despite receiving this dual glutamatergic-GABAergic input (Romo-Parra et al., 2003; Treviño et al., 2007). This is probably due to the differential release of glutamate and GABA on each target cell. Finally, the mossy fibers themselves have axonal $GABA_A$ and mGlu receptors that modulate the excitability of the fibers upon activation by GABA and glutamate released by parallel mossy fibers (Ruiz et al., 2003; Treviño and Gutiérrez, 2005), thereby reducing neurotransmitter output. This putative independent regulation of two signals (glutamate and GABA; Gutiérrez, 2003, 2005) could provide an additional degree of freedom in the transmission of information from the dentate gyrus to the hippocampus.

Another way by which two neurotransmitters that are released in a time-locked manner may regulate the time course of the post-synaptic conductances is by differentially activating ionotropic and metabotropic receptors. Many effects of neurotransmitters are dependent on the activation of members of the G protein-coupled receptor superfamily (GPCRs). Indeed, stimulation of particular GPCRs results in activation of signaling pathways that can subsequently interact with those activated by other GPCRs (Selbie and Hill, 1998), as well as those stimulated by ionotropic receptors.

In the case of ACh and glutamate, these two transmitters differ greatly in both the time-course and the mechanisms by which the resulting synaptic response is produced. While glutamate produces a fast ionotropic response, ACh produces a slow, G-protein-mediated effect via muscarinic receptors. Furthermore, and as shown for basal forebrain neurons that simultaneously release ACh and glutamate, their release is subject to powerful auto-inhibition mediated by Ach released through presynaptic M_2-muscarinic receptors (Allen et al., 2006).

Perhaps the best studied post-synaptic interaction between metabotropic and ionotropic receptors is that occuring in striatal GABergic projection neurons between dopamine (D_1- and D_2-like) receptors and glutamate ionotropic/metabotropic receptors. Although the inputs of the striatal GABAergic projection neurons do not co-release DA and glutamate, the following lines will help to define the functional consequences of such an interaction.

The interactions of dopamine and glutamate take place at different levels. The activation of D_1 receptors can increase the response to intrasomatic current injection and the increased response is at least partially mediated by enhanced opening of L-type Ca^{2+} channels (Hernández-López et al., 1997). Interestingly, the latter are anchored in spines near glutamatergic synapses (Olson et al., 2005). In addition, D_1 receptor activation enhances NMDA receptor-induced depolarization and NMDA-induced transient rises in intracellular Ca^{2+} (Liu et al., 2004). Also, D_1 receptor activation has been shown to increase the phosphorylation state of the NR1 subunit. Therefore, the ability of D_1 receptors to regulate NMDA receptor currents can be attributed to the synergistic increase in phosphorylation and the decrease in dephosphorylation of the NR1 subunit of the NMDA receptor. By contrast, D_2 receptor activation produces a negative modulation of L-type Ca^{2+} channels (Hernández-López et al., 2000) and activates K^+ channels providing an additional mechanism for such receptors to reduce the responsiveness of striatal neurones (Greif et al., 1995). Furthermore, D_2 receptor activation attenuates neuronal responses mediated by non-NMDA glutamate receptors (Cepeda et al., 1993; Herández-Echeagaray et al., 2004). With regard to non-NMDA glutamate receptors, D_1 receptor activation enhances AMPA receptor-mediated currents (Price et al., 1999).

There is also evidence that dopamine D_1- and D_2-like receptors regulate glutamate post-synaptic actions in striatal projection neurones in an opposite manner, by regulating receptor expression. D_1 receptor activation increases the insertion of NMDA receptors containing the NR2B subunit into the plasma membrane (Hallett et al., 2006; Snyder et al. 2000). Trafficking and localization might also be affected by a direct physical interaction between the dopamine D_1 and glutamate NMDA receptors and it results in decreased cell-surface expression of the NMDA receptors (reviewed by Cepeda and Levine, 2006). In terms of the D_2 receptors, their activation promotes trafficking of AMPA receptors out of the synaptic membrane (Sun et al., 2005; Håkansson et al., 2006).

Let us now consider the cross-talk between metabotropic glutamate receptors (mGluRs) and dopamine receptors. Group I mGluRs (mGluR1 and mGluR5) couple to the PLC/Ca^{2+}/PKC pathway (Conn et al., 2005). In addition to presynaptically modulating the release of dopamine by spillover (Zhang and Sulzer, 2003), the activation of striatal mGluR1/5 receptors also affects post-synaptic responses counteracting the effect of D_1 receptor stimulation in striatal neurones (Conn et al., 2005). Conversely, in the rat globus pallidus, where mGluR1 activation causes robust depolarization (probably due to calcium entry), D_1-like and D_2-like receptors act synergistically to control the signalling properties of mGluR1 by inhibiting PKA activity (Poisik et al., 2007). Thus, dopamine exerts complex modulatory effects on ionotropic and metabotropic glutamate receptor-mediated responses, in function of the excitatory amino acid receptor and the specific dopamine receptor subtype activated. The aforementioned interactions of DA and glutamate that occur in the striatum can be extended to other cell types and to other pairs of neurotransmitters acting on ionotropic and/or metabotropic receptors.

Finally, the co-release of classical neurotransmitters may play a role in maintaining the homeostasis of certain circuits. *Ex uno plures* implies divergence, so that the action of a given neuron on a target cell may be different from its effect on another target (depending on the receptors that each target cell expresses). However, it is also possible that a cell releases different transmitters in different target areas. Indeed, the release of chemicals from different sites or compartments of a neuron has been documented. For example, glutamate can be released from the dendrites of non-glutamatergic cells in the olfactory bulb, cerebellum and neocortex producing recurrent depolarization that may underlie recurrent excitation or that may potentiate GABA release (Harkany et al. 2004; Didier et al., 2001; Duguid and Smart, 2004). Some molecular mechanisms that contemplate such a phenomenon have been described. For example, the role of VGlut3 in the nervous system remains unknown as this transporter is structurally and functionally similar to VGlut1 and 2, yet it is expressed in terminals of cholinergic, serotonergic, GABAergic and glycinergic neurons (Fremeau et al., 2004). However, it has been proposed that VGlut3 supports glutamate release from these "non-glutamatergic" neurons and VGlut3 has been reported to mediate glutamate release from dendrites (Harkany et al., 2004). Indeed, it is clear that a complex intracellular machinery is able to selectively direct the necessary molecular apparatus to synthesize and accumulate transmitters into vesicles of different terminal branches of a neuron (Nishimaru et al., 2005).

As already recognized (Nishimaru et al., 2005), local recurrent positive and negative feedback loops are found in many areas of the brain, including the cortex, where they can balance membrane potentials close to the firing threshold to control the optimal conditions for persistent firing. Having both transmitters in the loops would stabilize activity, permitting the optimal integration of activity over the pool of connected neurons.

14.1 What Do We Still Need to Know?

The last question that we posed in chapter 2 was: does the postsynaptic target cell have a say in all this? While in this last chapter we have considered the position of the post-synaptic cells, some questions still need to be addressed.

Hence, does the post-synaptic cell communicate with the pre-synaptic cell in some way that can determine the neurotransmitter that the latter cell releases (in the case of it having more than one)? Several retrograde transmitters and their actions on synaptic plasticity have been described (e.g. NO, endocanabinoids, adenosine) but no studies have been made addressing their possible selective control of the release of either of the co-released neurotransmitters. Additionally, it has not been addressed whether the post-synaptic cell, by releasing these messengers, induces a change in the neurotransmitter phenotype of the pre-synaptic cell.

Is the mobility of the post-synaptic receptors a crucial factor in the activation of a mixed synapse? and how is this mobility commanded? Furthermore, do clustered receptors differ from the clusters of independent receptors? that is, do new interactions arise among them when they form clusters? How often and under what circumstances can transactivation of receptors occur?

We could continue to speculate about the almost unlimited possibilities of the combinations of classical neurotransmitters (and neuromodulators). It is maybe awkward to follow such an approach but on the other hand, this is a way to open doors to new avenues of research.

We are sure that while we prepare this book, more data regarding the co-existence of classical neurotransmitters and their receptors and transporters are appearing. This is a rapidly growing field and new and exciting discoveries will continue to shape the ongoing research. We need to integrate these findings as fast as they appear in order to reinforce the bridges being formed between cognitive functions, behavior and pathologies and the co-existence and co-release of classical neurotransmitters. Since the transmitter phenotype of neurons is highly plastic, so must we be.

Acknowledgement This work was supported by Consejo Nacional de Ciencia y Tecnología, México. We thank Dr. Jose Bargas and Dr. Peter Somogyi for critical review of this chapter

Reference

Allen TGJ, Abogadie FC, Brown DA (2006) Simultaneous release of glutamate and acetylcholine from single magnocellular "cholinergic" basal forebrain neurons. J Neurosci 26:1588–1595

Bekkers JM (2005) Presynaptically silent GABA synapses in hippocampus. J Neuroscience 25:4031– 4039

Ben-Ari Y, Khazipov R, Leinekugel X, Caillard O, Gaiarsa JL (1997) GABAA, NMDA and AMPA receptors: a developmentally regulated 'ménage a trois'. Trends Neurosci 20:523–529

Bergersen L, Ruiz, A, Bjaalie JG, Kullmann DM, Gundersen V (2003) GABA and GABAA receptors at hippocampal mossy fibre synapses. Eur J Neurosci 18:931–941

Burnstock G (1997) The past, present and future of purine nucleotides as signaling molecules. Neuropharmacology 36:1127–1139

Cepeda C, Buchwald NA, Levine MS (1993) Neuromodulatory actions of dopamine in the neostriatum are dependent upon the excitatory amino acid receptor subtypes activated. Proc Natl Acad Sci USA 90:9576–9580

Cepeda C, Levine MS (2006) Where do you think you are going? The NMDA-D1 receptor trap. Sci. STKE 333:pe20

Conn PJ, Battaglia G, Marino MJ, Nicoletti F (2005) Metabotropic glutamate receptors in the basal ganglia motor circuit. Nat Rev Neurosci 6:787–798

Chandler KE, Princivalle AP, Fabian-Fine R, Bowery NG, Kullmann DM, Walker MC (2003) Plasticity of GABA(B) receptor-mediated heterosynaptic interactions at mossy fibers after status epilepticus. J Neurosci. 23:11382–11391

Didier A, Carleton A, Bjaalie JG, Vincent J-D, Ottersen OP, Storm-Mathisen J, Lledo P-M (2001) A dendrodendritic reciprocal synapse provides a recurrent excitatory connection in the olfactory bulb. Proc Natl Acad Sci U S A 98:6441–6446

Duguid IC, Smart TG (2004) Retrograde activation of presynaptic NMDA receptors enhances GABA release at cerebellar interneuron-Purkinje cell synapses. Nat Neurosci 7:525–533

Fremeau RT Jr, Voglmaier S, Seal RP, Edwards RH (2004) VGLUTs define subsets of excitatory neurons and suggest novel roles for glutamate. Trends Neurosci 27:98–103

Gillespie DC, Kim G, Kandler K (2005) Inhibitory synapses in the developing auditory system are glutamatergic. Nat Neurosci 8:332–338

Greif GJ, Lin YJ, Liu JC, Freedman JE (1995) Dopamine-modulated potassium channels on rat striatal neurons: specific activation and cellular expression. J Neurosci 15:4533–4544

Gutiérrez R (2003) The GABAergic phenotype of the “glutamatergic” granule cells of the dentate gyrus. Prog Neurobiol 71:337–358

Gutiérrez R (2005) The dual glutamatergic-GABAergic phenotype of the hippocampal granule cells. Trends Neurosci 28:297–303

Håkansson K, Galdi S, Hendrick J, Snyder G, Greengard P, Fisone G (2006) Regulation of phosphorylation of the GluR1 AMPA receptor by dopamine D2 receptors. J Neurochem 96:482–488

Hallett PJ, Spoelgen R, Hyman BT, Standaert DG, Dunah AW (2006) Dopamine D1 activation potentiates striatal NMDA receptors by tyrosine phosphorylation-dependent subunit trafficking. J Neurosci 26:4690–4700

Harkany T, Holmgren C, Hartig W, Qureshi T, Chaudhry FA, Storm-Mathisen J, Dobszay MB, Berghuis P, Schulte G, Sousa KM Fremeau RT Jr, Edwards RH, Mackie K, Ernfors P, Zilberter Y (2004) Endocannabinoid-independent retrograde signaling at inhibitory synapses in layer 2/3 of neocortex: involvement of vesicular glutamate transporter 3. J Neurosci 24:4978–4988

Hernández-Echeagaray E, Starling AJ, Cepeda C, Levine MS (2004) Modulation of AMPA currents by D2 dopamine receptors in striatal medium-sized spiny neurons: are dendrites necessary? Eur J Neurosci 19:2455–2463

Hernández-López S, Bargas J, Surmeier DJ, Reyes A, Galarraga E (1997) D1 receptor activation enhances evoked discharge in neostriatal medium spiny neurons by modulating an L-type Ca2+ conductance. J Neurosci. 17:3334–3342

Hernández-López S, Tkatch T, Perez-Garci E, Galarraga E, Bargas J, Hamm H, Surmeier DJ (2000) D2 dopamine receptors in striatal medium spiny neurons reduce L-type Ca2+ currents and excitability via a novel PLCβ1-IP3-calcineurin-signaling cascade. J Neurosci 20:8987–8995

Jo YH, Schlichter R (1999) Synaptic corelease of ATP and GABA in cultured spinal neurons. Nature Neurosci 2:241–245

Jonas P, Bischofberger J, Sandkühler J (1998) Corelease of two fast neurotransmitters at a central synapse. Science 281:419–424

Li WC, Soffe S R, Roberts A (2004) Glutamate and acetylcholine corelease at developing synapses. Proc Natl Acad Sci USA 101:15488–15493

Liu JC, DeFazio RA, Espinosa-Jeffrey A, Cepeda C, de Vellis J, Levine MS (2004) Calcium modulates dopamine potentiation of N-methyl-D-aspartate responses: electrophysiological and imaging evidence. J Neurosci Res 76:315–322

Lu T, Rubio ME, Trussell LO (2008) Glycinergic transmission shaped by the corelease of GABA in a mammalian auditory synapse. Neuron 57:524–535

Nishimaru H, Restrepo CE, Ryge J, Yanagawa Y, Kiehn O (2005) Mammalian motor neurons corelease glutamate and acetylcholine at central synapses. Proc Natl Acad Sci USA 102:5245–5249

Olson PA, Tkatch T, Hernandez-Lopez S, Ulrich S, Ilijic E, Mugnaini E, Zhang H, Bezprozvanny I, Surmeier DJ (2005) G-protein-coupled receptor modulation of striatal CaV1.3 L-type Ca2+ channels is dependent on a Shank-binding domain. J Neurosci 25:1050–1062

Poisik OV, Smith Y, Conn PJ (2007) D1- and D2-like dopamine receptors regulate signaling properties of group I metabotropic glutamate receptors in the rat globus pallidus. Eur J Neurosci 26:852–862

Price CJ, Kim P, Raymond LA (1999) D1 dopamine receptor-induced cyclic AMP-dependent protein kinase phosphorylation and potentiation of striatal glutamate receptors. J Neurochem 73:2441–2446

Rao A, Cha EM, Craig AM (2000) Mismatched appositions of presynaptic and post-synaptic components in isolated hippocampal neurons. J Neurosci 20:8344–8353

Romo-Parra H, Treviño M, Heinemann U, Gutierrez R (2008) GABA actions in hippocampal area CA3 during post-natal development: differential shift from depolarizing to hyperpolarizing in somatic and dendritic compartments. J Neurophysiol. 99:1523–1534

Romo-Parra R, Vivar C, Maqueda J, Morales MA, Gutiérrez R (2003) Activity-dependent induction of multitransmitter signaling onto pyramidal cells and interneurons of area CA3 of the rat hippocampus. J Neurophysiol 89:3155–3167

Ruiz A, Fabian-Fine R, Scott R, Walker MC, Rusakov DA, Kullmann DM (2003) GABAA receptors at hippocampal mossy fibers. Neuron 39:961–973

Selbie LA, Hill SJ (1998) G protein-coupled receptorcross-talk: the fine-tuning of multiple receptor-signalling pathways. Trends Pharmacol Sci 19:87–93

Snyder GL, Allen PB, Fienberg AA, Valle CG, Huganir RL, Nairn AC, Greengard P (2000) Regulation of phosphorylation of the GluR1 AMPA receptor in the neostriatum by dopamine and psychostimulants in vivo. J Neurosci 20:4480–4488

Somogyi J (2006) Functional significance of co-localization of GABA and Glu in nerve terminals: a hypothesis. Curr Top Med Chem 6:969–973

Sun X, Zhao Y, Wolf ME (2005) Dopamine receptor stimulation modulates AMPA receptor synaptic insertion in prefrontal cortex neurons. J Neurosci 25:7342–7351

Treviño M, Gutierrez R (2005) The GABAergic projection of the dentate gyrus to hippocampal area CA3 of the rat: pre- and post-synaptic actions after seizures. J Physiol 567:939–949

Treviño M, Vivar C, Gutiérrez R (2007)β/γ oscillatory activity in the CA3 hippocampal area is depressed by aberrant GABAergic transmission from the dentate gyrus after seizures. J Neurosci 27:251–259

Zhang H, Sulzer D (2003) Glutamate spillover in the striatum depresses dopaminergic transmission by activating group I metabotropic glutamate receptors. J Neurosci 23:10585–10592

Zimmermann H (1994) Signalling via ATP in the nervous system. Trends Neurosci 17:420–426

Index

Color Plate 1

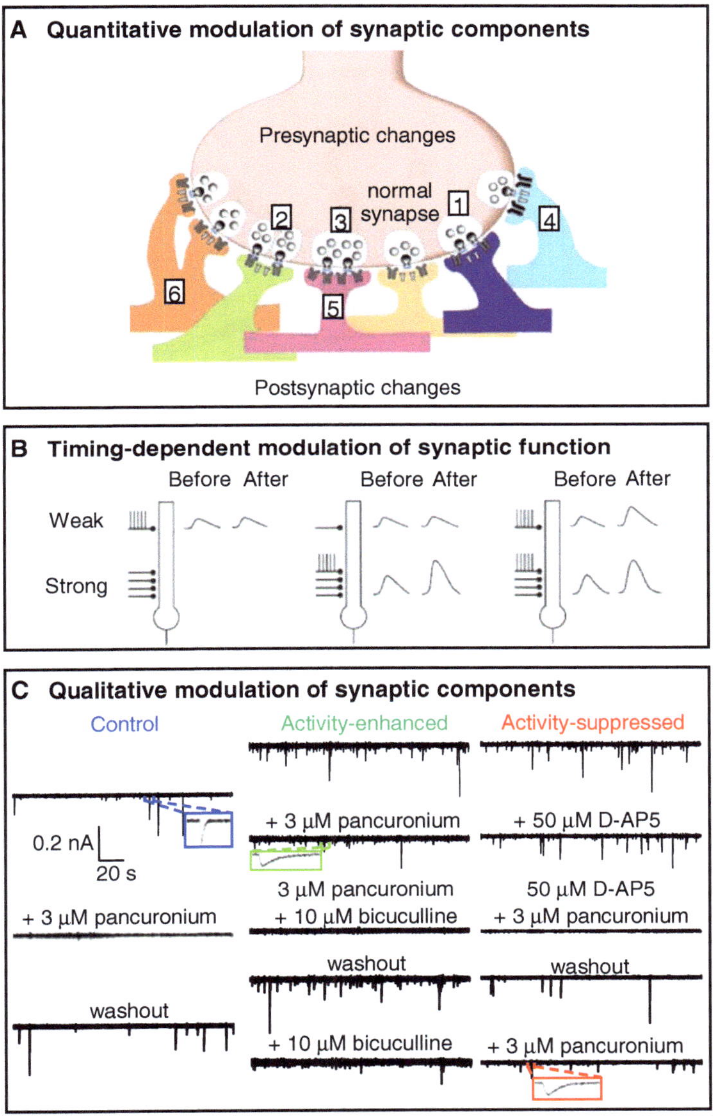

Fig. 3.1 Different levels of modulation of synaptic activity. (**a**) Synaptic activity can be enhanced by presynaptic (1–3), postsynaptic (4–5) or coordinated (6) changes: 1. Increase in release probability; 2. Increase in the number of release sites; 3. Increase in the number of vesicles available for release; 4. Increase in the sensitivity of preexisting neurotransmitter receptors; 5. Increase in the number of functional receptors; 6. Growth of new synaptic contacts. **(b)** Synchronization of stimuli at different synapses can cause changes in synaptic activity. Left, a tetanic stimulation of the weak input on a pyramidal cell does not cause

Color Plate 2

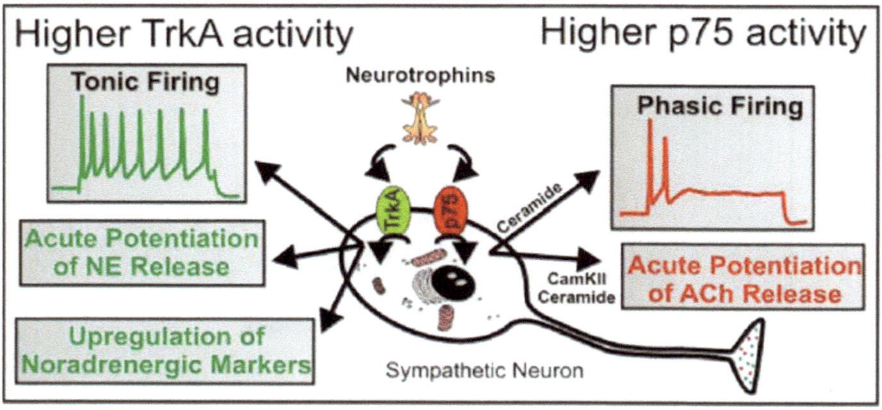

Fig. 4.1 Target-derived neurotrophins rapidly regulate functional neuronal properties differentially through p75 and TrkA signaling pathways. Sympathetic neurons express both TrkA and p75. When TrkA activation is higher than p75 activation the neurons tend to release norepinephrine and fire in a tonic pattern (left). When p75 activity predominates cholinergic transmission is potentiated and cells fire in a phasic pattern. Although the signaling pathways involved are incompletely understood, evidence shows that the second messenger molecule, ceramide, is involved in promotion of both phasic firing and potentiation of acetylcholine release. Calcium/calmoulin-dependent protein kinase II (CamKII) also has been implicated in promoting cholinergic transmission. (*See* page 47)

Fig. 3.1 (continued) long-term potentiation (LTP) in the pathway (same EPSPs); middle, tetanic stimulation of the strong input alone causes LTP in the strong pathway but not in the weak one; right, tetanic stimulation of both inputs together causes LTP in both pathways. **(c)** Perturbation of electrical activity induces co-transmission and homeostatic changes in types of neurotransmitter and receptors involved in the synapse. Recordings of skeletal muscle miniature postsynaptic currents (mpscs) are shown for control (left), activity-enhanced (middle) and activity-suppressed (right) *Xenopus* larvae. While in control embryos mpscs are completely blocked by a nicotinic acetylcholine receptor antagonist (pancuronium), in activity-suppressed and –enhanced embryos NMDAR- (D-AP5-sensitive) and $GABA_AR$-mediated currents (bicuculline-sensitive) can be recorded along with cholinergic mpscs (pancuronium-sensitive). (a) *Adapted from Wang et al., 1997 and Fundamental Neuroscience*, 2nd edition. (b) *Adapted from Nicoll et al., 1988*. (c) *Adapted from Borodinsky and Spitzer, 2007*. (*See* pp. 28–29)

Color Plate 3

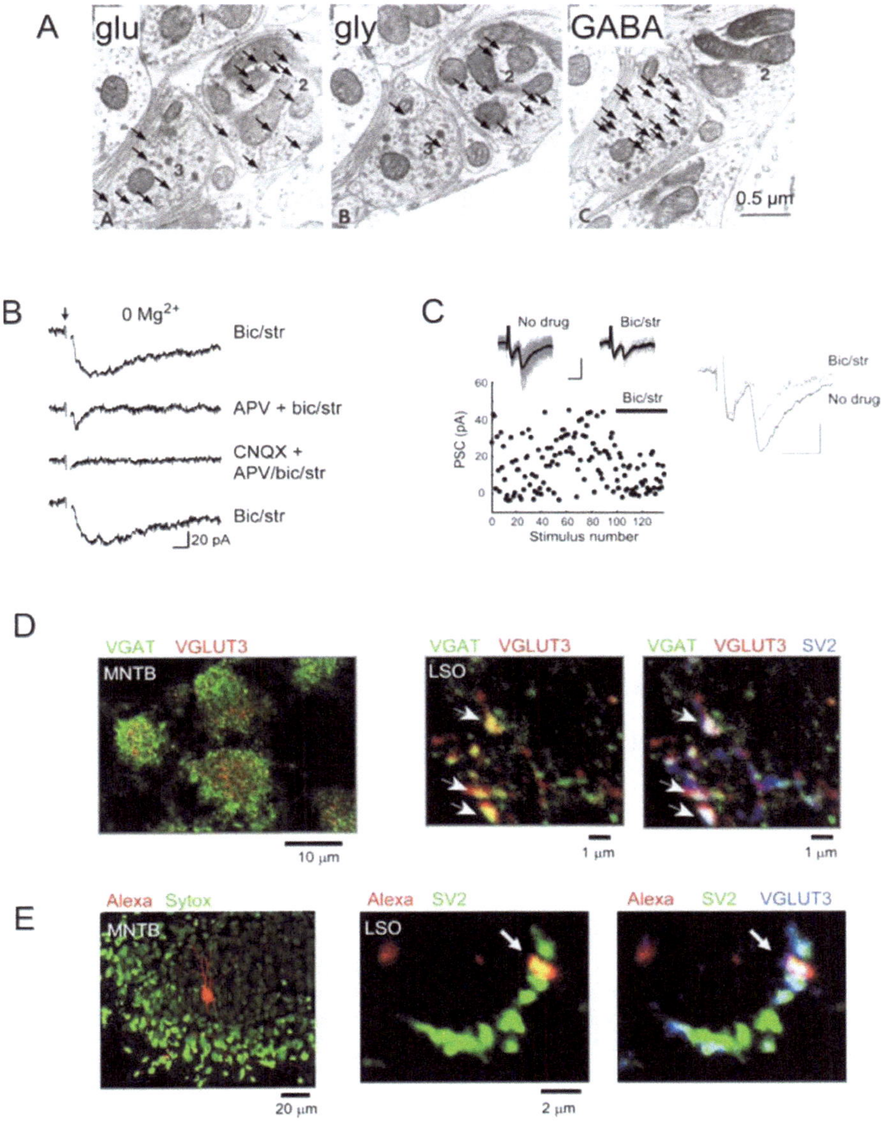

Fig. 5.3 Glutamate co-release at developing GABA/glycine auditory synapses. **(a)** Electron micrographs of an individual terminal from serial sections immunolabeled for glutamate, glycine, and GABA. Terminal 3 is immunoreactive for all three neurotransmitters, terminal 2 for glutamate and glycine. Arrows point to gold particles that tag immunopositive sites. **(b)** Blocking $GABA_A$Rs and GlyRs with bicuculline and strychnine uncovers a MNTB-elicited glutamatergic response in LSO neurons. In these recordings, magnesium was excluded from the bath to unblock NMDA receptors. Glutamatergic responses are mediated by

Color Plate 4

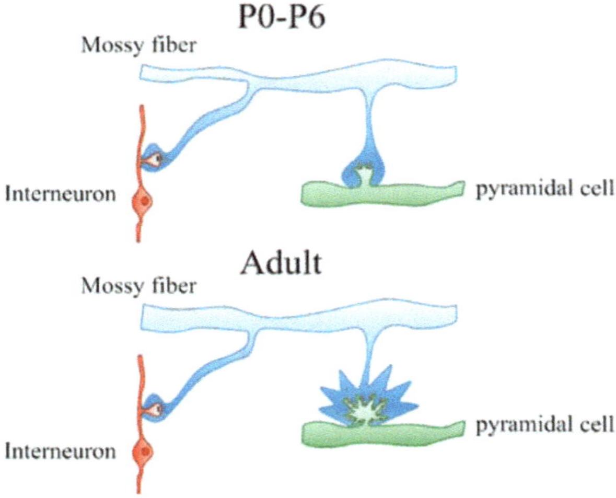

Fig. 6.1 Schematic representation of a mossy fiber (blue) making synaptic contacts with a GABAergic interneuron (red) and a pyramidal cell (green) during the first week of postnatal life (P0–P6) and in adulthood. (Modified from Amaral and Dent, 1981). (*See* page 84)

Fig. 5.3 (continued) NMDA and AMPA receptors, as they are partially blocked by the NMDAR antagonist APV and completely blocked by addition of the AMPAR antagonist CNQX. **(c)** Single GABA/glycinergic MNTB axons can co-release glutamate. Activation of single MNTB axons by minimal stimulation elicits postsynaptic currents in LSO neurons (black trace is average of grey, overlaid responses) that are only partially blocked by bicuculline and strychnine. **(d)** Expression of the vesicular glutamate transporter 3 (VGLUT3) in the MNTB-LSO pathway. MNTB neurons are immunopositive for both the vesicular GABA transporter (VGAT) and VGLUT3. In the LSO, VGAT co-labels with VGLUT3 in small clusters, which also label for the synaptic vesicle protein 2 (SV2). Arrows point to presumed presynaptic endings (SV2-positive) that also label with both VGLUT3 and VGAT. **(e)** Identified MNTB terminals in the LSO express VGLUT3. A single MNTB neuron was filled with the dye Alexa 568 (red). In the LSO an Alexa-filled terminal of this neuron is identified by expression of SV2 (yellow). This terminal also expresses VGLUT3 (white). *Adapted with permission from: A) Helfert et al.* 1992; ***B-E)*** *Gillespie et al.* 2005. (*See* pp. 67–68)

Color Plate 5

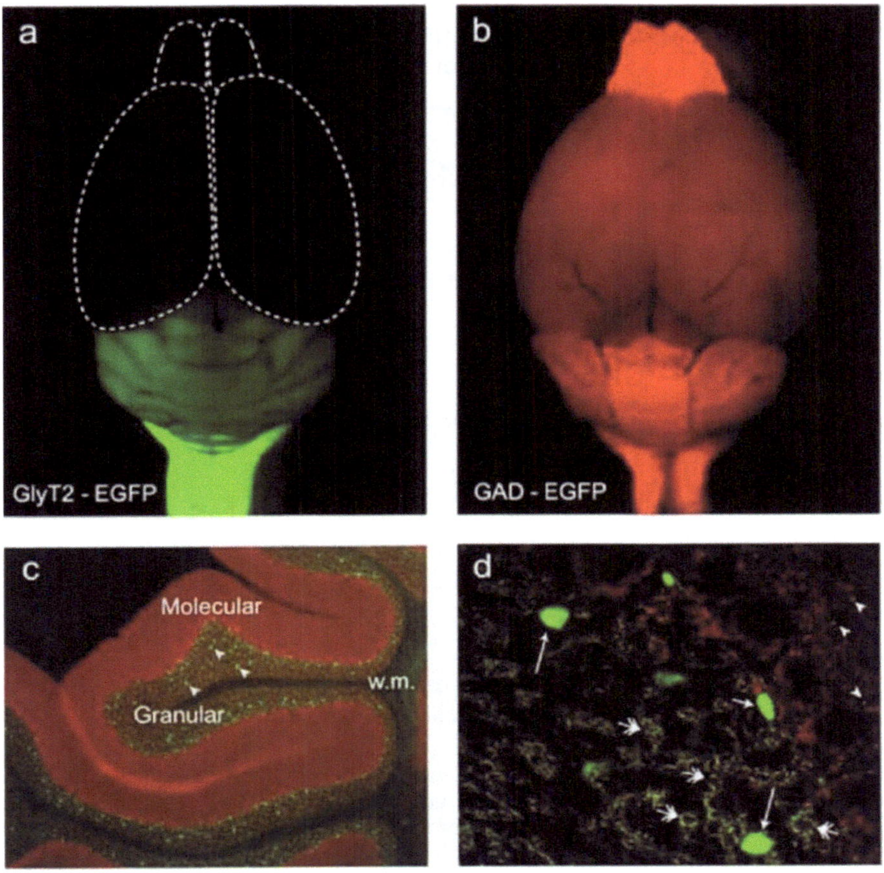

Fig. 7.1 GABAergic and glycinergic transmission in the central nervous system of rats. **(a)** A strong rostro-caudal gradient of glycinergic neuron density is evident in a transgenic mouse expressing GFP under the promoter of the neuronal glycine transporter GlyT2 (kind gift of H.U. Zeilhofer). **(b)** Similar view for a mouse expressing EGFP under the promoter of GAD65 that catalyzes the limiting step for GABA synthesis in GABAergic neurons (kind gift of G. Szabo) (Lopez-Bendito et al., Cereb Cortex, 2004). **(c)** Parasagittal section showing a lobule of the cerebellar cortex of the GlyT2-EGFP mouse. Immunostaining for GABA is shown in red. Note the prevalence of GABA in the molecular layer and the abundance of glycinergic interneurons in the granular layer (arrowheads). Most of the Golgi cell axons in the granular layer appear to co-stain for GABA (yellow background). GFP-positive axons, presumably of granular layer interneurons, are running in the white matter (w.m.). **(d)** Detail of a cerebellar lobule of the GlyT2-EGFP mouse stained for VIAAT, the vesicular cotransporter of GABA and glycine, in red. Golgi cell bodies (long arrows) and Lugaro cell bodies (short arrow) are clearly visible. Golgi cell axonal varicosities delineate the contour of cerebellar glomeruli (double arrows). Note that some of the VIAAT-stained profiles correspond to GFP-negative boutons from GABAergic Golgi cells. Lugaro cell varicosities, that co-release GABA and glycine, are visible in the molecular layer (arrowheads) as a minoritary population. (*See* page 101)

Color Plate 6

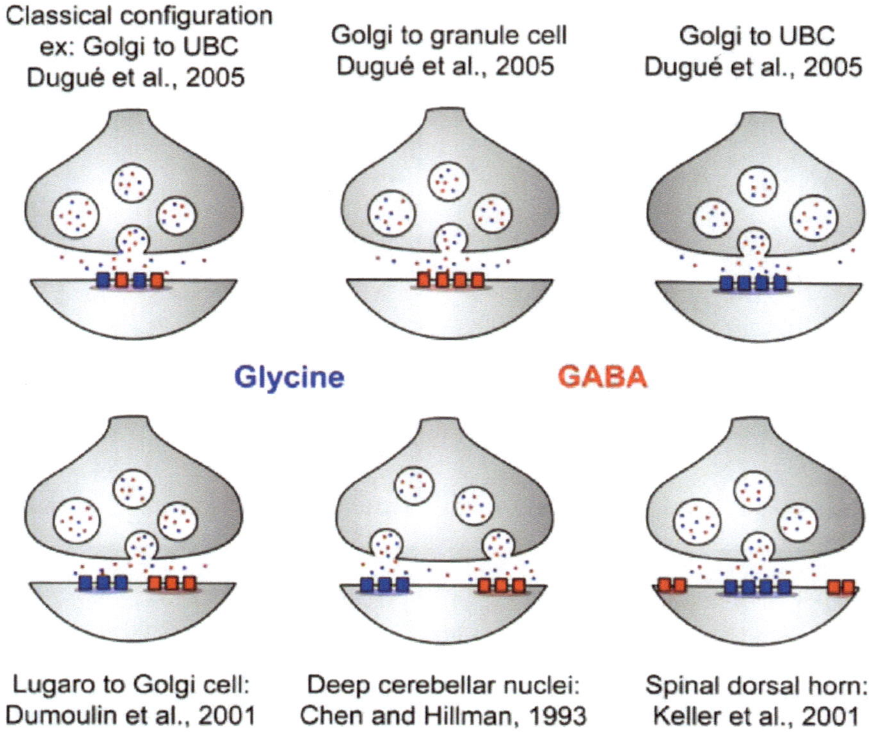

Fig. 7.2 Postsynaptic heterogeneity at synapses co-releasing GABA and glycine. We show here six different cases reported in the literature (see citations in the figure), where distinct postsynaptic configurations are present in front of presynaptic sites co-releasing GABA and glycine. Each instance is described in the text in detail. (*See* page 104)

Color Plate 7

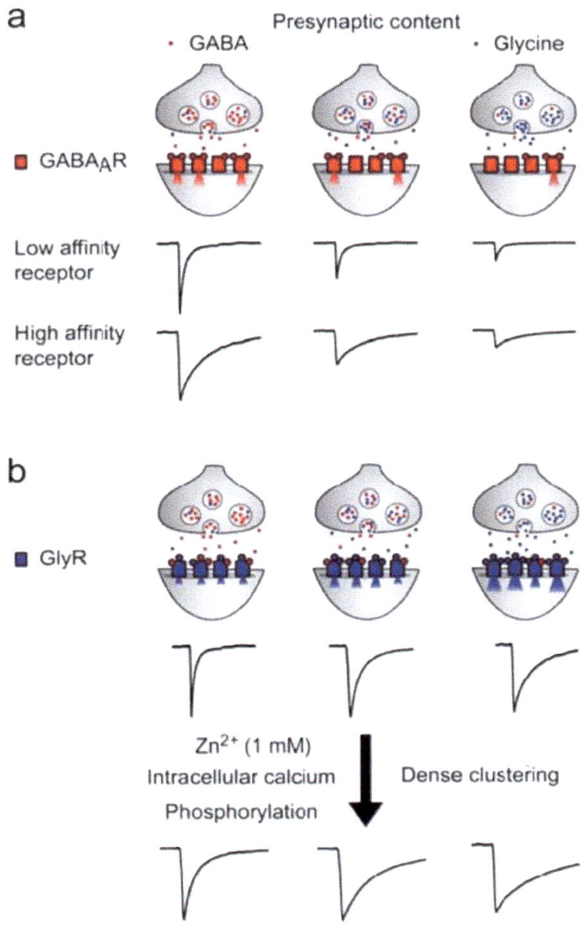

Fig. 7.3 Different mechanisms govern the amplitude and duration of GABAergic and glycinergic IPSCs at mixed synapses. **(a)** $GABA_A$ receptors activation at mixed inhibitory synapses. The amount of GABA released by mixed vesicles determines the amplitude of the GABAergic component of the IPSCs (left to right). The degree of occupancy of agonist binding sites defines the number of fully liganded receptors, which will open. The molecular composition of the $GABA_A$ receptors present at the synapse is the main determinant of the receptors affinity. It fixes the decay kinetics of the IPSCs. **(b)** Glycinergic receptors are allosteric devices with tunable decay kinetics. Partially liganded receptors display a low apparent affinity for their agonists and mediate fast IPSCs. Synaptic current should decay more slowly at glycine-rich synapses (right), where fully liganded receptors show a higher apparent affinity. In GABA-rich synapses, GABA can compete for glycine at the receptor binding sites. The low affinity and fast unbinding rate of GABA result in very brief glycinergic IPSCs. Allosteric transitions governing the gating kinetics of glycinergic receptors are modulated by a variety of environmental effectors that promote high apparent affinity states. (*See* page 112)

Color Plate 8

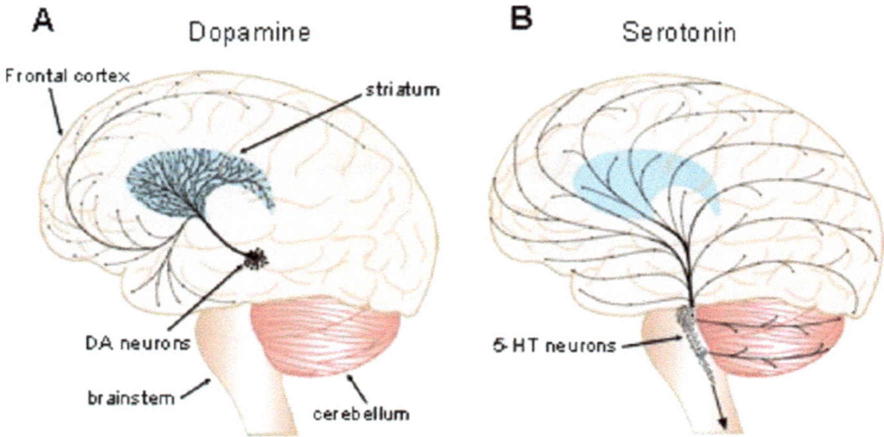

Fig. 9.1 A simplified representation of DA and 5-HT projections in the human brain. **(a)** The cell bodies of the mesoforebrain dopamine system are located mainly in the substantia nigra pars compacta (SNc, A9) and the ventral tegmental area (VTA, A10) of the midbrain. SNc dopamine neurons mainly project to the dorsal striatum, and VTA dopamine neurons mainly project to the mesocorticolimbic areas such as the nucleus accumbens, amygdala, and prefrontal cortex. **(b)** The cell bodies of the serotonin system are located in the raphe nuclei. These serotonin neurons project to virtually all brain areas in a relatively even manner, though some regional variation exists. *From Zhou and Dani*, 2006. (*See* page 146)

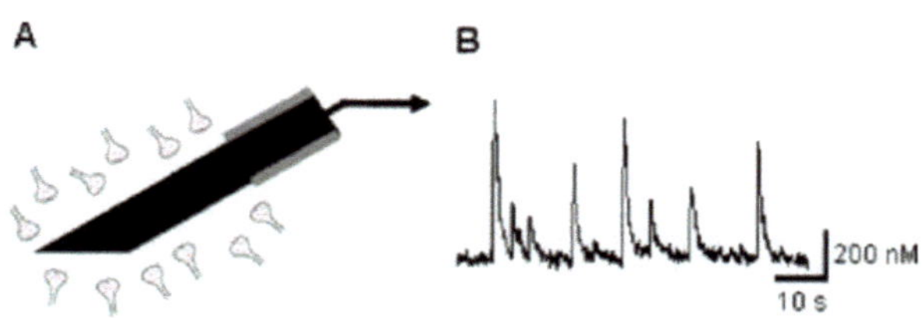

Fig. 9.2 DA released measured with carbon-fiber electrodes using fast-scan cyclic voltammetry. **(a)** Illustration of carbon-fiber electrode used for fast-scan cyclic voltammetry recordings of vesicular DA release in the striatum. Not drawn to scale. **(b)** The trace shows action potential-dependent spontaneous vesicular DA release recorded in the striatum using voltammetry. Action potential-independent single DA vesicle release events (minis) are below the detection sensitivity for voltammetry. Thus, these relatively large DA spikes arise from synchronized release of multiple axon terminals and varicosities originating from the same parent DA axon fibers in the vicinity of the carbon-fiber electrode. *Unpublished data of Zhou and Dani.* (*See* page 148)

Color Plate 9

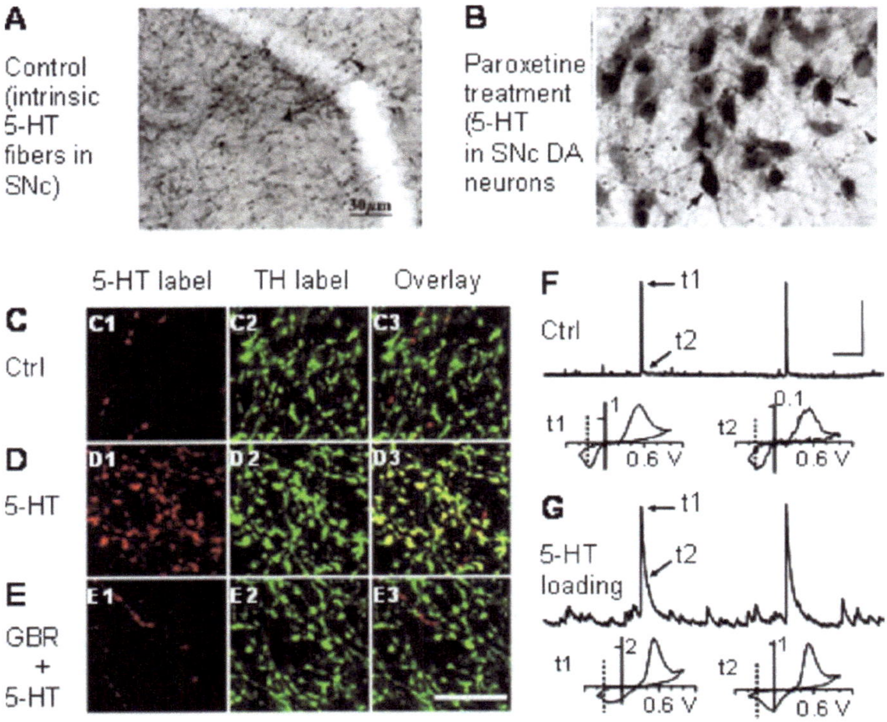

Fig. 9.4 5-HT enters DA neuron somata and axon terminals. **(a)** Under normal condition, 5-HT-immunostained fibers but not cell bodies are seen in the SNc. **(b)** Treatment of animals with paroxetine, a commonly used SSRI antidepressant, induces 5-HT accumulation in DA neurons in SNc in mice. From Zhou FC et al. (2002) with permission. **(c)** In striatal brain slices under control conditions, 5-HT axon terminals as detected by anti-5-HT antibody are relatively sparse (red, C1), and DA terminals as detected by anti-tyrosine hydroxylase (TH) antibody are dense (green, C2). The DA and 5-HT terminals generally do not overlap (absence of yellow, C3). **(d)** Incubation of striatal brain slices with 5-HT (2 μM) increases the number of 5-HT-positive terminals (red, D1), and DA terminals are relatively unchanged (green, D2). Most of the 5-HT-positive terminals coincide with DA terminals (yellow, D3), suggesting that DA terminals in this case also contain 5-HT. **(e)** Pretreatment with the DAT inhibitor, GBR12909 (1 μM), blocks the effect of 5-HT (2 μM). 5-HT terminals remain sparse (red, E1), DA terminals are dense and unchanged (green, E2), and the 5-HT-positive terminals rarely coincide with TH-positive terminals (yellow, E3). Scale bar = 10 μm. **(f)** A control trace showing small spontaneous voltammetric release events and two large electrically stimulated events. The scale bars are 1 μM DA and 30 s. Time t1 indicates the peak of the signal. At time t2, 4 s after the peak, the signal is small but still detectable. The two voltammograms (below) from t1 and t2 are typical DA voltammograms with the oxidation peak near 500 mV and the reduction peak near –230 mV (dotted vertical lines). The y-axis of the voltammograms varies, and is labeled for comparison in nA. **(g)** In the presence of 5-HT (2 μM) after subtracting off the 5-HT background, both the spontaneous and evoked events are larger and longer lasting than in the control case. The voltammogram from t1 is a mixed DA/5-HT voltammogram

Color Plate 10

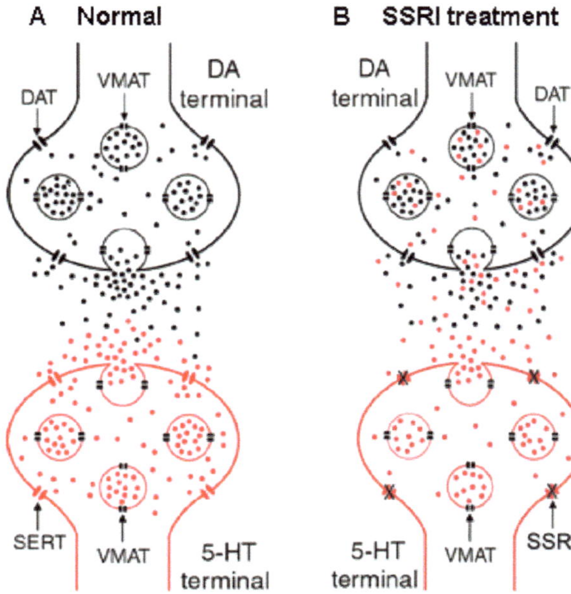

Fig. 9.5 Schematic, didactic representation of the hypothesized DAT-dependent 5-HT accumulation into striatal DA terminals after chronic SSRI antidepressant treatment. **(a)** Under normal conditions when the 5-HT transporters (SERTs) are functional, the released 5-HT molecules (red dots) are efficiently taken back into 5-HT terminals. Consequently, the extracellular 5-HT concentration remains low. DATs on DA terminals have little chance to uptake 5-HT. **(b)** DATs may uptake a portion of endogenously released 5-HT into DA terminals when the extracellular 5-HT concentration is elevated after the SERTs are inhibited by SSRIs. Once inside the DA terminals, 5-HT is further sequestered into DA terminals through the vesicular monoamine transporter (VMAT), which has similar affinity for DA and 5-HT. Subsequently, 5-HT partially displaces DA (black dots) from DA vesicles, and the two neurotransmitters are co-released. *From Zhou et al.* 2005 *with permission.* (*See* page 161)

Fig. 9.4 (continued) with a broad reduction peak from –230 mV (DA, dotted vertical line) to –30 mV (near 5-HT’s reduction peak), whereas the voltammogram from t2 is a predominantly 5-HT voltammogram with a reduction peak at –30 mV. (1) and (b) are from *Zhou FC et al.* 2002 *with permission.* (c-g) are from *Zhou et al.* 2005 *with permission.* (*See* page 158)

Color Plate 11

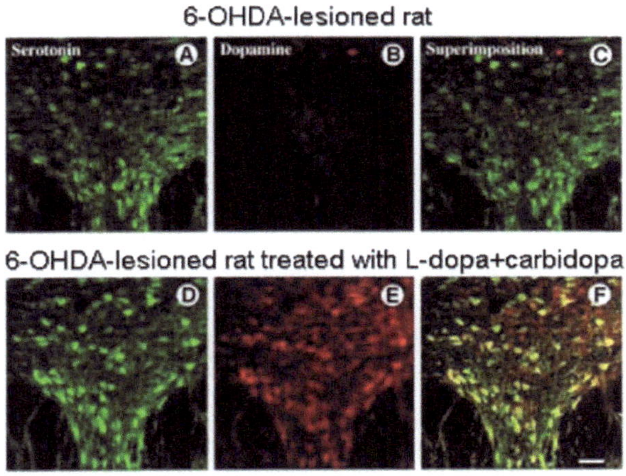

Fig. 9.6 DA in 5-HT neurons of the raphe. **(a)** Dorsal raphe 5-HT neurons immunostained with anti-5-HT antibody in a 6-hydroxydopamine (6-OHDA, a DA neuron toxin) lesioned rat. **(b)** Only a few DA terminals are revealed by an anti-DA antibody in the raphe area in the same section as in (A). **(c)** Superimposition of (A) and (B) shows no co-localization of 5-HT and DA-containing processes. **(d)** Dorsal raphe 5-HT neurons immunostained with anti-5-HT antibody in a 6-OHDA-lesioned rat treated with L-Dopa and carbidopa. **(e)** DA-containing neurons in the same section as in (D) revealed by an anti-DA antibody. **(f)** Superimposition of (D) and (E) shows that in addition to intrinsic DA terminals, DA co-localizes within raphe 5-HT neurons. The results indicate that L-dopa enters 5-HT neurons and, then, is converted to DA in 5-HT neurons. *From Yamada et al.* (2007) *with permission.* Scale bar = 10 μm. (*See* page 162)

Color Plate 12

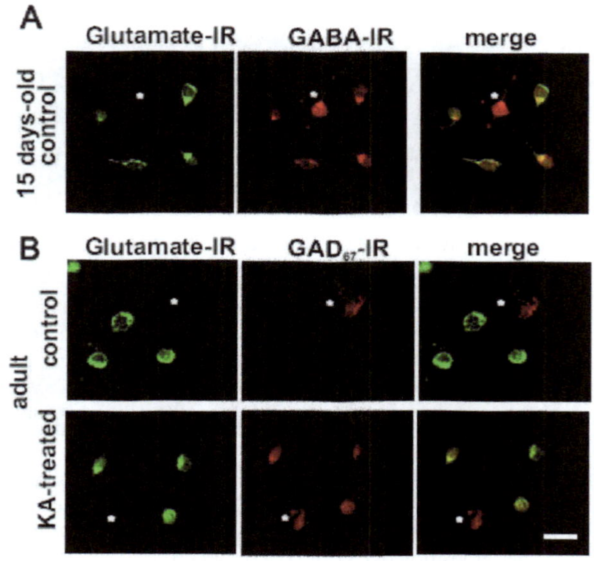

Fig. 10.2 Immunocytochemistry for glutamate, GAD_{67}, and GABA in cultured GCs. **(a)** Confocal images of short-term GC and interneuron cultures (asterisks) from a 15 day-old rat showing the co-expression of glutamate and GABA in GCs. **(b)** In cultures from adult rats, GCs express glutamate but not GABAergic markers unless they are exposed to KA or BDNF, whereupon GAD_{67} and glutamate-IR colocalize. Scale bar in B (15 μm) also applies to panel A. *From Gómez-Lira et al.*, 2005, *with permission*. (*See* page 185)

Color Plate 13

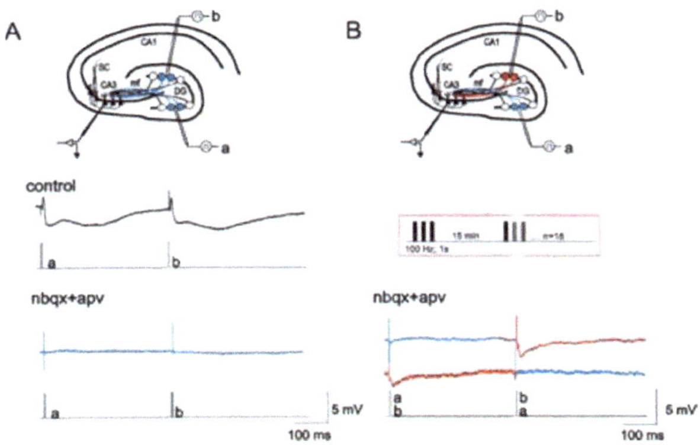

Fig. 10.5 Spatial specificity of the kindling-induced inhibitory responses in vitro. **(a)** Schematic representation of a hippocampus, showing two sites of stimulation in the DG (a,b); actual separation during the experiments 200–300 μm) and the recording electrode in CA3. Initially, test pulse stimulation at both sites of stimulation evoked EPSP/IPSP sequences in the pyramidal cell at a RMP of –64 mV (control), which were blocked by GluRAs (blue trace). **(b)** After applying 17 ± 1 kindling trials to site b (in red), test pulse stimulation of the kindled site (b) but not of the non-kindled site (a, in blue), provoked a fast IPSP in the pyramidal cell. Changing the order of stimulation inverted the order of the responses. All traces are an average of 10 responses. *From Gutiérrez* (2002), *with permission*. (*See* page 189)

Color Plate 14

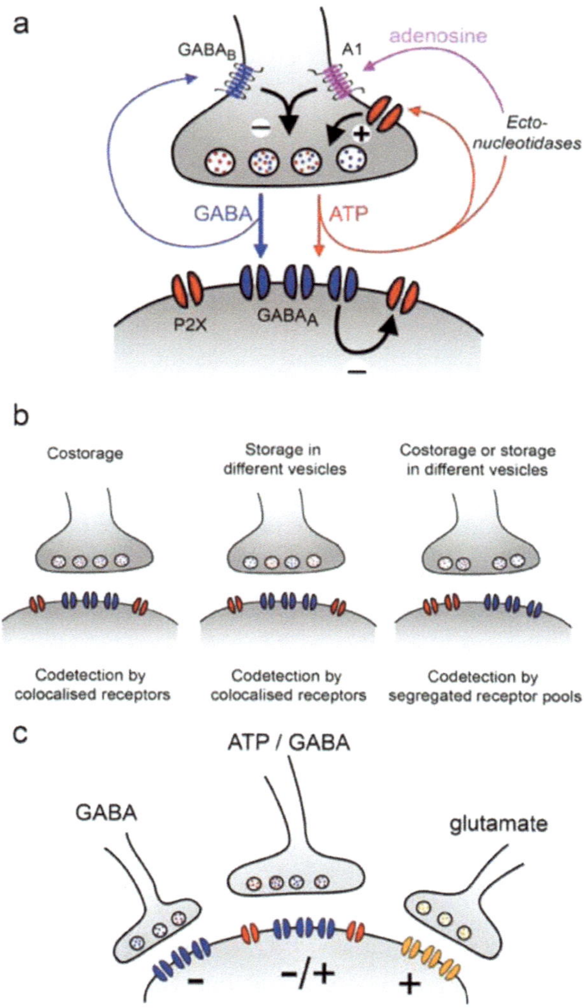

Fig 11.1 The ATP/GABA cotransmission. ATP and GABA are coreleased from the presynaptic terminal and act at postsynaptic ionotropic excitatory ATP receptors (P2X receptors) and inhibitory anionic $GABA_A$ receptors. **(a)** In the dorsal horn of the spinal cord, the synaptic corelease of ATP and GABA is modulated by presynaptic autoreceptors. ATP can facilitate GABA release by acting at presynaptic P2X receptors. Inhibition of corelease involves metabotropic $GABA_B$ receptors and A1 adenosine receptors which act by a partially convergent presynaptic mechanism. Adenosine is generated by the extracellular hydrolysis of ATP by ectonucleotidases. **(b)** Possible anatomical substrates and scenarios of ATP/GABA cotransmission. ATP and GABA might be costored in the same vesicle or stored in different vesicles. Postsynaptic P2X and $GABA_A$ receptors might be colocalized in front of the same release sites or segregated in different synapses. **(c)** The mixed excitatory/inhibitory ATP/GABA

Color Plate 15

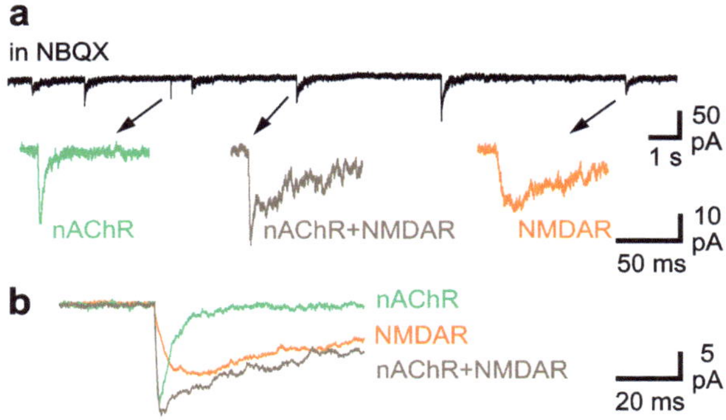

Fig. 12.2 Analysis of mEPSCs shows glutamate and ACh are co-released from single vesicles. **(a)** When glutamate AMPA receptor blocker NBQX was used, three types of mEPSCs were identified (arrows point to individual examples). **(b)** Averaged shape of the three types of mEPSCs. The time course of green and orange traces matched that of nAChR and NMDAR mEPSCs separated pharmacologically. The brown trace has a fast rise like nAChR and a fall like NMDAR currents, hence are most likely mixed nAChR + NMDAR responses. *Figure modified from Li et al. (2004a) with permission.* (*See* page 229)

Fig. 11.1 (continued) cotransmission can finely tune the balance of excitation and inhibition converging on the same neuron by shifting the net equilibrium toward inhibition or excitation, depending on the relative weight of the ATPergic and GABAergic components of the cotransmission. (*See* page 205)

Color Plate 16

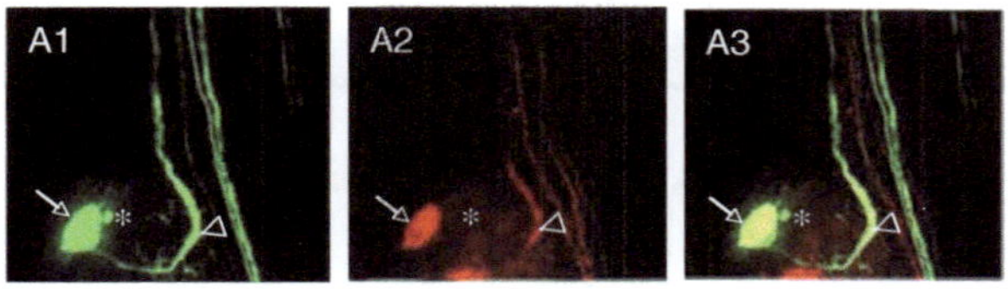

Fig. 13.1 Colocalization of tyrosine hydroxylase-like immunoreactivity and serotonin-like immunoreactivity in the L-cell of the blue crab, *Callinectes sapidus*. (A1) THli in the L-cell (*arrow*) and a second small cell body (*asterisk*; detected with a mouse monoclonal primary antibody and Alexa 488 goat anti-mouse secondary antibody). Note that the axon of the L-cell is constricted as it passes through the commissural ganglion, but that it then widens (*arrowhead*) after entering the circumesophageal connective to ascend to the brain (see text and Fort et al. 2004 for overall L-cell structure). (A2) 5HTli in the same preparation shown in *a1*. 5HTli was observed in the L-cell soma (*arrow*) and in the axon ascending in the connective (*arrowhead*; visualized with rabbit polyclonal antibody and Alexa 546 goat anti-rabbit secondary antibody). It was not detected in the initial segment of the L-cell. 5HTli was not seen in the small THli neuron (*asterisk*) or in other THli fibers, supporting the deduction that icolocalization of markers does not reflect an artifact of marker 'bleedthrough'. *Calibration bar* = 100 μm, applies to *A1–A3* (*See* page 249)

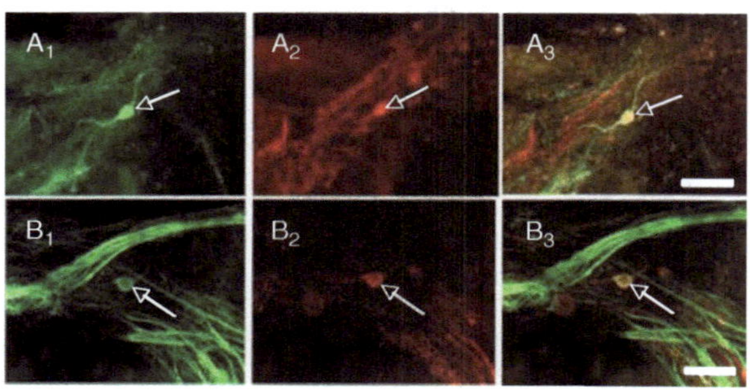

Fig. 13.4 Colocalization of TH-like immunoreactivity and GABA-like immunoreactivity in the buccal ganglion of *Dolabrifera dolabrifera*. (*A1*) A single neuron (*arrow*) on the rostral surface was marked with an antibody against tyrosine hydroxylase (mouse monoclonal; Alexa 488 goat anti-mouse secondary). (*A2*) When the same preparation was processed for GABA-like immunoreactivity (rabbit polyclonal; Alexa 546 goat anti-rabbit secondary), the same neuron (*arrow*) was marked. (*A3*) The labeled neuron (*arrow*) appears yellow in an overlay of panels *A1* and *A2*. *Calibration bar*: 100 μm applies to all *A* panels. (*B1*) A neuron (*arrow*) in the lateral region of the caudal surface of each buccal hemiganglion (only the right is shown) was marked with an antibody against TH. (*B2*) When the same preparation was processed for GABA-like immunoreactivity, the same neuron (*arrow*) was marked. (*B3*) The labeled neuron (*arrow*) appears yellow in an overlay of panels *B1* and *B2*. *Calibration bar*: 100 μm applies to all *B* panels (*See* page 253)

MIX
Papier aus verantwortungsvollen Quellen
Paper from responsible sources
FSC® C105338

If you have any concerns about our products,
you can contact us on
ProductSafety@springernature.com

In case Publisher is established outside the EU,
the EU authorized representative is:
Springer Nature Customer Service Center GmbH
Europaplatz 3, 69115 Heidelberg, Germany

Printed by Libri Plureos GmbH
in Hamburg, Germany